00-WSM-5

GLOBAL LESSONS FOR WATERSHED MANAGEMENT IN THE UNITED STATES

by:

James Goldstein

Annette Huber-Lee

Tellus Institute

2004

WERF

The Water Environment Research Foundation, a not-for-profit organization, funds and manages water quality research for its subscribers through a diverse public-private partnership between municipal utilities, corporations, academia, industry, and the federal government. WERF subscribers include municipal and regional water and wastewater utilities, industrial corporations, environmental engineering firms, and others that share a commitment to cost-effective water quality solutions. WERF is dedicated to advancing science and technology addressing water quality issues as they impact water resources, the atmosphere, the lands, and quality of life.

For more information, contact:
Water Environment Research Foundation
635 Slaters Lane, Suite 300
Alexandria, VA 22314-1177
Tel: (703) 684-2470
Fax: (703) 299-0742
www.werf.org
werf@werf.org

This report was co-published by the following organizations. For nonsubscriber sales information, contact:

Water Environment Federation
601 Wythe Street
Alexandria, VA 22314-1994
Tel: (800) 666-0206
Tel: (703) 684-2452
Fax: (703) 684-2492
www.wef.org
pubs@wef.org

IWA Publishing
Alliance House, 12 Caxton Street
London SW1H 0QS, United Kingdom
Tel: +44 (0) 20 7654 5500
Fax: +44 (0) 20 7654 5555
www.iwapublishing.com
publications@iwap.co.uk

Library of Congress Catalog Card Number: 2004109529

IWAP ISBN: 1-84339-712-9 WEF ISBN: 1-57278-227-7 WERF ISBN: 1-89366-485-6

Tellus Institute

The research on which this report is based was funded, in part, by the United States Environmental Protection Agency through Cooperative Agreement No. CR-827345-01 with the Water Environment Research Foundation (WERF). Unless an EPA logo appears on the cover, this report is a publication of WERF, not EPA. Funds awarded under the Cooperative Agreement cited above were not used for editorial services, reproduction, printing, or distribution.

ACKNOWLEDGMENTS

The authors would like to thank the Water Environment Research Foundation and the U.S. Environmental Protection Agency for providing financial support for this project. Special thanks are due Margaret Stewart, WERF Project Manager, for her guidance and patience throughout the project. Professor Kenneth Strzepek of the University of Colorado helped conceptualize the approach and provided early direction. We would also like to thank the watershed experts (listed in Appendix D) who participated in the project workshop and provided invaluable insight and practical experience on a broad range of watershed management issues. In particular, we would like to acknowledge the substantial contributions of the international participants, who provided data and feedback on their respective watersheds for in-depth case studies, which are essential elements of this report. Finally, we would like to thank Sara Donahue of Tellus Institute for her detailed efforts in finalizing the report.

Report Preparation

Principal Investigators:
James Goldstein
Annette Huber-Lee, Ph.D.
Tellus Institute

Project Team:
Stephen Bickel
Julia Bowden
Tellus Institute

Project Subcommittee

Robin L. Autenrieth, Ph.D., Chair
Texas A&M University

William Booty, Ph.D.
National Water Research Institute (Canada)

Gary S. Bowen, P.E.
Toronto and Region Conservation Authority

Dan Cloak, P.E.
Dan Cloak Environmental Consulting

Tom Feijtel, Ph.D.
The Procter and Gamble Company

Jim Pendergast
U.S. Environmental Protection Agency

Water Environment Research Foundation Staff

Director of Research:	Daniel M. Woltering, Ph.D.
Senior Project Manager:	Margaret Stewart

ABSTRACT AND BENEFITS

Abstract:

Water resource management in the United States is evolving in the face of continuing challenges to protect water quality, provide adequate quantities of water for competing uses, and protect habitat and other natural resources. In many jurisdictions and agencies this evolution is increasingly leading toward adoption of watershed management. This approach is characterized by planning and decision making on a watershed scale, integration of a variety of competing water resource priorities and goals, cooperation of multiple stakeholders and governmental agencies, and increased levels of public participation. While the number and diversity of watershed management initiatives underway in the United States is impressive, successful transition to this integrated approach remains challenging due to institutional, regulatory, and information barriers.

In certain respects—Geographic Information Systems mapping, wetlands regulations, citizen participation—U.S. watershed initiatives are highly advanced and serve as models for efforts around the world. In other respects—cross jurisdictional coordination and cooperation, agreements on the sharing of resources, habitat protection and restoration— innovative approaches implemented outside the U.S. show great promise and offer important lessons to U.S. decision makers.

This report identifies the most promising watershed planning and management approaches from around the world; evaluates how they operate, their benefits and limitations; and assesses the degree to which these approaches could be successfully adapted to the U.S. context. Drawing on this international experience, the report is intended to inform policy makers and practitioners and to promote the implementation of integrated watershed management approaches that are most likely to succeed.

Benefits:

- Provides a decision-making framework of watershed management efforts at all scales in the United States
- Evaluates past U.S. watershed management experience and identifies key characteristics for success as well as major challenges and opportunities for improving the watershed approach.
- Summarizes and evaluates international case studies where innovative watershed management techniques have been used.
- Identifies 10 key lessons for sustainable water management, including the role of water/wastewater utilities based on the experience of the international case study watersheds.

Keywords: integrated water resource management, decision-making framework, competing uses, international experience, watershed management

TABLE OF CONTENTS

LIST OF FIGURES

LIST OF ACRONYMS

ACAP	Atlantic Coastal Action Program (Canada/North America)
ACT	Australian Capital Territory
AECV	Atlantic Estuaries Cooperative Venture (Canada)
AMP	Asset Management Plan (Great Britain)
ARD	Acid Rock Drainage
ARW	Association of the Rhine Waterworks
AWBR	Association of Waterworks, Lake Constance/Rhine
AwwaRF	American Water Works Association Research Foundation
BCEN	British Columbia Environmental Network
BMP	Best Management Practices
BOD	biochemical oxygen demand
BOR	Bureau of Reclamation (U.S.)
CALFED	CALFED Bay Delta Program (U.S.)
CARP	Clean Annapolis River Project (Canada)
CBA	Cost Benefit Analysis
CEE	Central and Eastern Europe
CERCLA	Comprehensive Environmental Response, Compensation, and Liability Act (U.S.)
CLF	Dutch Clean Water Foundation
CMA	Catchment Management Authority
CMP	Catchment Management Plan
CNA	National Water Commission (Mexico)
COAG	Council of Australian Governments
CORE	Commission on Resources and Environment (Canada)
Corps	Army Corps of Engineers (U.S.)
CSE	Centro de Soporte Ecológico (Mexico)
CTIC	Conservation Technology Information Center (U.S.)
CWB	Central Water Board (Tanzania)
CWMA	Catchment Water Management Act (Australia)
DEFRA	Department of Environment, Fisheries and Rural Affairs
DFO	Department of Fisheries and Oceans (Canada)
DLAA	Department of Land Affairs and Agriculture (South Africa)
DOE	Department of Ecology (Washington State, U.S.)
DOE	Department of Environment (Canada)
DPIE	Department of Primary Industries and Energy (Australia)
DWAF	Department of Water Affairs and Forestry (South Africa)
EA	Environment Agency (Great Britain)
EC	electrical conductivity
EOEA	Executive Office of Environmental Affairs (Massachusetts)
EPDRB	Environmental Programme for the Danube River Basin
ESA	Endangered Species Act (U.S.)
EU	European Union
FBC	Fraser Basin Council (Canada)
FBMP	Fraser Basin Management Program (Canada)

FBMB	Fraser Basin Management Board (Canada)
FERC	Federal Energy Regulatory Commission (U.S.)
FRAP	Fraser River Action Plan (Canada)
FREMP	Fraser River Estuary Management Program (Canada)
FRPA	Fraser River Port Authority (Canada)
FWS	Fish and Wildlife Service (U.S.)
GBI	Georgia Basin Initiative (Canada/U.S.)
GEF	Global Environment Fund
GIS	Geographic Information Systems
GONW	Government Office for the North West (Great Britain)
GRCA	Grand River Conservation Authority (Canada)
GVRD	Greater Vancouver Regional District
HRCC	Herbert River Catchment Coordinating Committee (Australia)
HRSTS	Hunter River Salinity Trading Scheme (Australia)
IAWR	Internationale Arbeitsgemeinschaft der Wasserwerke im Rheineinzugsgebiet (Rhine)
ICAS	Interstate Council for the Aral Sea
ICI	(Great Britain)
ICM	integrated catchment management (Australia)
ICPR	International Commission for the Protection of the Rhine
IFAS	International Fund for the Aral Sea
IPES	individual parcel evaluation system
IWRM	integrated water resources management
LRMP	Land and Resource Management Planning (Canada)
MBC	Mersey Basin Campaign (Great Britain)
MDB	Murray-Darling Basin (Australia)
MDBC	Murray-Darling Basin Commission (Australia)
MDBI	Murray-Darling Basin Initiative (Australia)
MELP	Ministry of Environment, Land and Parks (Canada)
MNR	Ministry of Natural Resources (Ontario)
MRC	Mekong River Commission
MOU	memorandum of understanding
NBI	Nile Basin Initiative
NEEF	Nechako Environmental Enhancement Fund (Canada)
NEEF MC	Nechako Environmental Enhancement Fund Management Committee (Canada)
NEP	National Estuary Program (U.S.)
NFPA	North Fraser Port Authority (Canada)
NGO	non-governmental organization
NLP	National Landcare Program (Australia)
NPDES	National Pollutant Discharge Elimination System (U.S.)
NRCS	National Resources Conservation Service (U.S.)
NRMS	Natural Resources Management Strategy (Australia)
NSW	New South Wales
NWC	Nechako Watershed Council (Canada)
NWDA	North West Development Agency (Great Britain)
OFWAT	Office of Water Services (Great Britain)
PBWB	Pangani Basin Water Board (Tanzania)
PCB	polychlorinated biphenyl

PIAC	Principal International Alert Center (Danube)
PRC	People's Republic of China
RAP	Rhine Action Plan
RIWA	Association of River Waterworks (Rhine)
RMA	Resource Management Act (New Zealand)
RMC	River Murray Commission (Australia)
RMWA	River Murray Waters Agreement (Australia)
RVI	River Valley Initiative (England)
SA	South Africa
SAGE	Water Development and Management Scheme (France)
SAP	Strategic Action Plan (Danube)
SDAGE	Masterplan for Water Development and Management (France)
SII	Sustainability Indicators Initiative (Canada)
SIP	Strategic Implementation Plan (Danube)
SRF	state revolving fund (U.S.)
TDR	(U.S.)
TMDL	Total Maximum Daily Load (U.S.)
TVA	Tennessee Valley Authority (U.S.)
UNDP	United Nations Development Programme
USDA	United States Department of Agriculture
U.S. EPA	United States Environmental Protection Agency
USFS	United States Forest Service
USGS	United States Geological Survey
WA	(Australia)
WEFTEC	Water Environment Federation's Technical Exhibition and Conference
WERF	Water Environment Research Foundation
WFW	Working for Water (South Africa)
WMC	Water Management Committee (Australia)
WQI	water quality index
WSSD	World Summit on Sustainable Development
WUA	Water User Association
WWTP	wastewater treatment plant

Executive Summary

ES.1 Background and Project Purpose

Watershed management approaches are increasingly being adopted to address the complex challenges facing water resource managers, including balancing competing water supply needs, maintaining water quality, and protecting natural ecosystems. The success of these approaches, however, has been limited. The Water Environment Research Foundation (WERF) sponsored this project to compile and evaluate existing information and case studies from around the world to help identify the most effective approaches for enhancing watershed management in the United States.

The report first reviews the context in which watershed management efforts have developed in the United States and characterizes their diverse motivations, scales, institutional structures, and approaches. Based on this review, it describes opportunities and gaps where practices could be improved. Finally, drawing on the review of international experience, the report identifies innovative approaches and key lessons for U.S. water resource managers. While the report concludes that there is no standard approach or blueprint for effective watershed management, the guidance provided is intended to inform policy makers and practitioners about various techniques that have been successful and to promote their adoption and implementation in appropriate integrated watershed management efforts in the United States.

ES.2 Research Approach

This project had the following five primary research tasks:

1) Conduct a literature review on the theory and practice of watershed-based management in the United States and abroad.
2) Prepare a compendium of international watershed management experience.
3) Develop in-depth case studies of leading international watershed management initiatives.
4) Conduct a workshop with U.S. and international watershed experts and practitioners to identify case study lessons and appropriateness for U.S. application.
5) Prepare a final report summarizing the key findings of the project.

ES.3 Context of U.S. Watershed Experience

The specific motivations and driving forces behind individual watershed initiatives in the United States are as varied as the watersheds themselves. Several major driving forces can be identified, but even these have varied in prominence by region and have changed over the years.

Since the early 1900s, increasing population, especially in the arid West, has led to water scarcities that have dramatically highlighted the need for a system of managing water resources at a scale that takes all users and areas of contribution into account. The federal environmental regulations born out of the environmental movement that began in the 1960s, particularly the

Endangered Species Act (ESA) and Clean Water Act, have had and continue to have a profound effect on watershed initiatives. For example, the threat of ESA listing is an extremely important driving force for watershed activities in the Pacific Northwest and elsewhere and has focused attention on and increased support for watershed initiatives in that region. More recently, Total Maximum Daily Load (TMDL) requirements have emerged as an element of the Clean Water Act that has the potential to force greater integration of point and nonpoint source pollution regulation and integration of both into watershed initiatives.

The watershed approach has provided an obvious and useful framework for handling each of the critical environmental and regulatory issues described above, including water scarcity, ecosystem health, compliance with ESA regulations and creation of TMDLs. Each of these issues has been or is becoming a significant driving force behind many of today's watershed initiatives.

ES.4 Characteristics of the U.S. Watershed Experience

One of the most noteworthy aspects of watershed initiatives in the U.S. is their diversity. There are nearly as many catalysts for formation, development paths, and organizational arrangements, not to mention scales, as there are watershed initiatives. The diversity of approaches arises in part from differences in priority environmental issues (e.g., nutrient pollution from agriculture, insufficient in-stream flows to sustain fish, or bacterial contamination impacting recreational uses) and the scale at which these issues can be managed. Issues addressed by watershed initiatives also differ from region to region, most notably from east to west, according to differences in climate, politics, water law and regulatory frameworks. Other characteristics of watershed initiatives are related to the particular institutions and individuals involved in a given watershed. Watershed programs have been sponsored by a multitude of government agencies at all levels, with a general devolution of authority from federal government to state and local agencies over time. Finally, broad stakeholder involvement and collaborative decision making are generally identified as key elements of today's watershed initiatives.

ES.5 Watershed Management "Opportunities"

Although the variety of watershed management initiatives throughout the United States have achieved great success at helping to manage water supplies, improve water quality, and address ecological and other environmental concerns, improvements to the watershed approach can be made in several areas.

- **Need for Coordinated Authority**

  While many U.S. watershed initiatives have successfully implemented cross-jurisdictional coordination, fragmentation of authority remains a significant impediment to *efficient* watershed management. In spite of attempts at systematic coordination, federal and state agencies (including water quality regulators, wildlife, public health, land management, forestry, and agriculture agencies) often operate independently in fulfilling their missions, which at times may be cross-purposeful. At the local level, water and wastewater utilities are often separate from one another, and from parks, land use planning, public health and conservation commissions. Moreover, such agencies often operate with very limited budgets and over-extended staff, and are not provided the direction or supporting framework to coordinate their activities.

This fragmentation of authority often inhibits watershed initiatives from operating as efficiently as possible. For initiatives to be successful, the array of local, state, and federal agencies with jurisdiction over watershed resources require regular guidance and assistance from a coordinated process and institutional framework that has been accepted by the stakeholders. When these agencies do not plan or implement their activities or provide information and assistance on a coordinated basis, watershed initiatives may suffer.

- **Dearth of Watershed Initiatives at the Largest Scales**

The absence of large-scale watershed management initiatives in most of the nation's river basins is a serious deficiency. These scales (approximately U.S. Geological Survey (USGS) accounting region and USGS resource region) are the appropriate ones for managing key water quality problems such as sedimentation, salinity, or nutrient loading, and many, if not most, water allocation problems can only be effectively managed at this scale. In regions where these are the principal problems and there is no large river-basin scale initiative, the impacts of local watershed efforts on ecological conditions are extremely limited. Where large-scale initiatives do exist, such as in the Chesapeake Bay watershed, they face additional challenges not faced by smaller scale watersheds:

*Effectively involving the full spectrum of stakeholders***.** The number of stakeholders impacted by the decisions of large-scale watersheds is extremely large, far beyond the optimal number for a collaborative process. Yet the interests and opinions of these stakeholders must be meaningfully included and fairly represented in planning and implementation activities. The relatively few very large watershed initiatives currently operating have developed a variety of solutions, but they have not been fully tested.

Inequity in cost and benefits among stakeholders in different parts of the watershed. In larger watersheds it is common that management decisions that provide benefit for all or part of the watershed do not distribute the costs similarly. This dichotomy is often seen between upstream and downstream communities. Expensive upstream measures that only marginally change water quality or quantity in the immediate area may be needed to prevent cumulative downstream impacts. Also, larger scale efforts generally do not enjoy the same inherent sense of community engendered by smaller scale projects, making voluntary sacrifices that much more difficult.

- **Integration of Large- and Small-scale Efforts**

Watershed management initiatives have been initiated in the United States on many different scales, ranging from small grassroots watershed initiatives focusing on several locally important issues, to initiatives encompassing entire river systems, crossing state boundaries and focusing on an array of regional issues, such as sediment loading and water allocation problems. Integration of these small and large-scale efforts can yield significant benefits, including 1) facilitation of successful project planning and implementation; 2) more effective involvement of the broad spectrum of stakeholders; and 3) reduction in the inequity of costs and benefits within a watershed.

- **More Use of Economic Instruments**

  Watershed initiatives in the United States today face the constant challenge of accomplishing their goals with limited and inconsistent funding. Obtaining more stable and reliable funding from government and non-government sources can help watershed initiatives more effectively plan for and utilize limited funds. At the same time, increasing the utilization of economic instruments, such as effluent trading, can help achieve watershed management goals without significantly reducing funds allocated to watershed planning.

- **Integration of Point and Nonpoint Source Pollution Management**

  To understand the ecological and water quality impacts of pollution and to develop cost effective management plans, watershed initiatives need reliable information about point and nonpoint source pollution. Historically, point and nonpoint source pollution management has not been well integrated at the federal and state level or in local watershed initiatives. With the recent implementation of TMDLs, this is starting to change, at least in some watersheds facing significant water quality problems. However, watershed planning and programs for the most part are still not well connected to local land use planning. Thus, watershed protection priorities tend to have little impact on growth management and regulatory decisions such as zoning, building design, and development choices that profoundly impact the hydrodynamics of the watershed, impervious surface area, stormwater runoff, and nonpoint source pollution.

  With TMDL development and implementation and increasing awareness of the connections between land-use planning and water quality (e.g., smart growth literature), watershed initiatives are in a unique position to focus on the integration of point and nonpoint source pollution management.

- **Improved Monitoring and Measurement of Watershed Conditions**

  Science-based decision making requires reliable, complete, long-term data to understand the dynamics of the watershed, assess environmental health, develop plausible management plans, and evaluate the impacts of actions taken. Unfortunately, in most cases, reliable, representative monitoring data on water chemistry, hydrology, and ecological composition and conditions for the necessary biological, chemical, and physical parameters do not exist. Despite 30 years of significant investment in water quality research and monitoring, U.S. data are incomplete, fragmented, and often not useful to watershed managers. The focus of most past monitoring efforts was to determine effluent quality. Past data on toxic contaminants are lacking or badly compromised. Different agencies have targeted different parameters, and agencies often use different analytical techniques that may no longer be considered acceptable or that have poor quality control procedures. Moreover, much of the data that do exist are stored in inaccessible formats. Thus, watershed planning initiatives often must make ecological management decisions with very incomplete pictures of the history, nature and dynamics of their watershed and the impact of human activities within.

  Fortunately, these problems have been recognized and a national level effort is underway to develop uniform sampling and analytical protocols, improve field and laboratory quality control procedures, and make data available in standardized formats. Many watershed initiatives have transformed their water quality and habitat data gaps into a means of engaging and involving citizens and volunteers. Such data needs can provide an opportunity to educate children, students, and the public about the concept of the watershed, presenting resource issues, the work of the watershed initiative, the scientific method, and various scientific disciplines.

While there are important data gaps that should be addressed to refine watershed planning and management efforts—absence of high-quality monitoring data, limited understanding of aquatic ecosystems—it must be noted that, relative to other countries, the United States is relatively data rich. The issue is really what level of information and certainty is required to begin implementing programs on a watershed level, and whether uncertainty or lack of detailed data is sometimes used in the United States to justify inaction. Lessons from international experience may be useful in this regard.

- **Multi-objective Frameworks**

  Increasingly, agencies involved with watershed management look for ways to simultaneously address the complex environmental and social concerns, multi-jurisdictional agendas, and resource limitations into a more integrated multi-objective perspective that meets the needs of all the stakeholders in the watershed. There are an increasing number of tools available to support these frameworks, including GIS in combination with other decision support systems. However, a multi-objective framework that allows stakeholders to examine tradeoffs and synergies in various options has rarely been used for real decision making. This may be due in part to the first of the 'opportunities' identified—the need for coordinated authority at the river basin or watershed scale. Without this institutional framework, a full examination of the possible objectives in a basin is unlikely to take place. Clearly, such ambitious planning requires involvement of more than water professionals; a successful community-based, multi-objective watershed initiative must involve representatives from as many different aspects of the watershed as possible.

- **Funding Challenges**

  Less then $1 billion in federal funding ($20 million per state), is allocated to grant programs specifically designed to assist watershed initiatives. Much more money indirectly funds programs through federal agency efforts, but this is still far short of the resource needs of local watershed efforts. States contribute proportionately little funding for watershed initiatives. Other sources, such as foundation grants, corporate donations, membership dues, and contributions from utilities, are also available. Available funds are often narrowly focused, leaving some important needs unmet, such as support for overhead and general operations. There is clearly a need for more total dollars with more flexibility to allow watershed efforts at all levels to implement the highest priority projects. Moreover, funding for watershed initiatives is generally unstable, with few programs having dedicated funding sources such as from hydropower revenue or water/wastewater surcharges. Stability of funding from year to year is extremely important as watershed initiatives can lose the ability to support key staff, resulting in the loss of both continuity and institutional memory.

ES.6 Lessons and Guidance for Sustainable Watershed Management

Based on a review of international watershed management experience and informed by five detailed case studies undertaken for this project, it is clear that the watershed approach provides a very different framework from traditional water management, one that starts with a geographic or natural resources rather than an economic or jurisdictional imperative. The natural resource constraints form the baseline for analysis in a watershed approach. Furthermore, the watershed approach necessarily implies the creation of an entity or cooperation among entities either with the legal or the de facto authority to plan, design, and implement decisions on watershed management. These decisions may include allocations in the watershed and among competing uses, what mechanisms to use to reduce pollution loads to the watershed, the imposition of water pricing or effluent charges, and changes in land use policy. The watershed approach with the appropriate authority facilitates the ability to shift priorities for water uses (e.g., between agricultural, municipal, and ecological uses) and increases the ability to manage the exigencies of nature in the form of droughts and floods.

A review of the literature on non-U.S. watershed approaches shows a consistent pattern of three key principles that are indispensable for effective watershed management:

1. There must be a clear decision-making framework that looks across scales.
2. Water development and management should be adaptive and based on a participatory approach involving users, planners, and policy-makers at all levels, with the authority to make decisions at the lowest appropriate level.
3. Instruments used to achieve effective management should recognize the economic value, as well as the social and environmental considerations of water in all of its competing uses, whether for ecosystem protection or restoration, for fisheries, for agricultural, industrial, municipal, navigation and recreational users, or for power generation.

The specific implementation of these principles necessarily differs based on the climate, scale and use of the watershed, as well as the politics, culture and social values of the watershed stakeholders. There is no blueprint for success, but it is clear that a combination of frameworks, institutions and instruments based on some degree of local decision making, support and funding is essential for successful watershed management.

Several lessons can be drawn from international watershed experience that may be appropriate for U.S. applications. Perhaps the most important lesson is that watershed management is a dynamic and evolutionary process that combines increasing local awareness of watershed issues with increasing local responsibility for action, where "local" will vary across the scale of the basin. Whether initiated as a result of public pressure or a government initiative, international examples of effective watershed approaches involve local citizens. In addition, a range of watershed stakeholders are engaged in capacity building that provides an understanding of the physical processes of the watershed as well as the social and economic value of water in its competing uses. These successes in watershed management established institutions to facilitate discussions and subsequent actions across jurisdictions, and to provide opportunities for meaningful stakeholder participation.

Watershed management authorities in the international context that have been reviewed here have some degree of self-sufficiency in funding, through a combination of water and pollution charges or resource-associated levies on stakeholders. In return, the local users have influence over decisions that may affect their use and/or enjoyment of the watershed. Watershed stakeholders and institutions (which vary according to local conditions) have used a variety of instruments to promote effective watershed management. Among the most important are the uses of:

- Economic instruments as a means of involving stakeholders and finding better, more efficient solutions to watershed issues;

- Regulatory instruments to facilitate cooperation among agencies across jurisdictions, and give decision-making authority to the most locally appropriate decision makers;
- Information and communication instruments, which can be combined to promote a common ground for discussion among stakeholders, both for competing users and various agencies; and
- Technology instruments, which have the potential to provide win-win opportunities in cases of conflict.

Finally, place matters—whatever institutions or instruments have proven to be useful, there is no standard approach or blueprint for effective watershed management. Rather, these elements must be tailored to meet the local conditions, such as climate, demographics, natural resources, existing uses, institutional, political, cultural, and economic context.

Following are the key lessons drawn from the detailed international case studies conducted for this project—the Mersey River in northwest England; the Fraser River in British Columbia, Canada; the Rhine River flowing through nine European countries from Switzerland to the Netherlands; the Grand River in Ontario, Canada; and the Murray-Darling River in southeast Australia. The watersheds that were the primary source for each lesson are listed in parenthesis. Appendix A includes the full case studies.

Lesson 1: Utilities play a critical role. (Mersey and the Rhine)

Utility service territories rarely follow watershed boundaries. However, utilities' withdrawal/discharge practices and their interest in providing high quality water and wastewater services, make them critical players in watershed planning and management. In cases where utilities cover relatively small geographic areas, such as a single municipality or small region, groups of utilities might form partnerships defined by watershed boundaries as a platform for joint planning and management. Activities could then be coordinated with local and regional watershed management initiatives, with water and wastewater utilities serving as important sources of technical and financial resources for watershed management.

Lesson 2: Multi-stakeholder processes provide a forum for effectively managing watersheds, including jointly addressing point and nonpoint pollution sources. (All)

One of the consistent obstacles to effective watershed management in the United States has been the difficulty in successfully engaging stakeholders and gaining their support for new approaches to manage the watershed. The credibility and success of a watershed initiative depends largely on the degree to which key interests and stakeholders participate throughout the process. Involvement at the earliest stages is critical, as there are numerous examples of the high costs and lack of success that can occur when stakeholders are not engaged early in the process. If important stakeholders only become involved in later stages, important interests may not be addressed and opposition and delays are more likely to result. Moreover, meaningful participation and open communication throughout the process provides the time necessary to build the trust among stakeholders and the institutional commitments required for the initiative to succeed. This trust is also built by creating a common understanding of watershed issues and challenges by providing all parties access to key data concerning watershed conditions.

The problem of nonpoint source pollution will only be solved with the ability to address decisions about changes in land-use and application of fertilizers and other chemicals to agricultural and non-agricultural lands. Historically, this has been difficult partly because the institutions governing land and water often do not share the same jurisdictional boundaries, or common mandates regarding the management and preservation of the watershed. Watershed initiatives that involve the full range of parties—from national to local government, and from environmental to agricultural and economic non-governmental interests—help break down some of these institutional barriers.

Watershed initiatives often help bridge the gap between government and non-governmental organizations (NGOs). In addition, international experience demonstrates the effectiveness of institutional arrangements that can lead watershed initiatives with a measure of independence, while still holding them accountable and ensuring they operate efficiently.

Lesson 3: Large-scale watershed management can succeed by cultivating and integrating discrete sub-watershed and stakeholder initiatives. (Mersey, Murray-Darling, and the Rhine)

Improving the effectiveness of watershed management initiatives in very large basins requires special effort to involve the full spectrum of stakeholders. One promising method is to develop individual institutions to serve subsets of stakeholders and to represent those stakeholders' interests within the larger watershed planning and management process. This may require creating organizational capacity dedicated to managing the nested smaller watersheds to ensure two-way communications and integration of sub-watershed concerns and activities with the larger watershed initiative.

Lesson 4: Integrative "win-win" methods support the resolution of upstream-downstream and human versus nature conflicts. (Murray-Darling, Fraser, Mersey and the Rhine)

Upstream-downstream and human-ecosystem tensions exist in many U.S. watersheds, particularly large ones. Creation of watershed institutions or working committees that involve both upstream and downstream users can help avoid open conflict and work towards mutually beneficial solutions. A formal consultative body provides a forum to continue negotiations and to resolve substantive and political barriers. This structure also promotes exploration of innovative options, such as cost sharing among upstream and downstream parties in cases where regulatory authority is lacking, or the provision of financial incentives from downstream interests to encourage upstream parties to modify policies and/or practices that degrade resources downstream. Examples of this approach can be seen in the efforts of several large U.S. utilities (e.g., in New York and Boston) to meet the requirements of the Safe Drinking Water Act by investing in upstream watershed protection measures rather than constructing new high-cost filtration plants.

Lesson 5: An engaged civil society can provide authority that may be lacking in the watershed organization. (Fraser, Mersey and the Rhine)

Watershed initiatives rarely enjoy direct executive or regulatory powers. Rather, their role is often to convene, facilitate, plan, and assess to inform decision making about policy and project implementation. In many of the international cases reviewed herein, the initiators of watershed management efforts lacked formal authority. Nonetheless, by effectively engaging key stakeholders, especially NGOs (or "civil society") and the public, watershed initiatives often gain de facto authority by influencing government and other key players to plan and implement their priority projects and policies. A cooperative relationship between participating governments and NGOs through watershed initiatives also avoids lawsuits and promotes negotiations among parties to find broadly acceptable solutions. Where watershed organizations in the United States do not have formal regulatory powers, the involvement of NGOs and the larger community may be a useful tool to create political support and achieve the goal of sustainable water management.

Lesson 6: Institutional stability and a clear mandate for watershed management can reduce fragmentation of authority and result in more efficient planning and implementation. (Grand and Murray-Darling)

Too often, fragmentation of responsibilities among several different agencies involved in watershed management leads to duplication of efforts, unclear mandates, and conflicting missions. Where governments have established clear frameworks and mandates for comprehensive watershed management, stable watershed institutions and planning processes often develop. This institutional stability is an important contributor to the long-term success of watershed initiatives. In some cases, enabling legislation can provide this mandate by specifying the structure of the watershed management institution, its roles and responsibilities, as well as its jurisdiction, membership, and funding. In other locations, government initiatives can promote collaboration or consolidation of watershed management activities. Alternatively, government funding programs can include eligibility criteria that foster coordination among multiple agencies and key stakeholders.

Lesson 7: Instilling regulatory authority in a watershed-based institution can facilitate effective watershed protection across political boundaries. (Grand and Murray-Darling)

While virtually all states and municipalities have laws and regulations to protect water and other natural resources, their implementation and enforcement is rarely carried out on a watershed-wide basis or by watershed-focused organizations. Regional planning entities generally lack this authority, particularly in terms of land-use decision making. From the international arena, some laws or regulations provide direct authority to watershed organizations to operate across political boundaries. In such cases, traditional jurisdictional borders are irrelevant when considering the natural resource impacts of a proposed activity. With this authority, watershed-based organizations can protect the watershed by adopting regional environmental standards or issuing permits for development based on water quality and other environmental sensitivity criteria. This can result, for example, in the creation of greenways along riparian corridors, or the reduction of impervious surfaces and downstream flooding impacts. Although politically difficult to adopt in much of the United States due to local resistance, the benefits could be significant, resulting in vastly improved administration and enforcement of resource protection regulations. Municipal boundaries would no longer limit the view of costs and benefits. Mechanisms to ease adoption of this approach may exist. These may include establishing model watershed-based regulations for implementation at the local level and providing incentives for local adoption and implementation, such as preferred access to state revolving funds or other grants for planning and improvement projects.

Lesson 8: Explicit policies and guidance documents can be used to promote the integration of watershed and land-use planning at the local level. (Grand)

Issues of water supply, wastewater assimilation, groundwater recharge and local ecology are critical to the sustainability of localities and their population. While most municipalities have local zoning ordinances and periodically engage in planning, most tend to focus on land use and zoning, and are not integrated with watershed management plans. Even where a detailed watershed plan has been developed, essential pieces may not be referenced in the local land-use plan. Directly linking municipal land-use planning and watershed planning processes can be accomplished through an array of regulatory, policy, and educational efforts. Where watershed plans have been developed and approved by state or local agencies, local land use planners, planning boards, and boards of zoning appeal should be directed or encouraged to incorporate relevant elements into local planning documents. Similarly, local planning officials could commission a sub-watershed study prior to their review and permitting of large-scale development projects under local planning regulations. Because local approval of and commitment to the watershed plan and its recommendations are essential to its integration with land-use plans, local municipal officials should be directly involved in the watershed planning process. Clear guidance for the integration of watershed planning with local land-use planning could be provided by states, along with incentives for local follow through. Incentives could include state grants and/or assistance for projects in municipalities that have incorporated elements of approved watershed plans into municipal policies.

Lesson 9: A system of apportioning costs and benefits equitably across the watershed can help secure consistent municipal participation and funding. (Grand and Murray-Darling)

Because watershed management activities naturally take place at the local level, they frequently require the acceptance and involvement of municipal officials. Often, recommended watershed improvement projects, such as separation of combined storm-sewer systems, require substantial support from municipal agencies through long-term dedication of funds and disruptive, on-the-ground implementation efforts. For this reason, it is critical to involve municipal decision-makers in the process of watershed planning from the outset. One way to engage and secure municipal participation is to establish a system for equitable distribution of watershed management costs and benefits. A common example is downstream water users offering assistance to upstream communities in the adoption of water conservation and water quality protection measures. Regional utilities could play an instrumental role in the coordination and balancing of municipal contributions to watershed management projects and ensuring that the costs and benefits are distributed equitably.

Lesson 10: Watershed decision making at the lowest appropriate level is most effective. (Fraser and Murray-Darling)

Coordinated management of large-scale watersheds, such as the Mississippi or the Colorado, is notably lacking in the United States. One possible explanation is the concern at smaller scales of a loss of decision-making authority. To allay such concerns and to encourage decision-making by the most affected parties, decisions should be made at the lowest level appropriate for a given issue. For issues with basin-wide implications, policy decisions should be made at the basin-wide level. Specific implementation issues and concerns that impact a limited part of the basin should be made on a more local, sub-watershed basis.

Conclusions

These 10 lessons have potential for broad application in the U.S. While drawn from the five more detailed case studies, they are further supported by the examination of 16 additional watershed approaches reviewed in the compendium in Appendix C. While some of the lessons would require the creation of new institutions and/or granting new regulatory authority, several can be implemented within existing frameworks. Certainly utilities can be a driving force for positive change in any of the institutional and regulatory frameworks currently in place in the U.S.

CHAPTER 1.0

INTRODUCTION

1.1 Project Purpose

Water resource management in the United States is evolving in the face of continuing challenges to protect water quality, provide adequate quantities of water for competing uses, and protect habitat and other natural resources. In many jurisdictions and agencies around the United States, this evolution is increasingly leading towards adoption of watershed management—an approach characterized by planning and decision making on a watershed scale, integration of a variety of water resource priorities and goals, cooperation of multiple stakeholders and governmental agencies, and increased levels of public participation. While the number and diversity of watershed management initiatives currently underway in the United States is impressive, a successful transition to this integrated approach remains challenging due to institutional, regulatory, and information barriers.

In certain respects (e.g., Geographic Information Systems [GIS] mapping, wetlands regulations, citizen participation) U.S. watershed initiatives are highly advanced and serve as models for efforts around the world. In other respects (e.g., cross-jurisdictional coordination and cooperation, agreements on the sharing of resources, habitat protection and restoration) innovative approaches implemented outside the United States show great promise and offer important lessons to U.S. decision makers.

The purpose of this report is to identify the most promising watershed planning and management approaches from around the world; synthesize information about how they operate, their benefits and limitations; and assess the degree to which these approaches could be successfully adapted to the U.S. Drawing on this international experience, the report is intended to inform policy makers and practitioners, avoid unsuccessful efforts, and promote implementation of integrated watershed management approaches that are most likely to succeed. In addition to this report, broad dissemination of the findings from this project is intended to reach not only utility planning personnel, but also local and state agencies, and other stakeholders with an interest in improving watershed management practices.

Increasingly, water and wastewater utilities are recognizing the importance of integrated watershed planning and management approaches and the critical role they can play in helping to lead or shape them. As issues such as water supply limitations, threats to water quality, and habitat deterioration become the focus of multi-stakeholder watershed management efforts, utilities must look beyond water supply and wastewater management issues and engage in these broader concerns in order to successfully provide for their customers in the long term.

1.2 Research Approach

This project had five primary research tasks.

- *Conduct a literature review on the theory and practice of watershed-based management in the United States and abroad.* This is summarized in Chapter 2.0 and documents the progression of environmental regulation and natural resource policy towards adoption of a watershed approach and the status of the application of such approaches in the United States. It also identifies the challenges and information gaps hindering successful implementation of watershed management approaches in this country.
- *Prepare compendium of international watershed management experience.* Developed by the project team to provide a background on international experience and identify potential in-depth case studies, the compendium (Appendix C) reviews more than 20 watershed initiatives in more than 15 countries, including four multi-national cases.
- *Develop in-depth case studies of leading international watershed management initiatives.* Based on the research conducted for the preparation of the compendium as well as follow-up research and communications with key stakeholders, the project team conducted five in-depth studies on the following basins: the Rhine (a multi-national river basin in nine European countries), the Mersey (United Kingdom), the Fraser (British Colombia), the Grand (Ontario), and the Murray-Darling (Australia). Appendix A provides details of these case studies.
- *Conduct a workshop with United States and international watershed experts and practitioners to identify case study lessons and appropriateness for U.S. application.* In conjunction with the 2001 Annual Water Environment Federation's Technical Exhibition and Conference (WEFTEC), a workshop was held on October 14, 2001, in Atlanta, Georgia. Approximately two dozen participants discussed the status of watershed management practices in the United States and reviewed the lessons from the international case studies. Feedback from this workshop has been incorporated into this report.
- *Prepare a final report summarizing the key findings of the project.* Draft reports on each of the preceding tasks were circulated to the Water Environment Research Foundation (WERF) Project Advisory Committee and selected others for review. These reports and the feedback received have been synthesized into this report.

1.3 Background

Natural resource and environmental managers in the United States are increasingly embracing a watershed-based approach in their work. Regulatory agencies, including the U.S. Environmental Protection Agency (EPA) and the Natural Resources Conservation Service (NRCS) in particular, are leading efforts at the national level to encourage inter-agency cooperation in planning natural resource management at the watershed level. Numerous states have also made efforts to encourage resource management organized by watershed, some of which focus on the need for coordination among various jurisdictions and stakeholders.

What is the impetus for these efforts? Many regions and watersheds in the United States are facing increasing pressure from population growth, development, water use/wastewater generation, and ecological stress. Planners and resource managers have begun to recognize that current practices are not sustainable from either a water quantity or a water quality perspective. This recognition and the need for sustainable resource management has been widely accepted and documented, not only by WERF, the American Water Works Association Research Foundation (AwwaRF), the President's Council on Sustainable Development and countless other U.S. organizations. Leading international bodies, such as the World Commission on Environment and Development (in its 1987 report, *Our Common Future*, also known as the Brundtland report), the International Conference on Water and the Environment (culminating in the Dublin Principles in 1992), the U.N. Conference on Environment and Development (in its 1992 report, *Agenda 21*), the World Water Forum held in the Netherlands in the spring of 2000, and most recently, the World Summit on Sustainable Development (WSSD) at its 2002 meeting in Johannesburg, also advocated this approach.

An integrated approach to watershed planning and management is a prerequisite to effective and sustainable delivery of competing watershed services and protection of the environment. Despite the impressive progress in the United States over the past decade in terms of broad recognition of the importance and value of integrated watershed approaches, there continue to significant challenges to realizing the watershed agenda in practice. One issue is institutional fragmentation in planning and operations. It is common, for example, to find multiple water and wastewater utilities planning and operating in a single watershed without the benefit of an overall coordinated watershed or river basin planning framework. Similarly, notwithstanding recent federal stormwater regulations, many environmental regulators continue to focus on point source pollution, even though there is strong evidence that nonpoint sources are often more important. Finally, opportunities for meaningful stakeholder involvement in planning and resource management decision-making processes are often inadequate.

A successful integrated watershed approach faces resource management issues in a coordinated, multi-jurisdictional manner. In the U.S., the regulatory framework is evolving from a point source focused approach based strictly on federally mandated criteria, as discussed in Chapter 2.0. Full implementation and real success will require new tools to integrate a range of issues aside from water quality, such as habitat restoration, land use, biodiversity, and equitable use of shared resources. A successful watershed management effort will require input from regional stakeholders and a better articulated governance structure that integrates national and regional criteria with local interests. To enhance this shift, the United States will benefit from a better understanding of how similar challenges have been addressed in other countries that have overcome jurisdictional roadblocks or conflicting stakeholder interests. Some examples of international approaches are presented in Chapter 3.0, with a summarized guidance provided in Chapter 4.0.

CHAPTER 2.0

CONTEXT OF U.S. WATERSHED EXPERIENCE

2.1 Driving Forces for the Watershed Approach

Increasingly, management of human activities and natural resources based on watershed boundaries has become accepted practice among non-governmental organizations (NGOs), states, and the federal government. The specific motivations and driving forces behind individual watershed initiatives in the U.S. today, however, are as varied as the watersheds themselves. Several major driving forces can be identified, but even these forces have varied in prominence from region to region and have changed over the years.

For decades, increasing population, especially in the arid West, has led to water scarcities that have dramatically highlighted the need for a system of managing water resources at a scale that takes all users and areas of contribution into account. In the 1960s, the emergence of the environmental movement, combined with the creation of new federal and state programs and agencies designed to protect natural resources (most notably the U.S. EPA in 1970), further set the stage for the transformation of water resource management in the U.S. Indeed, federal environmental regulations born out of this period, particularly the Endangered Species Act (ESA) and Clean Water Act, have had a profound effect on watershed initiatives and are likely to continue to do so. For example, the threat of ESA listing is an important driving force for watershed activities in the Pacific Northwest and elsewhere, and has focused attention on and increased support for watershed initiatives in that region. More recently, total maximum daily load (TMDL) requirements have emerged as an element of the Clean Water Act that has the potential to force greater integration of point and nonpoint source pollution regulation and better integration of both into watershed initiatives.

The watershed approach has provided an obvious and useful framework for handling each of the critical environmental and regulatory issues described above, including water scarcity, ecosystem health, compliance with ESA regulations, and creation of TMDLs. Each of these issues has been or is becoming a significant driving force behind many of today's watershed initiatives and each is described in more detail below.

2.1.1 Increasing Scarcity of Water

Annual precipitation patterns have had perhaps the greatest influence on the development of watershed initiatives in the United States. Water supply and allocation issues have been at the forefront of water management initiatives in the arid West for over a century. Meanwhile, the Pacific Northwest, Great Lakes, and East Coast states have traditionally had an abundance of water and have been able to focus more directly on water quality issues. Until recently, water quantity in these regions was rarely an issue.

However, water consumption has increased significantly throughout the U.S., and water scarcity issues no longer belong solely to the arid West. Indeed, throughout "water rich" regions, there are examples of water use patterns that significantly reduce stream flow, lower water tables, impair water quality, and significantly impact local ecological systems and economies.

Washington, one of the most heavily watered states, has recently embraced significant watershed management legislation intended to help address water allocation conflict issues. Meanwhile, the more arid western states, such as Arizona, New Mexico, and Colorado, which have historically had water scarcity problems, face still greater concerns as they experience tremendous growth, with projections of a 30% increase in population by 2020. In these states and others, water supply and allocation issues have been and will continue to be core issues for watershed initiatives.

2.1.2 Increasing Importance of Ecosystem Concerns

The problems of toxic pollution and environmental degradation entered public awareness in the 1960s with the publication of Rachel Carson's *Silent Spring* and a number of high profile environmental disasters. Public concern translated into federal concern and led to the passage of a series of federal pollution control laws, including the Water Quality Act of 1965 and the Clean Water Act of 1972. The Clean Water Act was the first major piece of legislation to provide new power and financial resources to the U.S. EPA. It set a goal that all U.S. waters should be fishable and swimmable by 1985. States soon created parallel environmental agencies at the state level with similar missions and programs. Initially, these programs validated public concern by including natural resource degradation, water quality, and habitat preservation, including in-stream flows, as competing uses and providing some regulatory tools to protect these uses. Later, the ecological perspective led to recognition of the need for a more holistic ecosystem and watershed-based approach to management of this inherently integrated system.

The introduction of ecological concerns to water supply concerns in the 1960s and 1970s added a critical new dimension to integrated resource management and remains at the core of the modern watershed management approach. Concern for ecological health is evidenced today by the growing number of watershed initiatives that include land conservation, habitat preservation, wetland restoration, and management of invasive species among their chief activities. Watershed groups routinely apply for an array of state, federal and foundation grants available for ecosystem research, management and preservation. This integration of water quality, water supply, and ecological health within a broad watershed management framework was rare just a few decades ago, but has now achieved firm and growing support across the U.S.

2.1.3 Endangered Species Act Compliance

Endangered species listings and recovery plans have helped catalyze the formation of local and state watershed efforts, heightened recognition of existing watershed initiatives, and prompted state and federal governments to increase the resources available to these efforts. For example, the Southwest Oregon Salmon Recovery Initiative, motivated by the increasing likelihood of the listing of coho salmon as an endangered species, is organized on a watershed basis and is designed around the use of watershed councils as the "backbone of habitat recovery efforts." This local salmon recovery initiative has increased the profile and resources available to watershed initiatives throughout the state. In fact, it has been suggested that the threat or fear that a species will be listed as threatened or endangered may be more beneficial to watershed efforts than actual listing. Prior to the actual listing of a species on the endangered list, efforts to protect it may involve greater local control and discretion and participation by stakeholders and local watershed groups. Actual listing, it is argued, places the major planning and decision-making activities directly under federal control and allows less involvement at the local level. In both cases, however, species recovery efforts tend to support an integrated, watershed-based, multi-stakeholder approach to environmental management.

2.1.4 TMDL Requirements of the Clean Water Act

TMDL allocations, a newly prominent regulatory element of the Clean Water Act, are likely to have a strong influence on watershed initiatives. Section 303 of the Clean Water Act requires states to write plans for meeting TMDLs of the key pollutants in all impaired water bodies, in order to bring the waterbodies in line with state water-quality standards and use designations. Key pollutants can be chemical, biological, physical, or thermal. Thus far, more than half the states have been required to write TMDL implementation plans, and U.S. EPA intends to require all states to do so in the near future.

TMDL regulations have the potential to assist watershed initiatives by helping to overcome two of the most significant pollution management challenges: 1) the integration of point source permitting programs with watershed initiatives; and 2) the integration of point and nonpoint source pollution management programs. Because TMDLs look comprehensively at the daily load conditions of each water body and do not distinguish between different sources of the same pollutant, point source permitting programs may be forced to coordinate efforts with nonpoint source management programs. Watershed initiatives are an excellent vehicle for facilitating this coordination and, arguably, offer the most viable means of actually implementing TMDL plans.

2.2 Characteristics of the U.S. Watershed Experience

One of the most noteworthy aspects of watershed initiatives in the U.S. is their diversity. There are nearly as many catalysts for formation, development paths, and organizational arrangements, not to mention scales, as there are watershed initiatives. The diversity of approaches arises in part from differences in priority environmental issues (e.g., nutrient pollution from agriculture, insufficient in-stream flows to sustain fish, or bacterial contamination impacting recreational uses) and the scale at which these issues can be managed. Issues addressed by watershed initiatives also differ from region to region, most notably from east to west, according to differences in climate, politics, water law, and regulatory frameworks. Other characteristics of watershed initiatives are related to the particular institutions and individuals involved in a given watershed. Watershed programs have been sponsored by a multitude of government agencies at all levels, with a general devolution of authority from federal government to state and local agencies over time. Finally, broad stakeholder involvement and collaborative decision-making are generally identified as key elements of today's watershed initiatives.

As described above, the major characteristics of the U.S. experience in watershed management include: 1) scale and boundary conditions; 2) recognition of different climatic conditions and legal/regulatory frameworks in the east and west; 3) the multitude of government agencies involved; 4) a general devolution of authority from federal to state and local agencies; and 5) increasingly broad stakeholder participation. Each of these characteristics is described below.

2.2.1 Scale and Boundary Conditions

Just as watersheds come in different scales, so too do the initiatives to protect and manage their resources. Data on watershed management activity in the western states and collections of case studies and reports that include the eastern half of the country indicate that most watershed initiatives operate in basins of 500 square miles or smaller. Some focus on extremely small watersheds. Efforts to coordinate activities in watersheds of several thousand square miles are not uncommon, but there are only a small number working in regions of tens or hundreds of thousands of square miles in area.

Defining scale according to geographic size is somewhat misleading, however. Scale in terms of population and relative size of political units is an equally important factor in how watershed initiatives function. Smaller watershed initiatives in more densely populated watersheds or eastern states with smaller municipal units may face as complex a collection of stakeholders as collaborative efforts in much larger but more sparsely populated basins in western states with geographically larger towns and counties.

Nevertheless, the geographic scale of a watershed initiative does generally tend to influence its membership. Smaller scale initiatives are more likely to include active participation by citizens, individual landowners and small businesses. Watershed efforts at larger scales, on the other hand, are inherently more complex, encompassing many more political units at higher levels and larger stakeholder groups. As a result, large-scale efforts tend to have proportionately fewer local citizens and small business participants.

To some extent, scale also plays a role in determining the number and breadth of environmental issues that can be successfully addressed by watershed initiatives. Correspondingly, priority environmental issues may dictate the required scale for watershed management initiatives. For example, to control nutrient pollution in the Chesapeake Bay, the Chesapeake Bay Program has coordinated activities on a large scale, encompassing three states and the District of Columbia. However, to implement their nutrient pollution prevention plans, the Chesapeake Bay Program has enlisted the assistance of several smaller, community watershed initiatives working in the Chesapeake Bay's sub-watersheds. This integration of the efforts of small- and large-scale watershed initiatives within a single watershed is critical to successful watershed management planning and implementation. Unfortunately, such integration has been lacking in many regions.

2.2.2 Differences Between East and West: Climate, Politics and Law

Just as climates, landscapes, and human histories vary across the United States, so too do environmental issues. Flooding problems due to heavy rains and a history of poor development practices may be the primary watershed management issue in some regions, while cleaning up rivers heavily polluted by industry may be the focus in others. Some regions must frequently act to protect endangered species' habitats, while others may simply need to protect already healthy watersheds from the detrimental effects of increasing development. Regional differences in politics and law can also affect watershed management initiatives, influencing the way in which water and water-related resources are governed and managed.

Arguably the most distinct differences in watershed management strategies occur between the eastern and western states. Although these differences are related both to hydrologic (i.e., climate and rainfall) and legal/regulatory issues, they can best be described through a discussion of water rights.

In the United States, water has generally been treated as property, with the nature and allocation of water rights governed by state and local laws according to three basic doctrines: riparian rights, appropriation rights, and permitting. Generally speaking, riparian rights are found in the eastern and Great Lakes states and appropriation rights in the western states. Permitting is a relatively new innovation and generally applied in conjunction with either riparian or appropriation rights.

2.2.2.1 Riparian Rights

Riparian law has been adopted by states with a relative abundance of water, such as many of the eastern states, and functions most effectively under these conditions. Under the riparian system, anyone who owns land adjacent to a body of water has rights to "reasonable use" of that water. All users have equal rights and are prohibited from degrading the quality of the water. Additionally, water may not be transferred out of the basin, so that downstream users benefit from return flow. Water rights are generally transferred with the land title, but can sometimes be sold separately. The primary method for dealing with scarcity is to require all users to reduce consumption equally.

Increasing demand for water and the resulting scarcity in states applying riparian law has lead many to overlay permitting systems. Permits tend to restrict the riparian rights, allowing a determined allocation of rights to withdraw defined quantities of water. These permits can be written and policed to allow for additional restrictions during emergencies. Permitting systems may be considered favorable for watershed initiatives because they allow for rational and coordinated control of water withdrawals by state government, whose interests and objectives are often compatible with those of watershed initiatives.

2.2.2.2 Appropriation Rights

Historically, access to water in the western states was essential for economic survival. The first settlers were farmers, ranchers, and miners. All of these occupations depend on water (the first miners used a hydraulic technique to mine surface deposits trapped in soil). These settlers adopted a water rights system that gave the earliest settlers the greatest security. Under the appropriation rights system, water rights are established by making a public claim to a specific quantity of water for a "beneficial use." The date of the claim establishes seniority ("first in time, first in right"), and senior appropriators are permitted to withdraw their full allotment even if that means no water is left for more junior appropriators.

Several features of the appropriation rights system have important implications for watershed initiatives. Protecting habitat in many cases requires increasing in-stream flows and regulating their volume. When watershed initiatives in states operating under appropriation rights systems have this objective they must purchase the necessary water rights. While in theory appropriation water rights can be sold, in practice this is often very difficult.

A second feature of appropriation rights systems is that water can only be appropriated for an accepted "beneficial use," such as domestic use, agriculture, or industry. Many western states have not expanded their definition of beneficial use to include wildlife/habitat protection or in-stream flow. In these states, water rights cannot be purchased for these ecological uses even if a watershed initiative has a seller and the transaction is practicable.

Further restricting the options of watershed initiatives in western states, the appropriation system typically operates under a "use it or lose it" principal, such that appropriators who do not use all of their allotment in a given year will forfeit the unused rights. This, of course, encourages consumption, discourages conservation, and creates even less flexibility for apportioning water for fish and other habitat purposes.

2.2.3 Multitude of Government Agencies Involved

An important characteristic of watershed management initiatives in the United States is the complex network of federal, state, and local government agencies involved. Indeed, all recent reviews of watershed management activities in the United States highlight that federal and state agency involvement is critical to the success of local watershed initiatives. In addition, federal and state government participation can be viewed as a form of enlightened self interest, in that federal and state agencies have discovered that watershed initiatives offer an efficient means of fulfilling their mandates. As a result, an array of both federal and state programs has sprung up around the country to help further the goals of watershed management initiatives.

2.2.3.1 Federal

The federal government's substantial authority to regulate water-related activities, founded on constitutionally based mandates to oversee navigability, hydroelectric power generation, defense, disposal and management of federal lands, and provision for the general welfare, has spawned 19 federal agencies that oversee some dimension of water management and are therefore directly relevant to watershed management initiatives. The most important of these agencies are the Army Corps of Engineers (navigation, flood control, and more recently, habitat restoration), the Fish and Wildlife Service (species and habitat protection), the U.S. EPA (water quality), the Natural Resources Conservation Service (NRCS; flood and erosion control, natural resource protection), the Bureau of Reclamation (water supply for drinking and irrigation), the Federal Energy Regulatory Commission (hydroelectric dam construction and power generation), and the National Oceanic and Atmospheric Administration (coastal zone management). The U.S. Geological Survey (USGS; hydrologic and data collection and analysis) is an important non-regulatory agency.

In the simplest terms, these agencies offer watershed planning and management initiatives four important things: money, permits, information, and interactive technical support. Watershed management activities in the United States take place within the context of thousands of pages of regulations generated by these federal agencies. The authority to accomplish one or more objectives of a watershed initiative frequently lies with one or more of these agencies and/or a state counterpart. This extensive federal involvement in regulating water and water-related resources means that federal agencies frequently participate in watershed management efforts.

Of course, all agencies are not relevant to all watershed initiatives. Some are only important in certain geographic areas or for certain issues. For example, the Bureau of Reclamation manages lands and projects west of the Mississippi River, and the Federal Power Administration is relevant only on rivers with hydropower installations. The NRCS is involved only when issues are related to commercial agricultural, range, or forest lands.

Overall, federal agencies and watershed initiatives have found cooperation mutually beneficial. Watershed initiatives have turned out to be extremely efficient vehicles for fulfilling agency missions, and agencies have provided watershed initiatives with resources and institutional support to implement their plans.

2.2.3.2 State

States have generally created sets of regulatory agencies with parallel functions to many federal agencies. These state agencies then influence watershed initiatives in a similar fashion to their federal counterparts. Federal agencies in some cases have delegated authority to enforce particular laws and regulations, such as the National Pollutant Discharge Elimination System (NPDES) regulations of the Clean Water Act, to the state agency counterparts. State governments have a variety of responsibilities over water-related resources, including conducting studies, devising TMDLs, monitoring water quality, issuing and managing water quality permits for point sources of pollution, addressing nonpoint source pollution, regulating water flows, managing fisheries and wildlife habitats, insuring safe drinking water, and dealing with water quantity allocation issues (the latter, primarily in the west, but increasingly in eastern states as well).

In addition, state regulations may provide watershed organizations with additional regulatory tools and funding mechanisms for pursuing their objectives. For example, some states have taken advantage of their license to enact environmental legislation that is more stringent than the federal government's and created regulations and programs that empower them to regulate nonpoint source pollution.

Unfortunately, the unique nature of how these state agencies are structured and their differing authorities often make innovations in watershed approaches more difficult to transfer between states. It may also be difficult to coordinate management of a single watershed that bridges two or more states with different state agency structures and authorities. Since rivers often serve as the border between states, there are many such watersheds in the United States that may encounter this problem.

2.2.4 Devolution of Authority

While federal policies and programs have played and will continue to play an important role in water resource management and the evolution of a watershed management approach, the greatest innovations in recent years have been at the state and local levels. As federal powers have generally devolved to the states over the past 20 years, states have developed the institutional and technical capacity to carry out their growing responsibilities. The gradual shift in authority has also prompted state-level experimentation and innovation in watershed management.

State experiments in integrating and promoting watershed approaches are wide ranging. A number of states have formally adopted a watershed approach, while others have not. Massachusetts, Washington, and Oregon are at the forefront of refocusing their land and resource management programs within a watershed framework. Doing so requires establishing the watershed as the organizational framework for one or more aspects of water resource management, including integrating or coordinating formally independent agency policies; regulatory activities relating to point and nonpoint source pollution control and related permitting; fish and wildlife conservation activities; forest management; and parks, soil conservation, agriculture, and transportation activities.

Leading states have provided additional and targeted funding for watershed efforts and have relocated agency staff to watersheds, often assigning them responsibility for leading or supporting watershed partnerships. The applicability of the successful models to other states is partly dependent on underlying legal/regulatory frameworks, climates, and political conditions, but individual elements of these models are likely to have broad transferability.

Massachusetts is regarded as having made the most significant institutional changes and has introduced a number of significant innovations. Massachusetts began with a somewhat advantageous institutional structure in which all state environmental agencies, including the departments of environmental protection, environmental management, and fisheries and wildlife, were overseen by a single entity, the Executive Office of Environmental Affairs (EOEA). The state has restructured the EOEA and its departments to better accommodate the watershed-based management approach. The state has created an Office of Watershed Management, which coordinates the Massachusetts Watershed Initiative and the activities of basin teams within the state's 27 USGS cataloging unit scale watersheds. Basin teams are led by full-time team leaders and include federal and state agency personnel, municipal representatives, businesses, local watershed association members, and other relevant stakeholders. The teams strive to develop broad, consensus-based decisions regarding basin planning, implementation, and assessment, as well as foster local support and capacity for watershed management activities. Moreover, the state coordinates its permitting and grant funding programs on a watershed basis.

In virtually all watershed initiatives, local and regional governments and other stakeholders are recognized as being closest to the issues facing watershed management and therefore play significant roles.

2.2.5 Stakeholder Participation

One of the defining characteristics of the modern watershed approach and one of the essential ingredients for success is broad stakeholder participation. The number of stakeholder groups actively participating in decision making has practical limitations, although watershed initiatives with 50 to 150 partner entities, such as the Bronx River Working Group and the Illinois River Watershed are not unheard of.

The many stakeholders involved in watershed initiatives tend to come with many different agendas, requiring an effective strategy for collaboration and decision making. This is especially noticeable with regard to competing uses for limited water supplies. The National Academy of Public Administration's review of case study reports identified five common elements of effective collaborative watershed processes: crisis, inclusion, rules, communication, and time. The academy found that:

- Collaborative processes work best when the situation is perceived as a crisis.
- Including of all stakeholders, and giving them meaningful opportunities to effect decisions, is essential.
- Processes need clear rules to operate smoothly.
- Participants need an explicit strategy to maintain communication with the organizations they represent.
- Building trust and mutual understanding usually takes a great deal of time.

Achieving broad stakeholder participation that is perceived as meaningful by participants is a considerable task for many watershed initiatives, especially for those operating in large watersheds. However, the evidence of its value is plentiful. Large watershed initiatives can learn from the examples of others. For example, the CALFED Bay Delta Program (CALFED) process in California uses numerous sub-committees and holds frequent public meetings to manage input from an array of stakeholders. While resource intensive, this strategy is reported to be working well.

In addition to variation in the number of participating stakeholder groups, the types of stakeholder groups also vary with project phase, location, and priority issues. Three quarters of western watershed initiatives surveyed included at least one federal agency representative. Eighty-five percent of the time, that representative was a member of the NRCS. State agencies, local governments, water districts, utilities, businesses, non-profits, foundations, universities, academics, citizens and others are all found in any number of combinations in U.S. watershed initiatives. CALFED and the Phalen Chain of Lakes Watershed Project illustrate the variability of stakeholder groups in watershed programs.

2.2.5.1 CALFED Bay Delta Program

CALFED is a partnership of federal and state natural resource agencies. It was established in 1995, following resolution of long-running disagreements between state and federal agencies on an array of issues arising out of inherent conflicts between protecting resources such as water quality, quantity, and endangered species. Established via a memorandum of agreement, the program now has direct authority to regulate, enforce, or make funding decisions. In practice, participating agencies accept its recommendations and act on its behalf. The program has its own staff and borrows staff from participating agencies. The program seeks to meet multiple objectives: providing good water quality for all beneficial uses, improving and increasing aquatic and terrestrial habitats, reducing the mismatch between water supplies and beneficial uses, and reducing damage to human and natural resources from levee failure. The program has undertaken a comprehensive, watershed-based resource assessment and analysis and employs coordinated planning process, including numerous task forces and advisory committees and hundreds of public meetings, designed to include all relevant stakeholders.

2.2.5.2 Phalen Chain of Lakes Watershed Project

The Phalen Chain of Lakes Watershed Project is governed by a citizen-based steering committee composed of local elected officials and commissioners, lakeshore owners, business interests, environmental organizations, and neighborhood representatives. Other project partners include the Ramsey-Washington Metro Watershed District, the Minnesota Department of Natural Resources, the City of St. Paul, the University of Minnesota Department of Landscape Architecture, seven other city governments, and two county governments. In addition, the project includes a technical advisory committee composed of city and county resource agency staff.

2.3 Watershed Management "Opportunities"

Although the variety of watershed management initiatives throughout the United States have achieved great success at helping to manage water supplies, improve water quality, and address ecological and other environmental concerns, improvements to the watershed approach can be made. The following sections describe some of the most important opportunities for improvement.

2.3.1 The Need for Coordinated Authority

While many U.S. watershed initiatives have successfully implemented cross-jurisdictional coordination, fragmentation of authority remains a significant impediment to *efficient* watershed management. In spite of attempts at systematic coordination, federal and state agencies (including water quality regulators, wildlife, public health, land management, forestry, and agriculture agencies) often operate independently in fulfilling their missions, which at times may be cross-purposeful. At the local level, water and wastewater utilities are often separate from one another as well as from parks, land use planning, public health, and conservation commissions. Moreover, such agencies often must operate with very limited budgets and over-extended staff. They are not provided the direction or supporting framework to coordinate their activities.

This fragmentation of authority often inhibits watershed initiatives from operating as efficiently as possible. For initiatives to be successful, the array of local, state, and federal agencies with jurisdiction over watershed resources requires regular guidance and assistance from a coordinated process and institutional framework that has been accepted by the stakeholders. When these agencies do not plan or implement their activities or provide information and assistance on a coordinated basis, the efforts of watershed initiatives may suffer.

2.3.2 Integration of Large- and Small-scale Efforts

As previously discussed, various approaches to watershed management have been initiated in the United States on many different scales. These range from small grassroots watershed initiatives focusing on several locally important issues to initiatives encompassing entire river systems, crossing state boundaries, and focusing on an array of regional issues, such as sediment loading and water allocation problems. Integration of these small- and large-scale efforts can yield significant benefits. These include:

- Facilitation of successful project planning and implementation,
- More effective involvement of the broad spectrum of stakeholders; and
- Reduction in the inequity of costs and benefits within a watershed.

2.3.2.1 Project Planning and Implementation

As one can imagine, there are far more small-scale watershed initiatives in the United States today than large-scale ones. However, large-scale watershed planning efforts commonly identify projects that must be implemented at the local level. Effective coordination with local small-scale watershed groups plays a critical role in the support, implementation, and monitoring of such projects. Conversely, small-scale watershed groups can provide important data and background information to assist in watershed planning at larger scales. Ideally, therefore, the small-scale initiatives within a river basin should communicate and coordinate efforts with each other and with a large-scale initiative covering the entire basin. This integration of large- and small-scale efforts can lead to a more comprehensive and successful watershed management strategy.

2.3.2.2 Effective Involvement of the Full Spectrum of Stakeholders

The number of stakeholders impacted by the decisions of large-scale watersheds is extremely large, far beyond the optimal number for a collaborative process. Yet the interests and opinions of these stakeholders must somehow be meaningfully included and fairly represented in planning and implementation activities. Likewise, the activities of small-scale watershed groups may have impacts on downstream stakeholders or others that could only be identified when viewing the watershed on a larger scale. Integration of large- and small-scale initiatives can provide a means of involving a great many more stakeholders in overall watershed planning.

The absence of large-scale watershed management initiatives in most of the nation's river basins is a serious deficiency. These scales (approximately USGS Water Resource Regions and Planning Regions) are the appropriate ones for managing key water quality problems, such as sedimentation, salinity, or nutrient loading, and many, if not most, water allocation problems can only be effectively managed at this scale. In regions where these are the principal problems and there is no large river-basin scale initiative, the impacts of local watershed efforts on ecological conditions are extremely limited.

2.3.2.3 Inequity of Costs and Benefits

Another benefit of integration could be a reduction in the inequity of costs and benefits among stakeholders in different parts of the watershed. In larger watersheds it is much more common that the management decisions that provide benefit for all or part of the watershed do not distribute the costs equitably. This inequity is often seen between headwaters communities and those farther downstream. Expensive upstream measures that only marginally change water quality or quantity may be needed to prevent cumulative downstream impacts. Also, larger scale efforts generally do not enjoy the same inherent sense of community engendered by smaller scale projects, making voluntary sacrifices that much more difficult. An integration of small- and large-scale watershed planning and management efforts within a single basin could help solve this problem.

2.3.3 More Use of Economic Instruments

Watershed initiatives in the United States face the constant challenge of accomplishing their goals with limited and inconsistent funding. Obtaining more stable and reliable funding from government sources can help watershed initiatives more effectively plan for and utilize these limited funds. At the same time, increasing the utilization of economic instruments, such as effluent trading, can help achieve watershed management goals without significantly reducing funds allocated to watershed planning.

2.3.3.1 State and Federal Funding

Less then $1 billion in federal funding ($20 million per state) is allocated to grant programs specifically designed to assist watershed initiatives. Much more money indirectly funds programs through federal agency efforts, but this is still far short of the resource needs of watershed efforts at the local level. States contribute proportionately little funding for watershed initiatives. Other sources such as foundation grants, corporate donations, membership dues, and contributions from utilities are also available. Available funds are often narrowly focused and leave some important needs unmet, such as support for overhead and general operations.

There is a need for more total dollars with more flexibility to allow watershed efforts at all levels to implement their highest priority projects. For example, insufficient federal funds have hampered USGS water monitoring efforts. Moreover, funding for watershed initiatives is generally unstable, with few programs having dedicated funding sources such as from hydropower revenue or water/sewer surcharges. Stability of funding from year to year is extremely important for sustained efforts at project planning, implementation, and monitoring. Additionally, without consistent funding, watershed initiatives can lose the ability to support key staff, resulting in the loss of important institutional memory.

2.3.3.2 Market-based Approaches

Market or quasi-market based approaches to increasing the efficiency of water pollution control are relatively rare. The Colorado Salinity Control Program's competitive bid system is one example. U.S. EPA has for sometime been promoting an effluent trading system, following on the heels of the successful sulfur dioxide emissions trading system. The phase-in of TMDLs also provides an excellent opportunity for the introduction of effluent trading systems.

Under an effluent trading system, a pollution source such as a wastewater treatment plant (WWTP) can sell or trade pollution reduction credits to another source that can only reduce its pollutants at greater cost. In the context of TMDLs, point and nonpoint sources could trade credits initially distributed based on the TMDL value for the watershed. For example, rather than installing expensive new equipment, a WWTP might pay a local dairy farmer to plant vegetated buffer strips along ditches and streams to reduce his nitrogen, phosphorous, and coliform bacteria loading. Trading systems such as this in other arenas have proved to be efficient at producing cost-effective yet flexible pollution control. For such systems to operate well, however, it is important that appropriate regulatory and monitoring systems are in place to ensure accountability and prevent market distortions.

2.3.4 Better Integration of Point and Nonpoint Source Pollution Management

Pollutant loading to the watershed and the resulting water quality is the sum of both point and nonpoint source pollution. To understand the ecological and water quality impacts of pollution and to develop cost effective management plans, watershed initiatives need reliable information about both sources. Historically, however, point and nonpoint source pollution management has not been well integrated at the federal and state level, or in local watershed initiatives. With the recent implementation of TMDLs, this is starting to change, at least in some watersheds facing significant water quality problems. However, watershed planning and programs for the most part are still not well connected to local land-use planning. Thus, watershed protection priorities tend to have little impact on growth management and regulatory decisions such as zoning, building design, and development choices that profoundly impact nonpoint source pollution (e.g., stormwater) and the hydrodynamics of the watershed (e.g., impervious surface area).

With the advent of TMDL development and implementation and increasing awareness of the connections between land-use planning and water quality (e.g., the smart growth literature), watershed initiatives are in a unique position to focus on the integration of point and nonpoint source pollution management. As described above, by definition watershed-based planning and management includes a broad range of stakeholders (e.g., local land use planners and advocates, water utility personnel, wastewater dischargers, state permitting agency staff) and focuses on comprehensive approaches to multiple issues. With these interests represented and communicating in a structured process, issues such as land use, stormwater management, and point and nonpoint source pollution can be addressed in an integrated fashion.

2.3.5 Improved Monitoring and Measurement of Watershed Conditions

Science-based decision making requires reliable, complete, long-term data to understand the dynamics of the watershed, assess environmental health, develop plausible management plans, and evaluate the impacts of actions taken.

Unfortunately, reliable, representative monitoring data on water chemistry, hydrology, and ecological composition and conditions for the necessary biological, chemical, and physical parameters do not exist in most cases. Many watershed efforts must make ecological management decisions with very incomplete pictures of the history, nature, and dynamics of their watershed and the impact of human activities within.

Despite 30 years of significant investment in water quality research and monitoring, data in the United States are incomplete, fragmented, and often not useful to watershed managers. The focus of most monitoring efforts in the past was determination of effluent quality. Past data on toxic contaminants are lacking or badly compromised. Different agencies have targeted different parameters, and agencies often use different analytical techniques that may no longer be considered acceptable or that had poor quality control procedures. Moreover, much of the data that do exist are stored in inaccessible formats.

Fortunately, these problems have been recognized and a national effort is underway to develop uniform sampling and analytical protocols, improve field and laboratory quality control procedures, and make data available in standardized formats. Research done in the Willow Creek Watershed in Colorado is an example of a watershed partnership that uses scientific data to inform remediation and management activities. In the Willow Creek Watershed, heavy metal contamination from abandoned mines was believed to be the primary source of water quality degradation. Fortunately, the watershed partnership took a measured scientific approach to the problem and made no premature conclusions. The U.S. Forest Service (USFS) and USGS have been tracing contamination in area streams to identify specific sources. The first dye-tracing phase suggests that the contamination in Willow Creek may be confined to a limited area, and remediation and management of the area financially and technically realistic.

Many watershed initiatives have transformed their water quality and habitat data gaps into a means of engaging and involving citizens and volunteers. Such data needs can provide an opportunity to educate children, students, and the general public about the concept of the watershed, present resource issues, demonstrate the work of the watershed initiative, and explain the scientific method and various scientific disciplines. Some watershed initiatives have found that engaging students and the public in data gathering activities (e.g., daily or weekly water quality monitoring, transect surveys of the riparian corridor, riffle pool sequence measurements, shellfish bed surveys) instills greater appreciation and personal connection to the natural life of the watershed.

2.3.5.1 Scientific Models and Decision-support Tools

GIS are fast becoming ubiquitous tools for the display, modeling, analysis, and interpretation of watershed-related data. The development of GIS capabilities and databases are significantly enhancing watershed research and planning efforts, especially as dynamic computer hydrological and other models have been integrated, enabling the water transport and flow to be simulated and viewed visually.

The rapid pace of adoption and utilization of GIS by watershed initiatives is outstripped by the growth rate of complex natural resource simulation models that are often linked to GIS. These models have so many adjustable interacting parameters that making a decision is difficult without a systematic method for evaluating the outputs. The National Research Council sees an urgent need to facilitate the development of decision support system software. It envisions a type of expert system that addresses: 1) problem identification; 2) selection of decision criteria; 3) selection of feasible management system or design alternatives; 4) evaluation of the alternatives by simulation models and/or historic data; and 5) recommendations for a decision.

2.3.6 Multi-objective Frameworks

Increasingly agencies involved with watershed management look for ways of simultaneously addressing the complex environmental and social concerns, multi-jurisdictional agendas, and resource limitations into a more integrated multi-objective perspective that meets the needs of all of the watershed's stakeholders. There are an increasing number of tools available to support these frameworks, including GIS in combination with other decision support systems. However, a multi-objective framework that allows stakeholders to examine tradeoffs and synergies in various options has rarely been used for real decision-making. This may partly be due to the first of the 'opportunities' identified—the need for coordinated authority at the river basin or watershed scale. Without this institutional framework, a full examination of the possible objectives in a basin is unlikely to take place. As described above, such ambitious planning requires involvement of more than the water professionals. A successful community-based, multi-objective watershed initiative must involve as many different aspects of the watershed as possible.

CHAPTER 3.0

LESSONS AND GUIDANCE FOR SUSTAINABLE WATERSHED MANAGEMENT

3.1 Introduction

Much of the popular literature on water resources discusses how rapid population and economic growth, in combination with increasing urbanization and emerging water uses such as the role of ecosystem services, are severely stressing natural resources and increasing the global perception that water is scarce (Postel, 1999; Postel, 2001). In this context, conflict over the distribution of water is inevitable, whether the water resource is shared between countries, states, districts, or competing uses. Traditionally, most water management policies focused on surface water quantity, ignoring the relationship with groundwater or land use and the importance of water quality. Moreover, such policies were both fragmented and often decided at the national level based largely on politics, with an almost exclusive focus on increasing the supply of water in order to meet municipal, power, and agricultural demands. Little attention was paid to other water resource uses, particularly ecosystem preservation and the human reliance on the services ecosystems provide. Traditional approaches have proven to be less effective as pressures on natural resources have grown (Rogers and Priscoli, 1993; Gleick, 2001).

The shortcomings of traditional water management have been recognized globally since at least 1977 (United Nations, 1977), but were not addressed seriously until the 1990s, when an international consultative process, culminating in the International Conference on Water and the Environment in Dublin in 1992, developed what has come to be known as the Dublin Principles. These principles, given below, continue to play a prominent role in framing water resources management around the world.

1. Fresh water is a finite and vulnerable resource, essential to sustain life, development, and the environment.
2. Water development and management should be based on a participatory approach, involving users, planners, and policy makers at all levels.
3. Women play a central part in the provision, management, and safeguarding of water. (This is particularly important in developing countries.)
4. Water has an economic value in all its competing uses and should be recognized as an economic good.

The Dublin Principles were subsequently endorsed at the United Nations Conference on Environment and Development in Rio de Janeiro, 1992. These principles have since been restated and elaborated at numerous major international water conferences, including those in

Harare, Paris, and the "Rio+5" follow-up meeting, all in 1998, and more recently at the World Water Forum, held in the Netherlands in the spring of 2000 (World Water Forum, 2000).

In the international context, the term integrated water resources management (IWRM) is often used interchangeably with the "watershed" or "river basin" approach. Largely based on the Dublin Principles, the Global Water Partnership (Global Water Partnership, 2000) defined IWRM as "a process that promotes the coordinated development and management of water, land and related resources, in order to maximize the resultant economic and social welfare in an equitable manner without compromising the sustainability of vital ecosystems." This definition, while similar to the U.S. EPA's definition of a watershed approach, is somewhat less prescriptive, and specifically mentions the economic or social aspects of watershed approaches.

Regardless of terminology, the watershed approach provides a very different framework from traditional water management, one that starts with a geographic or natural resources imperative rather than an economic or jurisdictional imperative. The natural resource constraints form the baseline for analysis in a watershed approach. Furthermore, the watershed approach necessarily implies the creation of an entity or cooperation among entities with the authority, legal or de facto (e.g., granted to the entity by popular backing), to plan, design, and implement decisions for watershed management. These decisions may include allocations in the watershed and among competing uses, what mechanisms to use to reduce pollution loads to the watershed, the imposition of water pricing or effluent charges, and changes in land use policy. The watershed approach with the appropriate authority facilitates the ability to shift priorities for water uses (e.g., between agricultural, municipal, and ecosystem uses) and increases the ability to manage the exigencies of nature in the form of droughts and floods. Furthermore, successful management requires an *adaptive management structure*, or one with the ability to change watershed priorities based on public input or new information and technologies.

A review of the literature on non-U.S. watershed or IWRM approaches shows a consistent pattern of three principles that are indispensable for effective watershed management:

1. There must be a clear decision-making framework that looks across scales.
2. Water development and management should be adaptive and based on a participatory approach involving users, planners, and policymakers at all levels, with the authority to make decisions at the lowest appropriate level.
3. Instruments used to achieve effective management should recognize the economic value, as well as the social and environmental considerations of water in all of its competing uses, whether for ecosystem protection or restoration; fisheries; agricultural, industrial, municipal, navigation, or recreational users; or power generation.

There is a fourth principle that is necessary, particularly in developing countries: there must be an enabling environment—one without war or a political system that prevents participation at all levels. This report assumes that this is not the case in the United States. Therefore, it is not included in the discussion. Note that this may be defined more broadly to include an enabling legal environment, which may be relevant in some U.S. cases.

The specific implementation of these principles necessarily differs based on the climate, scale, and use of the watershed, as well as the politics, culture, and social values of the watershed stakeholders. Described below are specific non-U.S. examples of watershed decision-making frameworks, institutions, and instruments that highlight the variety of approaches that have been effective. Each example highlights one element of watershed management; however, other equally important elements are also described to give a full picture of what is necessary for effective management. There is no blueprint for success, but a combination of frameworks, institutions, and instruments based on some degree of local decision making, support, and funding is essential for successful watershed management.

3.2 Decision-making Framework

The need for a decision-making framework for planning and evaluation for sustainable watershed management has been well articulated (e.g., Stakhiv, 1996; Clements et al., 1996). Both Stakhiv and Clements et al. state that coordination among various agencies is not sufficient to enable effective management. There must be clearly defined goals against which projects and programs can be evaluated, and there must be clear roles for the decision-making process. Stakhiv, in particular, urges that scale must be taken into account in such a framework through a hierarchy of nested planning levels, as was suggested in the Water Resources Planning Act of 1965 (now defunct). The top level (A) would address planning for large river basins, such as the Mississippi or Colorado Rivers, or large systems, such as the Great Lakes, and would look strategically at what is needed for sustainable development in those basins or systems. The middle level (B) would be for smaller river basins (at the USGS 2-digit scale) guided by the planning at Level A, and again would set strategic plans and projects for the basin. The lowest level (C), guided by the planning processes at Levels A and B, would involve looking at project feasibility at the sub-basin or watershed scale.

While planning at Level C happens regularly in the United States, the evaluation of projects tends to be reactive and without a systematic framework on large-scale impacts or sustainability that would be in place if planning also occurred at Levels A and B. Two international examples that have clear decision-making frameworks are France and the Murray-Darling River Basin in Australia. Each approach to a decision-making framework is briefly described below. The Murray-Darling River Basin is described in more detail as a case study in Appendix A2.

3.2.1 French Basin Agencies

One example of a clear framework for decision-making is in France, which created River Basin Financing Agencies in 1964 for each of its six major river basins. Under this approach, basin committees, made up of representatives from the state and local authorities and various users, are responsible for preparing and implementing water resource management policies. Associated with each basin is a water agency, under the supervision of the basin committee and the national-level Ministries of Environment and Finance. The water agency is responsible for the implementation of policies dictated by the basin committees. Figure 3-1 illustrates this structure.

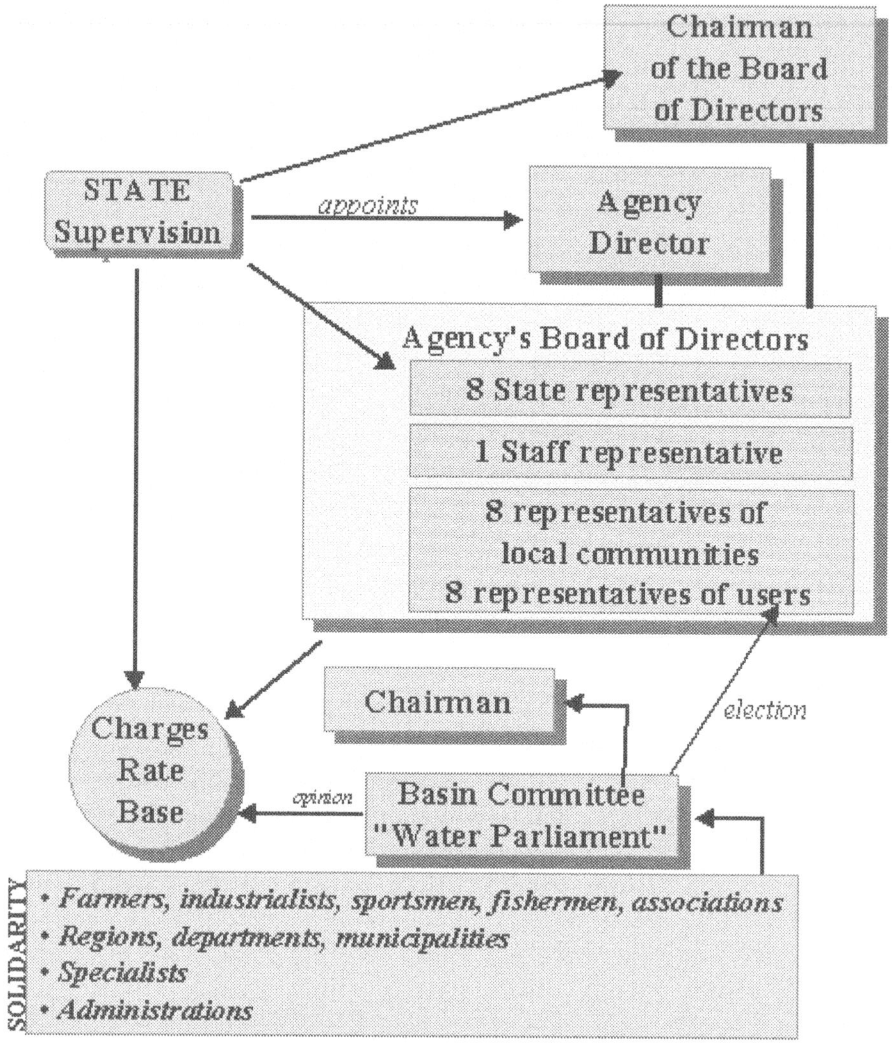

Figure 3-1. Organization of Water Management in France: Organization of the River Basin (International Office for Water).

At the national level, a member of parliament chairs a water committee responsible for giving guidance. A 1992 water law established by the parliament emphasized a more integrated approach to water rather than the more sector-oriented logic used in the past (see the Paris example Appendix C1, which promotes the integration of point and nonpoint source pollution). The water law established clear goals and priorities from which projects can be evaluated. Each basin is required to establish a master plan based on two main principles: 1) progression from "water management" to "water environment management" and, 2) giving priority to collective interests. These plans, developed at the basin level, attempt to achieve balanced water resource management in the form of ecosystem conservation and management, use reconciliation, and water resource protection and development.

3.2.2 Murray-Darling

The development of planning at the river basin scale happened quite differently in Australia, where historically the emphasis has been on state rather than federal control over watershed management. Two federal laws passed in the 1990s furthered control at smaller scales for local watershed management. The Catchment Water Management Act of 1995 created local catchment management boards with responsibility for crucial aspects of planning and management, including flood control, recreation, and preserving and promoting environmental quality. The second important piece of legislation, the Water Resources Act of 1997, emphasized the integration of ecological, environmental, and economic considerations on a geographic basis. For example, local boards generally retain authority over local resource management, subject to standards and regulations set at the state and federal level.

However, there was growing recognition that for problems on the scale of a large river basin, the institutions involved needed to be appropriately scaled. For example, salinity is very problematic in the Murray-Darling Basin (MDB), which covers an area of approximately 1 million square kilometers. To address the problems of the MDB, the federal government established the Murray-Darling Basin Agreement, which formed a consortium among catchment management boards and state and federal officials.

The agreement created three decision-making bodies: a ministerial council, a commission, and a community advisory committee. The responsibilities of each are as follows:

- MDB Ministerial Council: Ministers from each of the signatory governments make up the council, which is the decision-making forum for basin-wide policies and programs.
- MDB Commission: Comprising senior executives from each partner government with an independent president, the council gives advice to the ministerial council and carries out the decisions made by the council.
- Community Advisory Committee: Comprises representatives from the basin's catchments, four key special interest organizations, and an Indigenous representative, the committee provides a two-way communication channel between the ministerial council and the community. In addition, the committee provides the council with advice from the basin-wide community.

The strategic and philosophical framework for achieving the purpose of the agreement is the *Natural Resources Management Strategy*, which was approved by the ministerial council in 1990. The strategy provides the broad charter for a community-government partnership to develop plans for the integrated management of the basin's water, land, and other environmental resources on a catchment basis. The *Natural Resources Management Strategy* established two fundamental pillars for handling natural resource management in the basin. The first pillar is the philosophy of integrated catchment management (ICM), recognizing the links between various biophysical processes that affect or are affected by water, its movement, and its uses. The second pillar is the community/government partnership, recognizing that neither party working in isolation can protect the basin's natural resources.

In order to improve the performance and accountability of work undertaken under the *Natural Resources Management Strategy*, the ministerial council put in place the *Basin Sustainability Plan* in 1996. Following consultation with key groups, the objectives of the *Basin Sustainability Plan* were reviewed and updated in 1999 to ensure their relevance to priority natural resource management issues across the basin. The plan provides the framework for the co-ordination of planning, monitoring, evaluating, and reporting for the *Natural Resources Management Strategy*, and covers all government and community investment for sustainable natural resources management in the basin. The *Natural Resources Management Strategy* and *Basin Sustainability Plan* together are the foundation of the commission's basin-wide planning processes for natural resource management.

The *Basin Sustainability Plan* contains long-term productivity and resource condition objectives for sustainable agriculture, water quality, nature conservation, and cultural heritage. For each of these priority thematic areas, specific objectives apply to irrigated and dryland regions of the basin and its riverine environments. Reporting against these objectives is designed to show short-term achievements (empowerment), medium-term achievements (implementation), and long-term outcomes (resource condition).

The plan also contains direction setting and management implementation objectives, which focus on people and management arrangements. These objectives are designed to help ensure that the arrangements for natural resource management enhance the partnership between community and government, and help the managers of land and water protect the basin's catchments.

3.2.3 Summary

Both France and Australia have evolved to form agencies/institutions with decision-making authority at the scale of large river basins. These institutions have clear objectives and mechanisms for review. The processes established allow for dynamic evaluation and change around water management, in other words, systems of adaptive management. Decision-making frameworks and institutional structure are necessarily interconnected.

3.3 Water Institutions

The formation of water institutions is almost always done in a broader social context and rarely fits with rational analytic models. Water planning has as much to do with flexibility and managing uncertainty as it does with discerning deterministic trends. Effective institutions must be adaptable to these changing conditions, with an ability to address cross-jurisdictional issues in their legal and political context, while providing adequate voice and power to the divergent views of the range of stakeholders in the watershed. Cross-jurisdictional cooperation and stakeholder empowerment are crucial and require persistent effort to be achieved. The most effective examples of watershed management are those with institutions that have evolved (adapted) over time.

We focus on these two institutional components of watershed approaches below. The first component, discussed in Section 3.3.1, addresses the need for integration of national goals with local priorities and for coordination of agencies across jurisdictions. The second component, discussed in Section 3.3.2, addresses approaches to improve meaningful stakeholder involvement—those that allow stakeholders to influence decisions in the basin. For examples of participatory approaches that have failed in the international context, see *Participatory Watershed Research and Management: Where the Shadow Falls* (Rhoades, 1998). Clearly the decision-making framework discussed in Section 3.2 is inherently critical to how well institutions function. Together, these components support the development of an institutional framework capable of integrating human systems (economic, social, and political) with natural systems.

3.3.1 Cross-jurisdictional Cooperation

Comprehensive watershed management necessitates coordination among the range of human activities that create the demand for water, determine land uses, and generate water-borne waste products. Creating a water-sensitive human system requires coordinating policy making at all levels, from national agencies to community-based organizations. Delineation of jurisdictional boundaries must be clear for watershed management to function well. This is the case for the two watershed approaches described below: Germany and the Grand River in Ontario, Canada, in which national, provincial, municipal, and local authorities have clearly defined roles, functions, and responsibilities. The Murray-Darling Initiative, described in Section, 3.2.2, is also a good example of an effective cross-jurisdictional institutional arrangement.

3.3.1.1 German Ruhrverband

One of the earliest examples of successful river basin management was the Ruhr Basin Water Association, also known as the Ruhrverband. The association was founded in 1913 as a self-governing body subject to legal supervision from the state. The Ruhrverband established several key principles. First, all stakeholders (all users and polluters of water, including communities, districts, and trade and industrial enterprises) would be members, and policies would be made by a "water parliament" or assembly of stakeholders, as well as a supervisory board and an executive board. Some positions are elected, some appointed, and some are based on representation from various water users/polluters. Second, economic instruments, such as water charges and pollution fees, would provide the finances for investments and other management activities of the Ruhrverband. Third, water quality objectives were to consider both the costs and the benefits of improvements. Fourth, municipalities along the Ruhr River would maintain the collection of wastewater and distribution of drinking water in their jurisdiction, while the Ruhrverband would manage the major infrastructure, such as the design, construction and operation of reservoirs and waste treatment facilities.

The Ruhrverband continues to be the basin water authority in charge of operating a system of reservoirs to protect drinking water supply while minimizing the risk of flooding. In addition, the Ruhrverband is responsible for the water quality management of the Ruhr. In that context, the Ruhrverband plans, constructs, and operates WWTPs and works to reduce nonpoint sources of pollution to enhance the ecosystem quality of the Ruhr basin.

The funding for the agency is founded on the application of the "user-polluter-pays" principle. The water agencies levy water charges on withdrawals and discharges from all users who affect water quality and/or modify the flow regime. The charges to industry are calculated by using various parameters appropriate to each sub-basin, and the amount of pollution produced by each establishment. Domestic charges are calculated for each community according to both permanent and seasonal populations, and are collected from individual users based on the quantity of water used as measured by a flow meter. The rates and charges for the basin are based on the cost of maintaining the existing system as well as investments in improvements. The water agencies are, therefore, self-sustaining with total charges equal to total budget requirements.

3.3.1.2 Grand River

Prior to World War II, renewable natural resources in Ontario were exploited to encourage industrial expansion and economic growth. As a result of public concern about the state of the environment, the government of Ontario passed the Conservation Authorities Act in

1946 to provide guidelines for the establishment and functioning of conservation authorities in Ontario. The act was based on three principles:

- The watershed is the best level to coordinate all conservation work dealing with renewable resources.
- Local people must provide the initiative for the establishment and support of a conservation authority with power to carry out conservation works within a watershed.
- In response to demonstrated local initiative and support, the Ontario government would be prepared to provide technical advice and financial assistance in the form of grants.

Under the terms of the act, the Grand River Conservation Authority (GRCA) was formed in 1948. The GRCA membership reflects a strong municipal partnership. The GRCA has representation from 34 watershed municipalities in nine regions and counties, and provides the means by which they work together, deal with cross-boundary issues, and influence each other's activities for mutual benefit. This governance model has proved to be highly successful.

Resource management issues and needs throughout the watershed are identified at the local level and endorsed by the authority membership. Member watershed municipalities participate in the administration and operation of the authority through their appointed representatives and support the authority financially through general municipal levies and special levies for projects that directly benefit a municipality or group of municipalities.

3.3.2 Stakeholder Involvement

Some level of stakeholder influence on decision making is a key component of effective watershed management. Since virtually everyone is a water user, and therefore a watershed stakeholder, participation should be broad and include representatives from government and NGOs from the national to the local level. Stakeholder involvement is needed to increase awareness of the integrated nature of water, and the many possibilities to effect improvements in water management. A participatory approach is necessary to achieve long-lasting consensus and common agreement. Effective stakeholder involvement processes in watershed management require participants to recognize the effect of their actions on other water users and ecosystems, and to accept the need to improve the efficiency of water use and allow the sustainable development and use of the resource. Although the watershed approach in the United States tends to focus water quality issues, the international literature on integrated water resources management addresses water use in terms of quality and quantity, and the possibilities of conservation and/or alternative allocations.

Two examples of how stakeholder involvement varies with different watershed institutional arrangements and issues are presented below. The Mersey Basin Initiative in the United Kingdom provides an example with more focus on water quality issues. The initiative is based on voluntary partnerships motivated by strong public support to improve the conditions of the basin. The Fraser River example illustrates the potential for a non-governmental and non-regulatory institution to play a critical role in providing a forum for the full range of stakeholder voices to be heard in a large watershed.

3.3.2.1 Mersey River

During the 1950s and early 1960s the United Kingdom's government introduced a number of acts of parliament to control the discharge of harmful waste into surface waters. While this led to a steady improvement in the quality of rivers in the 1970s and 1980s, many remained heavily polluted, particularly in the eyes of the general public. In the early 1980s, strong public support led to the suggestion for a project based on a partnership approach for the Mersey Basin, located in northwest England. The Mersey Basin Campaign (MBC), funded by the Department of Transport Environment and Regions, promotes, manages, and supports the campaign partnership. The campaign is a flexible partnership between organizations and individuals, helping and empowering communities to act for themselves. It enjoys the backing of the expertise and resources of its partners, which include local authorities, businesses, voluntary organizations, and government-sponsored agencies.

A notable element of the MBC is the involvement of businesses. In 1987 ICI, Shell, and Unilever agreed to support campaign-related projects. With the increasing number of businesses actively supporting the campaign, the Mersey Basin Business Foundation was launched in January 1992. The original three members of the foundation have increased to more than 20 and cover a wide range of business sectors. The principle aim of the business foundation is to accelerate the progress of the MBC by encouraging greater business participation in partnership with the public and voluntary sectors, contributing to the strategic development of the campaign, and providing sponsorship.

Already, many projects have been implemented, with some support from the Department of Transport Environment and Regions, the Environment Agency, and primary support from other campaign partners. The main project themes include grant schemes for voluntary groups, campaign publicity and promotion, environmental education for schools, developing water quality monitoring systems, encouraging sponsorship for River Valley Initiatives (i.e., specific sub-basin projects), award schemes and competitions, together with "help-in-kind" through various services. The MBC was awarded the first International River Prize for Best River Management Initiative at the Second International River Symposium in Brisbane, Australia.

3.3.2.2 Fraser River

The Fraser River Basin is the fifth largest river basin in Canada, spanning an area of 238,000 km^2, an area comparable to the state of Oregon. The basin supports 48% of British Columbia's commercial forest area, 60% of metal mine production, and 44% of agricultural land. The Fraser and its tributaries are important shipping and transportation links for these resource-based industries. Over 60% of the province's population of two million lives in the basin and depends on the river as a source of drinking water. The combination of industrial and municipal water use results in annual discharges to the river on the order of 900 million cubic meters. The basin is also a key attraction for the recreation and tourist industries, which are valued at approximately $300 million annually.

In an attempt to address the deterioration of the Fraser River and its salmon stocks, the Government of Canada initiated the Fraser River Action Plan (FRAP) in June 1991. The FRAP was conceived as a six-year, $100 million green plan initiative to be undertaken jointly by the Department of Environment (DOE) and Department of Fisheries and Oceans (DFO), each of whom was to administer half of the program's funding. In developing its Green Plan, the federal government concluded that building partnerships would be critical to its success. However, both agencies had for some time been developing their own plans for the FRAP. Severe restrictions had been placed on discussion of these plans outside of their organizations before they were to be announced. DOE and DFO intended to remedy this situation by initiating a collaborative process of building partnerships.

An agreement was reached in 1992 to initiate the Fraser Basin Management Program (FBMP), to be led by a multi-interest committee called the Fraser Basin Management Board (FBMB). Comprising multiple governments and NGOs, the board is required to make decisions by consensus and to encourage consensus-based decision making in all Basin activities as well as to facilitate the development of decision-making processes that are locally based.

No formal basin-wide organization existed prior to the FBMB. The board was given five years to accomplish this goal. At the end of the five years, a new institution was established—the Fraser Basin Council (FBC). It is clear that the institutional arrangement of the council greatly benefited from the experiences gained during the life of the board. The FBMP and its board set a path, with much public involvement, toward a very different kind of public-private partnership dedicated to balancing the components of a sustainable future for the Fraser Basin.

The council is composed of governmental, non-governmental, private, and civil sectors and facilitates the emergence of common ground in large partnership processes toward sustainable solutions. Senior involvement from each of these sectors enables the council to address multi-jurisdictional issues collectively, and then through the use of specific organizational mandates, implement actions to help resolve those issues. By design, the council is not an authority and has no legislated powers. It functions as an impartial facilitator, catalyst, jurisdiction and conflict resolution agent, and educator. As a result of the council, more decisions are being made at the community level with senior governments providing support and participating whenever possible. The FBC is renowned as a leading example of best practice river basin management. It also assists the emergence of community-based governance models in Russia, Brazil, and the Philippines.

3.3.3 Summary

Institutional arrangements for watershed management around the world vary considerably—with formal government-led frameworks such as the Ruhrverband, the Grand, and the Murray-Darling, to semi- or non-government frameworks, such as the Mersey and Fraser basins. In all cases, cross-jurisdictional issues are addressed up front, and stakeholder processes are formally established. Broad representation is consistently the rule in these basins, cutting across scales within the government and between government and civil society. This combination of cross-jurisdictional frameworks with broad stakeholder representation is consistently an effective combination for watershed management.

3.4 Instruments

Watershed management instruments can be viewed as a set of tools that enable decision makers to make informed choices among alternative actions that will promote the sustainable and equitable use of natural resources. Instruments are defined in a variety of ways in the literature, but may be grouped into the following four categories: information and communication instruments, regulatory instruments, economic instruments, and technology instruments. Each is described below with examples of their application in international river basins.

3.4.1 Communication and Information

Communication instruments enhance stakeholder involvement by providing timely and relevant information to all concerned parties in a form that can be understood. Typical instruments include public information sessions, expert panel hearings, publications, citizen juries, and posting information in the local newspaper or on the internet. Information should be communicated in an open and transparent format. Information instruments provide decision makers with an understanding of the watershed, from the perspective of both human and ecosystem interactions. Examples of such instruments may include assessments and/or models of relevant water resources, environmental impacts, demands, and risks, as well as monitoring systems. Also, the mandatory disclosure of information, for example about water use and pollutant loads, allows the public to push for improvements, or make decisions to purchase other brands. The labeling of products to indicate the "quality" or "promotion of environmental sustainability" in the production of a good or service is also an effective instrument. The Rhine and the Mersey River Basins provide good examples of how these instruments have been used successfully in watershed management.

3.4.1.1 Rhine River Basin

The Rhine River Basin offers an example of how the utility industry in particular, but other stakeholders as well, used information and communication instruments to convince governments to consider the watershed as an integrated whole.

There is a long history of stakeholder involvement in the Rhine, although this originally occurred absent a formal institutional structure. Some of the earliest influences came from the Dutch waterworks organization, Association of River Waterworks (RIWA) in the 1930s that formed in response to deteriorating water quality. Soon after its formation, RIWA contacted the Dutch government with proposals for action. Furthermore, to improve contacts with other governments RIWA joined forces with fellow organizations in the Rhine catchment area to form the Internationale Arbeitsgemeinschaft der Wasserwerke im Rheineinzugsgebiet (IAWR). In its various Rhine memorandums the IAWR has proposed extensive demands on the governments of the riparian states along the Rhine. In addition, the waterworks were also able to influence policies because representatives of the drinking-water companies held administrative and political jobs in each of the riparian states.

The environmental NGOs in the region initiated fact-finding activities about pollution sources, which in turn influenced perceptions in the catchment area. While the IAWR provided the states with data concerning the pollution levels of the Rhine, environmental organizations like the Dutch Clean Water Foundation (Stichting Reinwater or CLF) and Greenpeace started monitoring campaigns on the river to clearly identify specific point sources of pollution. Both succeeded in influencing perceptions concerning the quality of the Rhine.

NGO-initiatives proved effective in reducing some political barriers to change. For example, the environmental NGOs ensured that upstream governments and industry were kept under continual pressure, for instance, through legal proceedings against polluters. In 1974, CLF and horticulturalists in the region initiated legal proceedings against the French potassium mines because their cucumbers and other products were suffering from high salt loads. After 14 years of proceedings the legal action was successful, resulting in an out-of-court settlement whereby the affected horticulturalists received financial compensation for the damage they suffered from the discharges from the mines. These successes at court inspired Dutch drinking-water companies and other victims of pollution problems to start other successful proceedings in France in administrative and criminal courts.

Finally, negotiations concerning the ecological recovery of the catchment were also influenced by NGO activities. The successes in the civil proceedings of CLF improved the position of the Dutch government in its negotiations with Germany over problems of contamination from industrial discharges. The Dutch delegation threatened to support the municipality of Rotterdam, which wanted to start proceedings as well. Rotterdam Harbor was polluted with contaminated sludge. Upstream discharging industries were held responsible for the costs to Rotterdam for removal and storage of this sludge. This made the Germans more eager to accept the Dutch proposals.

The openness of the political system in the Rhine catchment area made governments receptive to the ideas of the NGOs. An open dialogue exists between NGOs and the governments. In the Netherlands, this openness is probably the most far-reaching, with the Dutch government in some cases supporting organizations critical of their own activities. Environmental organizations like CLF are partly funded by the government (Van Leeuwen 1992). Recently, NGOs obtained an observer status at the International Commission for the Protection of the Rhine (ICPR). With the creation of the European Union (EU), the national political systems are more integrated. Apart from the contacts within the ICPR, many other contacts between the riparian states exist on political, business, and cultural levels.

NGOs, including the IAWR and RIWA, have also had an impact on chemical pollution. Publicity campaigns have influenced environmental awareness of industries, and governments and have mobilized public opinion in the catchment area. Through buying shares in polluting companies, the Dutch CLF tried to influence the decision making at these companies. Major polluters of the Rhine were confronted with the results of monitoring campaigns in the so-called water tribunal, a non-official court in which they were condemned. CLF also published its arguments in newspapers, conducted grassroots education of schoolchildren, and had contacts with ministers and political parties. These activities resulted in a growing willingness by industrial dischargers to intensify treatment of their wastewater to meet environmental standards. In response to these efforts, concentrations of heavy metals on the Dutch border have decreased and an international agreement on specific standards was more easily reached.

3.4.1.2 Mersey Basin

Management of the Mersey Basin on a watershed basis grew out of the insight that the region's problems of industrial decline and decay and poor water quality were linked and required an approach that could accommodate both issues simultaneously. In a 1981 visit to Liverpool, Michael Heseltine, then government minister for Merseyside, made a public statement describing the Mersey River as, "the dirtiest river in Europe…and an affront to civilized society." The indictment brought high-level and focused attention to the dismal state of the watershed and, through the continued efforts of Heseltine (first as government minister and later as secretary of state for the environment) and others, led to the establishment of the 25-year government-backed MBC.

Three organizations, each focusing on a particular community of stakeholders, evolved to form a coordinated effort under the MBC: the Mersey Basin Campaign Administration Ltd., the Mersey Basin Trust, and the Mersey Basin Business Foundation. Formed two years after the launch of the campaign, the Mersey Basin Trust is a registered charity whose mission is to educate, coordinate, and assist the voluntary, community, and educational sectors in support of the MBC's goals and to represent the interests of these sectors within the MBC. The trust has become an increasingly important player in the MBC, as development of a framework for informal intervention, local initiative, and the fostering of community involvement has become a prime objective of the campaign (Wood and Handley, 1999). All types of organizations are welcomed, with most falling into one of five categories: wildlife and open-space conservation groups, recreational organizations, community groups, heritage preservation organizations, and schools. Membership is free, and the trust now claims over 600 member organizations. The trust oversees the operation of campaign projects, including small grant programs, sub-watershed management groups, educational curricula for the public schools, and other educational and assistance programs. A 24-member board of trustees elected by member organizations oversees the trust.

Through the Mersey Basin Trust, the campaign has cleverly sought to reach the broader public by partnering with existing local organizations and institutions. The trust primarily seeks organizations as members, not individuals. The trust maintains and strengthens these partnerships by providing valuable resources such as improving access to financing, technical assistance, educational materials, and logistical guidance as exemplified in programs such as:

- *Water Watch:* targets the problems of litter in and around rivers and canals
- *Stream Care*: supports voluntary action to clean up and care for streams and rivers
- *Grants and Awards:* small grants to fund projects and awards to recognize achievement and dedication
- *Water Detectives:* river-focused curricula and support for schools and teachers
- *Mersey Basin Weekend*: a weekend offering over 100 conservation and recreational activities on and around watercourses in the Mersey Basin.

These services, combined with clearly defined objectives and the regional focus of the campaign, in turn create a framework for independent initiative and actions that further the objectives of the campaign.

3.4.2 Regulatory Instruments

Regulatory instruments can be powerful tools to achieve goals of watershed sustainability and have historically been used extensively in water resource management in the United States. Both direct and indirect regulatory instruments can be used. For example, direct controls can be made on which organization has control over what aspect of the resource, such as allocations of quantities of water or pollution allowed into the resource, land use and zoning, and water utility pricing and operation. Water rights may also be directly regulated. Standards and guidelines may be established with more or less flexibility than direct controls, although the United States has primarily relied upon direct controls. For example, a direct control may specify the technology to be used to meet pollution standards, while an indirect, or more flexible requirement allows the polluter to choose how the standard is met, whether through alternative technologies or processes, or through pollution trading.

The international experience in using regulatory instruments encompasses the breadth of direct and indirect tools. As the United States has extensive experience in direct control, two examples of indirect or flexible regulatory instruments are provided below. While the Grand River Conservation Authority has regulatory authority over land-use planning, it is done with extensive consultation with municipalities, and thus flexibly to meet local planning needs. The Murray-Darling has placed a firm cap on any new withdrawals of water from the basin, but states have control over how they each meet their respective caps.

3.4.2.1 Grand River

Perhaps the most important lesson that the Grand River can offer U.S. watershed managers involves the integration of watershed planning and local land-use planning. In the

Grand watershed and throughout Ontario, municipal land-use planning and watershed planning processes are seen as inextricably linked through conservation authorities. Conservation authorities have jurisdiction over all aspects of natural resource conservation in one or more watersheds, as approved by the Ministry of Natural Resources. The authorities have power to undertake an array of conservation projects within their jurisdiction, as well as to establish and enforce watershed-wide regulations as necessary to protect life, property, water quality, and ecosystem health. Such regulations may include restrictions on floodplain development, filling of wetlands, and development on steep slopes.

Guidance from provincial agencies in the form of legislation and Memoranda of Understanding has led to an established process by which:

1) Conservation authorities produce watershed plans.
2) Watershed plan recommendations are incorporated into municipal official plans.
3) Sub-watershed plans are developed when necessary to guide impending land-use changes at the local level.
4) Official plans are then amended to incorporate the more detailed recommendations of the sub-watershed plans.
5) Specific site development plans are tailored to meet the requirements of the newly amended official plans.

This straightforward process, clearly outlined by the provincial government and implemented locally, fosters cooperation between planners and watershed managers and leads to development that meets the needs of the watershed as a whole. Without this integration, some of the most important watershed management recommendations might not be implemented at the local level.

The process outlined by the government of Ontario for integration of watershed plans with municipal official plans could be applied to the United States For instance, where watershed plans have been developed and approved by state or local agencies, local land-use planners, planning boards and boards of zoning appeal could be directed or encouraged to incorporate relevant elements into local planning documents. When a large-scale development proposal is anticipated, local planning officials could commission a sub-watershed study prior to their review and permitting of the project under local planning regulations. However, one reason why the process works so well in Ontario is that municipal representatives, through the conservation authority, are themselves directly involved in development of the watershed plans. In the U.S., municipalities may be unwilling to adopt land-use recommendations from a watershed plan that they had no part in developing. Likewise, at the sub-watershed level, local officials may be reluctant to allow a sub-watershed study that crosses municipal boundaries and was prepared by an outside organization to restrict local development initiatives. Local approval of and commitment to the watershed plan and its recommendations are essential to its integration with land-use plans. This again highlights the importance of municipal officials being directly involved in cooperative watershed planning, such as in the Grand River Conservation Authority model.

3.4.2.2 Murray-Darling

Possibly the most significant management decision taken by the Murray-Darling Basin Initiative (MDBI) has been to cap water diversions, in part to preserve aquatic ecosystems. The decision-making process for the cap is closely related to salinity management issues in the basin, particularly the amount of water committed in yet-to-be-used licenses for water diversions. In the context of the working to address salinity problems, the effect of growth in diversions was explored, which revealed that the over commitment of surface water in the basin was a serious problem that required political intervention. This was raised at a meeting of the Ministerial Council in 1993, where it was decided that the Murray-Darling Basin Commission (MDBC) should prepare a report on the scope for continued growth in diversions in the Basin. The MDBC prepared a report, *Limits to Surface Water Diversion in the Murray-Darling* in 1994 that concluded that the then current licensing and allocation system served only to limit water diversion during droughts. During periods of more plentiful supply, various practices in the Basin encouraged additional diversions. Without a change to the existing system, diversions would certainly increase over time. At a subsequent council meeting in 1995, it was agreed that there was a need for an appropriate balance to be achieved between water diversions and in-stream water use in the basin. It was decided that water diversions must be capped, and an immediate moratorium was placed on further diversions of water from the rivers of the basin (the 'cap' on diversions). A number of working groups were then established to examine the various issues around the cap, including determining the appropriate level for the cap.

The cap was a very significant decision since in effect it represented a distinct modification in how water was allocated among the states. Water use by the different states was to be determined based on the level of development that existed in the basin in 1994 rather than solely by the formula agreed upon by the states several decades before (that had been accepted by the Murray-Darling Basin Agreement).

Since the introduction of the cap, there has been an annual water audit monitoring its implementation in order to ascertain whether the states are meeting their commitments. Each state has been required to develop reliable monitoring and reporting systems.

The cap has drawn recognition to the importance of striking a balance between consumptive and in-stream (ecosystem) demands, but the cap has also generated interest in a market for the trading water. In response, the MDBC established a Water Market Reform Working Group to fast track the process of advancing water markets and the trading environment. The commission is piloting permanent interstate water trading in the southern part of the basin.

3.4.3 Economic Instruments

Economic instruments can provide both incentives for behaviors consistent with sustainable watershed plans, as well as funding for the institutions that do the planning and management for sustainable watersheds. The lines between the two can be blurred, and thus are presented jointly in this section. Adequate and sustained funding especially to cover core operational costs such as rent and administrative staff support is one of the most commonly cited weaknesses in U.S. watershed management. State and local government funds that are available are often restricted to specific types of programmatic activities, such as wetland and erosion protection, and rarely provide more than a year or two worth of support for the designated activities. The lessons learned from the international basins are that while sustained financial support is necessary for a stable institutional setting for watershed planning and management, the funding can come from a range of public and private sources, and while all successful examples had some level of local funding, state and national level funding sources played a role as well.

Economic incentives may include water pricing, tariffs, and subsidies. Any of these instruments may be used either as incentives or disincentives, and may apply both to water quantity and quality issues. The money raised from these incentives can be used to provide core funding for watershed management agencies, as well as funding for infrastructure or programs for more sustainable water management, as discussed below in the examples of the Fraser River, Grand River, and the Mersey River, respectively. Water and pollution markets are a possible set of economic instruments, but require appropriate regulatory and institutional environments to account for market imperfections and other externalities, as illustrated in the examples of the Hunter River in New South Wales (NSW) and the Murray-Darling at the end of this section.

3.4.3.1 Fraser River

The FBC was created in 1997 as a not-for-profit NGO with a mandate to implement the Charter for Sustainability, the strategic plan for the Fraser Basin. To carry out its assignment, the FBC does not have any regulatory authority to control activities, nor funds to allocate, but it is provided a budget funded by each of the local government signatories contributing 10 cents per capita (totaling approximately $200,000) and by the federal and provincial governments providing the balance equally between them. This broad-based funding mechanism combined with a highly integrated institutional structure gives the full range of stakeholders in the basin some voice in the decision-making processes. The FBC's structure has provided a forum for discussions among the variety of government and NGOs, from local to national scales. The current set of initiatives operating through the FBC is extensive, ranging in scale from basin-wide to specific sub-regions of the basin. Each of these initiatives has input from the full breadth of stakeholders, from the federal to the local level, and from environmental to agricultural and economic interests. A number of successes have been achieved within these projects, both in building partnerships and in developing cooperative solutions.

3.4.3.2 Grand River

A basic system of funding for conservation authorities, including the GRCA was established in the Conservation Authorities Act. The act specifies that funding for the administration and operation of the authority is to be collected through levies from the participating municipalities. Estimated administrative costs are to be divided equally among municipalities and collected through levies. Costs for special projects are born by municipalities also through levies, according to the following system outlined in Section 27 (2) of the act:

> "…after determining the approximate maintenance costs for the succeeding year, the authority shall apportion the costs to the participating municipalities according to the benefit derived or to be derived by each municipality, and the amount apportioned to each such municipality shall be levied against the municipality."

This cost sharing system provides an equitable means of raising municipal funds for watershed improvement projects, where those benefiting most directly contribute a proportionately higher amount. Provincial governments also fund the authority through transfer payments from the Ministry of Natural Resources. In the past, the province has provided 50% or more of the authority's project costs in this way. Recently, however, it has limited its funding support to flood control and flood warning programs. Provincial funding is no longer provided for development of watershed plans, capital works, recreation programs, and environmental improvement programs, such as soil and water conservation and stream rehabilitation. Figure 3-2 shows the changes in funding levels from different sources over the past 10 years. Government grants have dropped significantly from 40% to 9% of the total budget. User fees, rentals, and donations have increased to 61%. Municipal funding has remained approximately the same at 30%.

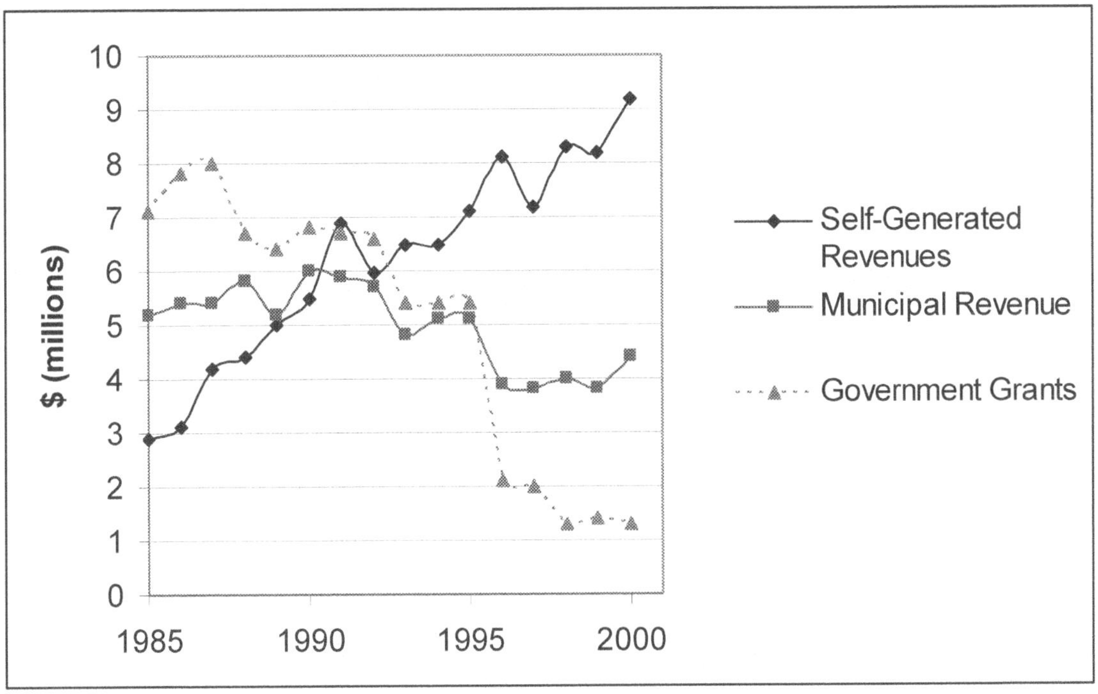

Figure 3-2 Grand River Conservation Authority 2001 Budget (©Grand River Conservation Authority).

3.4.3.3 Mersey Basin

In 1985 the British government committed to providing core-funding support for the MBP for 25 years. The MBC is, in its own words, "a government-backed partnership, which brings together local authorities, businesses, voluntary organizations, and government agencies to deliver water quality improvements and waterside regeneration throughout the Mersey Basin River System." The organization provides the implied functions of forum, focus, and agent of change through a collection of institutions and programs that assist, coordinate, and guide the activities of its many partners. While that funding commitment only covers a small portion of administrative costs today, it accomplished three things:

- It created a sense of confidence in the long-term stability of the nascent institution.
- It provided genuine financial stability during the first several years of operation.
- It enabled the organization to concentrate on strategic and programmatic development.

One of the three branches of the MBC, the Mersey Basin Business Foundation was formed in 1992, with the principal aims of encouraging business participation in campaign activities in collaboration with the public and voluntary sectors, facilitating business input into strategic development of the campaign, providing project support through sponsorship and/or in kind contributions, and improving their own environmental performance. The foundation grew from the initial interest and commitment in 1987 of three large companies—ICI, Shell and Unilever—to support campaign-related projects. To join the foundation a business must make a minimum commitment of £30,000 (~$50,000) over three years. Some corporations make more substantial contributions, such as Shell UK Ltd., who dedicated funds for a senior manager to fill the position of executive director from 1992 through 1998. Although the foundation is technically open to businesses of all sizes, the minimum contribution requirement appears to select for relatively large, often multinational businesses.

The success of these programs and strategies allowed for the observed robust institutional growth by broadening the partner base and improving the campaign's competitiveness in receiving grant awards.

3.4.3.4 New South Wales

The NSW government established Water Management Committees (WMCs) in 1997 to help create management plans for major river catchments throughout the state. These plans provide the basis for water use and trading by industry and communities, river care, and other environmental programs for a period of five years. The minister for land and water conservation and agriculture and the minister for the environment jointly approve the WMC plans.

The WMCs have successfully developed and implemented a variety of creative and innovative approaches that rely on government partnership with the community. The committees typically include representatives of all relevant government agencies, the general community, the farming and fishing industries, indigenous Australian people, and nature conservation interests. Their approaches include developing a strategic planning framework and process to carry out the necessary rapid research and preparation of the plans, and educating the multi-dimensional group of representatives, particularly in conflict management, consensus decision making, and water-based resource management. Indigenous and other cultural elements are brought into committee activities and planning through art, songs, and storytelling.

Another innovative program was initiated in 1996 in the South Creek area of the Hawkesbury-Nepean River. This small, self-contained, emissions trading scheme involves economic and regulatory measures. This bubble scheme allows the three participating sewage treatment systems to adjust their individual discharges, provided the total pollutant load limit for the scheme is not exceeded. This enables efforts to reduce pollution to be focused where the costs are lowest. The regulatory load limits mandated under the bubble scheme require an 83% reduction in total phosphorus and a 50% reduction in total nitrogen by 2004 compared to a "business as usual" scenario.

3.4.3.5 Murray-Darling

The management of salinity problems in the MDB has been one of the major focuses of the initiative throughout its operation. In fact, the issue of irrigation and salinity was considered by the ministerial council at its inaugural meeting, and it was agreed then to establish a number of working groups to study various salinity and related issues in the basin, in addition to the salinity investigations already being carried out by the MDBC's predecessor, the River Murray Commission. A framework for cooperative action by the participating governments to manage salinity problems in the basin was subsequently developed, called the Salinity and Drainage Strategy. The council formally adopted the strategy in 1988. This inter-governmental agreement imposes rights and responsibilities on each government member of the commission, and holds them accountable for all future actions impacting river salinity.

The strategy's objective is to effectively manage the pressing problems of water logging and land salinization in the irrigation districts of the Murray Valley in NSW and Victoria, and the river salinity in the lower Murray River. The essential features of the Salinity and Drainage Strategy are:

- River salinity levels over the period 1975 to 1985 are adopted as the benchmark for attributing impacts of all future actions that affect river salinity.
- The rules for river operation are changed to provide for improved dilution of salinity at critical times.
- Each state is responsible for its actions after 1988 that affect river salinity.
- No actions that increase salinity are allowed unless they are offset by ameliorating works.
- These works are authorized with limited salt disposal rights to the river.
- These rights were negotiated on the understanding that the upper states would provide financial support for salinity reduction works further downstream.
- The rights have become known as salinity credits; they are a tradable pollution right.

In the words of Senator Peter Cook, chairman of the MDB Ministerial Council at the time the agreement was reached on the strategy:

> "For the first time in our history, the states have agreed to tackle a major environmental problem through a common effort across their borders. The strategy paves the way for further interstate cooperation in promoting the sustained use of our land, water and environmental resources."

Trading in Salinity Credits

A salinity reduction of 80 electrical conductivity (EC) is one of the stated goals of the strategy, and the states are allocated prescribed "credits" or "debits" toward this target. Each state has the right to handle the credit or debit in its own territory from that point on, including the right to trade credits and debits with other states. The three main states involved, Victoria, NSW, and South Australia, have each taken a different approach in allocating the credits/debits. In Victoria and NSW, the state pays for the interception schemes that generate credits and then allocates these credits to land and water salinity management plans at the time of their approval. However, the beneficiaries of these credits are expected to bear some of the operation and maintenance costs of these large schemes. Further, the beneficiaries do not own the credits unless they also contributed to the capital costs of the interception schemes. In Victoria, the allocation of salinity credits is based on established cost sharing arrangements and considerations of public and private goods. In NSW, the allocation is based on an assessment of whether a work or measure is in the public interest and has an acceptable benefit–cost ratio. South Australia has taken a different approach altogether, allocating its share of the credits to improve the river salinity, and is committed to remain salinity neutral in its developments. To achieve this, South Australia rigorously imposes policies designed to improve the irrigation efficiency of existing developments and minimize the salinity impact of new developments.

Since 1988 the strategy has achieved a 61 EC reduction against the 80 EC target through the construction of five new or enhanced salinity control schemes. These works cost $44 million in capital investment and require $1.8 million per year for operation and maintenance. The combined pre- and post-strategy schemes prevent 222,000 tons of salt per year from reaching the Murray.

3.4.4 Technology Instruments

Technology instruments may include the research and development of new technologies that better promote the sustainability of natural resources, such as water-saving technologies, and monitoring, but also can contribute to a better understanding of aquatic ecosystem health. Encouraging an assessment of appropriate technologies may also lead to more cost-effective and sustainable technology choices. These instruments are frequently combined to achieve watershed goals. The examples given below highlight the instruments used to promote the integration of point and nonpoint source pollution, including innovative measures, such as effluent trading, and the instruments successfully used to promote ecosystem/habitat protection and restoration.

3.4.4.1 New South Wales

The Hunter River Water Management Committee created a unique and innovative pollution control program called the Hunter River Salinity Trading Scheme (HRSTS), believed to be the world's first 24-hour online credit exchange for a water emissions trading scheme. HRSTS is the only emissions trading scheme based on real-time environmental conditions rather than modeled predictions. It manages discharges of saline water from coalmines and electricity generators to the Hunter River so that river salinity does not exceed levels that are detrimental to agricultural productivity or environmental quality downstream. This is achieved by:

- Extensive and continuous real-time monitoring of environmental conditions and discharges,
- Scheduling saline discharges to complement high river flow rates and low background salinity levels so that salinity targets are not exceeded, and
- Sharing the total allowable discharge according to dischargers' holdings of tradable salinity credits.

By maximizing discharge opportunities in response to river and weather conditions, the scheme also saves the participating discharging industries millions of dollars by reducing the need for large storage dams and desalination programs.

3.4.4.2 Ara River, Japan

The Ara River flows from the northern mountains to Tokyo. Extensive development has lead to ecological deterioration in the river. The Japanese adopted the European approach of neo-natural river reconstruction to address the degradation issues of the Ara River and its tributaries.

In the early 1980s, Germany was cutting forests at an accelerated rate to create space for urban development and agricultural land. There was a collapse in the ecological balance of plants and animals in rivers that were once considered clean, but now lined with concrete. In 1984, the idea of neo-natural river reconstruction emerged from Germany with the intent to address ecological deterioration in the river systems. "Neo-natural river reconstruction consists of the introduction of various species of flora and fauna into a river environment, not with the intent of recreating the organic ecosystem that has deteriorated at the hand of man, but creating a new ecosystem which protects the surviving inhabitants, restores ecological balance and improves the living situation of animal and plant life." Some refer to these locations of restored nature as biotops (meaning "a place for life"). Original policy elements of neo-natural river reconstruction include the following:

- A river basin must be regarded as an ecological complex.
- Pre-existing species of flora and fauna should be preserved to the greatest possible extent, and species diversity in the ecosystem should be enhanced.
- The characteristics of each species' habitat should be developed in harmony with the natural dynamic alterations that take place in a river, such as erosion, sedimentation, and floods.
- A balance of natural and human influences according to the condition of the surroundings should develop the characteristic of each river.
- Consistent maintenance of the natural habitat and its unique characteristics is a priority.
- First preserve what remains of the original biostructure.
- Repair and maintenance should be done in harmony with ecological rhythms and cycles.
- Incorporate new agricultural technology.
- Provide sufficient space as dictated by the ecological system requirements.

Canals were dug into the riverbed where there used to be flat barley fields. The canals vary in width and depth and are intentionally crooked to provide a better medium for species variety. The terrain surrounding the river was made uneven with small mounds and gaps, creating shelter and protection for small animals. Kingfishers were provided wooden posts for resting spots, and the banks of the river were given a gentle slope, allowing for plant growth and fish spawning habitat. The restoration did require the removal of several recreational facilities. In 1994, the Japanese pursued the idea that connecting two or more biotops would create a better environment for a greater number of animals and plants. Using yashiki groves and paddy fields as a stepping-stone network, tablelands and forests are connected.

After only five months of biotop construction, willows were sprouting and kingfishers were breeding. Foxes, raccoon dogs, and various types of ducks now inhabit the area. The Ara River Basin is now home to an extensive wildlife population with a variety of water animals and plants. The basin is also home to some species that are on the Japanese endangered species list. While the neo-natural river restoration concept is beginning to spread, the outreach has not been extensive. Valuable information for future projects will come from the monitoring of wildlife in current biotops.

CHAPTER 4.0

SUMMARY GUIDANCE

4.1 Summary Guidance

Several lessons can be drawn from the examples above that may be appropriate for U.S. applications. Perhaps the most important lesson is that watershed management is a dynamic and evolutionary process that combines increasing local awareness of watershed issues with increasing local responsibility for action. "Local" will vary across the scale of the basin. International examples of effective watershed approaches were initiated either because of pressure from the local public (e.g., Fraser River Estuary, the Rhine, and the Mersey Basin) or involved the local public through a state or national initiative (e.g., Australia, Mexico, Germany, and NSW). In either case, the local public and the watershed stakeholders began a process of capacity building that allowed for the development of an understanding of the physical processes of the watershed, and the social and economic value of water in its competing uses. Also, these successes in watershed management established institutions to facilitate discussions and subsequent actions across jurisdictions, and to provide opportunities for meaningful stakeholder participation.

Watershed management authorities in the international context that have been reviewed here have some degree of self sufficiency in funding, through a combination of water and pollution charges or resource-associated levies on stakeholders. In return, the local users have influence over decisions that may affect their use and/or enjoyment of the watershed. Watershed stakeholders and institutions (which vary according to local conditions) have used a variety of instruments to promote effective watershed management, as discussed in Chapter 3.0. Among the most important are the uses of:

- Economic instruments as a means of involving stakeholders and finding better, more efficient solutions to watershed issues;
- Regulatory instruments to allow cooperation among agencies across jurisdictions, and give decision-making authority to the most locally appropriate decision makers;
- Information and communication instruments, which can be combined to promote a common ground for discussion among stakeholders, both for competing users and various agencies; and
- Technology instruments, which have the potential to provide win-win opportunities in cases of conflict.

Finally, place matters—whatever institutions or instruments have proven to be useful, there is no standard approach or blueprint for effective watershed management. Rather, these elements must be tailored to meet the local conditions, such as climate, demographics, natural resources, existing uses, institutional, political, cultural, and economic context.

4.2 Lessons

Watershed management practices in the United States can be improved based on the following lessons from global experience.

Lesson 1: Utilities play a critical role.

Utility service territories rarely follow watershed boundaries. However, utilities' withdrawal/discharge practices and their interest in providing high quality water and wastewater services, make them critical players in watershed planning and management. In cases where utilities cover relatively small geographic areas, such as a single municipality or small region, groups of utilities might form partnerships defined by watershed boundaries as a platform for joint planning and management. Activities could then be coordinated with local and regional watershed management initiatives, with water and wastewater utilities serving as important sources of technical and financial resources.

In the Mersey, for example, utility involvement was necessary to effect the dramatic improvements in water quality, but alone would not have been sufficient. Certainly the geographic scale of the utilities' territory and the coordinated careful planning process under which improvements are carried out have also been of great importance. United Utilities serves all of the Mersey River Basin and has worked cooperatively with other key actors in the MBC to develop the long-term improvement plan.

While U.S. utilities have made significant investments in wastewater infrastructure since the 1970s, often in response to regulatory requirements of the Clean Water Act (e.g., the Massachusetts Water Resources Authority's efforts to facilitate the clean up of Boston Harbor), it is rare to find the confluence of factors found in the Mersey:

- A large watershed that falls within the jurisdiction of a single water/wastewater utility,
- The ability and willingness of that utility to make substantial monetary investments,
- Cooperative regulators and a regulatory process that encourages strategic long- and middle-range planning, and
- A well organized watershed initiative with deep voluntary and business support.

Nevertheless, the leading role played by United Utilities has played in the Mersey Basin is something water/wastewater utilities and/or watershed managers in the United States might emulate.

The situation in the Rhine is more typical of the United States: many jurisdictions and at least historically little coordination among them. There is a long history of stakeholder involvement in the Rhine, although this originally occurred absent a formal institutional structure. Some of the earliest influences came from the Dutch waterworks organization RIWA in the 1930s that formed in response to deteriorating water quality. Soon after its formation, RIWA contacted the Dutch government with proposals for action. Furthermore, to improve contacts with other governments, RIWA joined forces with fellow organizations in the Rhine catchment area to form IAWR. In its various Rhine memorandums the IAWR has proposed extensive demands on the governments of the riparian states along the Rhine. Water utilities in the Rhine, as represented by IAWR, strongly and successfully promoted the innovative concept that waterworks should ideally be able to rely on simple treatment only to supply good drinking water quality. This can only happen with high quality source water. The ICPR ultimately bought into this cutting-edge philosophy and adopted it in the bylaws of the commission.

Lesson 2: Multi-stakeholder processes provide a forum for effectively managing watersheds, including jointly addressing point and nonpoint pollution sources.

One of the consistent obstacles to effective watershed management in the United States has been the difficulty in successfully engaging stakeholders and gaining their support for new approaches to manage the watershed. The credibility and success of a watershed initiative depends largely on the degree to which key interests and stakeholders participate throughout the process. Involvement at the earliest stages is critical, as there are numerous examples of the high costs and lack of success that can occur when stakeholders are not engaged early in the process. If important stakeholders only become involved in later stages, important interests may not be addressed, and opposition and delays are more likely to result.

The history of watershed management on the Rhine, for example, demonstrates the need for balance among stakeholders. For much of the late 19th and early 20th centuries, the fishing industry lacked a voice in the river's management. The result was the complete elimination of the most valuable fish on the Rhine, the salmon, as well as decimation of the fishing industry. In the latter part of the 20th century, however, the ICPR was able to achieve a unique balance among stakeholders in the basin. This is a balance not only between upstream and downstream interests, but also among industrial, agricultural, hydropower, and ecological interests. Such balance is possible in the United States (and undoubtedly occurs in some watersheds), but the equal footing given to ecological protection and restoration in the Rhine remains the exception rather than the rule.

Moreover, meaningful participation and open communication throughout the process provides the time necessary to build the trust among stakeholders and the institutional commitments required for the initiative to succeed. This trust is also built by creating a common understanding of watershed issues and challenges by providing all parties access to key data concerning watershed conditions. Watershed initiatives often help bridge the gap between government and NGOs. In addition, international experience demonstrates the effectiveness of institutional arrangements that can lead watershed initiatives with a measure of independence, while still being accountable and having the capacity to operate efficiently.

The initiative in the Fraser Basin shows not only the potential and essential role of collaborative multi-stakeholder processes in pursuing sustainability but also the challenges of putting them into practice in watershed management. One of the reasons that the FBC has been successful in raising awareness of the Fraser watershed is its ability to provide a middle ground between government and NGOs and interests. The credibility of the council depends on it operating, and being seen to operate, with a measure of independence while still being accountable and efficient.

The Fraser Basin experience illustrates a common lesson with broad applicability: the need for all stakeholders to be present, to understand, and to be committed to the intent and language of the management approach and plan. The rapid drafting of an initial agreement in the early 1990s and its swift passage through bureaucratic and political approvals of three levels of government were misleadingly impressive. It could have been disastrous, and it took years following the agreement to build trust among the stakeholders so that they could work jointly towards the shared goals of sustainability, in the environment and society.

The problem of nonpoint pollution will only be solved with the ability to address decisions about changes in land use and application of fertilizers and other chemicals to agricultural and non-agricultural lands. Historically, this has proved difficult both in Canada and the United States, where the governing institutions of land and water often do not share the same jurisdictional boundaries, or common mandates regarding the management and preservation of the watershed.

Watershed initiatives that involve the full range of parties—from national to local on the government side, and from environmental to agricultural and economic interests on the non-government side—help break down some of these institutional barriers. The Fraser Basin in British Columbia is an example where such barriers are being overcome, and as described in the Grand River case study in Chapter 3.0, the establishment of Conservation Authorities in Ontario represents a successful institutional forum established in watersheds throughout the province where point and nonpoint sources of pollution are addressed on an integrated basis.

Lesson 3: Large-scale watershed management can succeed by cultivating and integrating discrete subwatershed and stakeholder initiatives.

A serious deficit in the U.S. watershed management movement is the absence of large-scale watershed management initiatives in most of the nation's river basins. Effectively involving the full spectrum of stakeholders (something the MBC, for example, does extremely well) is one of the obstacles to the establishment of such initiatives and a challenge to the few that do exist.

To improve the effectiveness of watershed management initiatives in very large basins requires special effort to involve the full spectrum of stakeholders. One promising method is to develop individual institutions to serve subsets of stakeholders and to represent those stakeholders' interests within the larger watershed planning and management process. This may require creating organizational capacity dedicated to managing the nested smaller watersheds to ensure two-way communications and integration of sub-watershed concerns and activities with the larger watershed initiative. The campaign's River Valley Initiatives represent a possible model, one with inherent flexibility and emphasis on local goal setting and control that is likely to have appeal in the United States. Moreover, the campaign's linking of economic and environmental improvement contributes to its success, attracting business and local governments in the economically depressed region that might be hostile to an exclusively environmental message.

Lesson 4: Integrative "win-win" methods support the resolution of upstream-downstream and human versus nature conflicts.

Tensions between upstream and downstream regions and human versus ecosystem interests exist in many U.S. watersheds, particularly large ones. Creation of watershed institutions or working committees that involve both upstream and downstream users can help avoid open conflict and work towards mutually beneficial solutions. A formal consultative body provides a forum to continue negotiations and to resolve substantive and political barriers. This structure also promotes exploration of innovative options such as the use of cost sharing among upstream and downstream parties in cases where regulatory authority is lacking, or the provision of financial incentives from downstream interests to encourage upstream parties to modify policies and/or practices that degrade resources downstream. Examples of this approach can be seen in the efforts by several large U.S. utilities (e.g., New York and Boston) to meet the requirements of the Safe Drinking Water Act by investing in upstream watershed protection measures rather than constructing new high-cost filtration plants.

This was the case in the MDBI in which a salinity and drainage strategy was developed with an explicit statement that "upper states would provide financial support for salinity reduction works further downstream." Another example can be found in the Rhine River Basin where the Netherlands, the most downstream country, consistently adopted strategies that avoided open conflict with upstream stakeholders. Using diplomatic channels, the Dutch promoted the creation of the ICPR, and then used this body to work with the country's upstream neighbors or stakeholders. The Dutch went further, with the consistent use of financial incentives to upstream parties to invest in pollution reduction activities, such as its funding for measures to reduce the amount of salt discharged to the Rhine from French mines. From an institutional and legal perspective, the complex of institutions revolving around water in the riparian nations of the Rhine are at least as complicated as river basins in the United States that cross county, state, and national borders. Yet, the parties in conflict have been able to find solutions that are satisfactory to all.

Lesson 5: An engaged civil society can provide authority that may be lacking in the watershed organization.

Watershed initiatives rarely enjoy direct executive or regulatory powers. Rather, their role is often one of convening, facilitation, planning, and assessment to inform decision making about policy and project implementation. In many of the international cases reviewed herein, the initiators of watershed management efforts lacked formal authority. Nonetheless, by effectively engaging the key stakeholders, especially NGOs (or "civil society") and the public, watershed initiatives often gain de facto authority by influencing government and other key players to plan and implement their priority projects and policies. Cooperative relationships between participating governments and NGOs through watershed initiatives also avoid lawsuits and promote negotiations among parties to find broadly acceptable solutions.

For example, the ICPR does not have any executive or coercive power to speak of, such as the authority to order the construction of new facilities or the imposition of standards. Its role is to carry out research into the nature, extent, and origin of the pollution in the Rhine; to propose measures to improve conditions; and to prepare international agreements to help implement these measures. The commission identifies environmental projects for the Rhine but, if adopted, they are funded and implemented by each member country. The commission simply monitors their application along the river's entire length. The non-regulatory approach in the Rhine is partly successful because of the role taken by NGOs, such as the waterworks lobby and the environmental movement, who put pressure on both governments and industries on a national and an international level. It should be noted, however, that judicial proceedings by NGOs have also been a powerful force in the Rhine, and have likely increased the responsiveness of governments and industry to the commission.

Similarly, the MBC essentially went "over the heads" of local government leaders with a direct public appeal and message that environmental improvement and economic development were inextricably linked.

Where watershed organizations in the United States do not have formal regulatory powers, the involvement of NGOs and the larger community may be a useful tool to create political support and achieve the goal of sustainable water management.

Lesson 6: Institutional stability and a clear mandate for watershed management can reduce fragmentation of authority and result in more efficient planning and implementation.

One weakness in watershed management in the United States is the lack of strong watershed-based institutions or agencies with authorities and expertise that are well understood and accepted by the public and municipalities. Too often, fragmentation of responsibilities among several different agencies involved in watershed management leads to duplication of efforts, unclear mandates and conflicting missions. Where governments have established clear frameworks and mandates for comprehensive watershed management, stable watershed institutions and planning processes often develop. This institutional stability is an important contributor to the long-term success of watershed initiatives. In some cases, enabling legislation can provide this mandate, by specifying the structure of the watershed management institution, its roles and responsibilities, as well as its jurisdiction, membership, and funding. For example, the institutional structure of the GRCA has not substantially changed since it was created in 1948 under provisions of the Conservation Authorities Act. This institutional stability, along with its consistency across the Province of Ontario in the 38 other authorities, has been a major factor in its success. The authority's clear mandate supplies legitimacy for important actions, such as promulgation and enforcement of watershed-wide regulations and integration of watershed management with local land-use planning.

In other instances, government initiatives can promote collaboration or consolidation of watershed management activities. Alternatively, government funding programs can include eligibility criteria that foster coordination among multiple agencies and key stakeholders.

The Tennessee Valley Authority and the CALFED Bay-Delta Program are two U.S. examples of strong federal (and federal/state) institutions with clear mandates for comprehensive large-scale watershed protection. However, past U.S. experience shows that existing federal agencies have traditionally been reluctant to yield control or jurisdiction to new watershed-based institutions. This may be more likely to occur at the state level. For example, Washington has initiated a state-wide process by which local governments and water utilities can choose to form an entity for watershed planning and protection with funding from the state. Legislation has given these local watershed entities legitimate authority to direct the management of state lands within the watershed. The success of this initiative has yet to be evaluated, but it provides an example of how this lesson could be applied in the U.S.

Lesson 7: Instilling regulatory authority in a watershed-based institution can facilitate effective watershed protection across political boundaries.

While virtually all states and municipalities have laws and regulations to protect water and other natural resources, their implementation and enforcement is rarely carried out on a watershed-wide basis or by watershed-focused organizations. The result is that, often, development is permitted and/or poorly monitored in one municipality without noticeable adverse effects; meanwhile, such development may be contributing to severe flooding or water supply problems for other municipalities downstream. Regional planning entities generally lack this authority as well, particularly in terms of land-use decision making.

From the international arena, there are cases in which laws or regulations provide direct authority to watershed organizations to operate across political boundaries. In such cases, traditional jurisdictional borders are irrelevant when considering the natural resource impacts of proposed activities. With this authority, watershed-based organizations can adopt regional environmental standards or issue permits for development, based on water quality and other environmental criteria to protect the watershed. For example, the Conservation Authorities Act in Ontario provides watershed-based conservation authorities the power to establish regulations applicable to their area of jurisdiction, such as restrictions on development or placing of fill in wetlands and floodplains, as well as restrictions on alteration of watercourses. This has allowed conservation authorities to monitor and regulate local development activities with watershed management principles in mind. This has resulted in the creation of greenways along riparian corridors, as well as in reduction of impervious surfaces and downstream flooding impacts.

Although politically difficult to adopt in much of the United States due to resistance by local governments, the benefits of watershed-based institutions with real authority could be significant. Such obstacles may be more easily overcome in regions with strong county systems, where certain land-use decisions may already be taking place at a regional level. Removing the constraints imposed by municipal boundaries on consideration of costs and benefits of various activities can often result in vastly improved administration and enforcement of resource protection regulations. There may be mechanisms to ease adoption of this approach. At the state level, for example, existing watershed initiatives or inter-agency forums for watershed planning could establish model watershed-based regulations for implementation at the local level. Incentives for municipal adoption and implementation could include access to state revolving funds or other grants for planning and improvement projects.

Lesson 8: Explicit policies and guidance documents can be used to promote the integration of watershed and land-use planning at the local level.

Issues of water supply, wastewater assimilation, groundwater recharge, and local ecology are critical to the sustainability of localities and their population. While most municipalities have local zoning ordinances and periodically engage in planning, most tend to focus on land use and zoning, and are not integrated with watershed management plans. Even where a detailed watershed plan has been developed, essential pieces may not be referenced in the local land-use plan. Directly linking municipal land-use planning and watershed planning processes can be accomplished through an array of regulatory, policy, and educational efforts. Where watershed plans have been developed and approved by state or local agencies, local land-use planners, planning boards, and boards of zoning appeal should be directed or encouraged to incorporate relevant elements into local planning documents. Similarly, local planning officials could commission a sub-watershed study prior to their review and permitting of large-scale development projects under local planning regulations. Because local approval of and commitment to the watershed plan and its recommendations are essential to its integration with land-use plans, local municipal officials should be directly involved in the watershed planning process. Clear guidance for the integration of watershed planning with local land-use planning could be provided by states, along with incentives for local follow-through. Incentives could include state grants and/or assistance for projects in municipalities that have incorporated elements of approved watershed plans into municipal policies.

This integration is precisely what occurs in the Grand River Basin and other watersheds throughout Ontario, where municipal land-use planning and watershed planning processes are seen as inextricably linked. Guidance from provincial agencies in the form of legislation and memoranda of understanding has led to establishment of a process by which 1) conservation authorities produce watershed plans; 2) watershed plan recommendations are incorporated into municipal official plans; 3) sub-watershed plans are developed when necessary to guide impending land-use changes at the local level; 4) official plans are then amended to incorporate the more detailed recommendations of the subwatershed plans; and finally, 5) specific site development plans are tailored to meet the requirements of the newly amended official plans. This straightforward process, clearly outlined by the provincial government and implemented locally, fosters cooperation between planners and watershed managers and leads to development that meets the needs of the watershed as a whole. Without this integration, some of the most important watershed management recommendations might not be implemented at the local level.

Lesson 9: A system of apportioning costs and benefits equitably across the watershed can help secure consistent municipal participation and funding.

Because watershed management activities naturally take place at the local level, they frequently require the acceptance and involvement of municipal officials. Often, recommended watershed improvement projects, such as separation of combined storm-sewer systems, require substantial support from municipal agencies through long-term dedication of funds as well as disruptive, on-the-ground implementation efforts. For this reason, it is critical to involve municipal decision makers in the process of watershed planning from the outset. One way to engage and secure municipal participation is to establish a system for equitable distribution of watershed management costs and benefits. A common example is of downstream water users offering assistance to upstream communities in the adoption of water conservation and water quality protection measures. Regional utilities could play an instrumental role in the coordination and balancing of municipal contributions to watershed management projects and ensuring that the costs and benefits are being distributed equitably.

The GRCA example goes beyond just involving municipalities in the watershed management process, and actually makes municipal membership the foundation of the watershed organization. Further, municipalities agree up front to share in the cost of the organization's operation and implementation of special projects. Member municipalities find this acceptable because the costs and benefits of watershed management are distributed equitably. Participating municipalities receive monetary assistance for improvement projects of direct benefit to them, while they contribute to other projects only in the amount proportional to the indirect benefit they receive.

The FBC was created in 1997 as a not-for-profit NGO with a mandate of implementing the Charter for Sustainability, the strategic plan for the Fraser Basin. To carry out its assignment, the FBC does not have any regulatory authority to control activities, nor funds to allocate, but it is provided a budget funded by each of the local government signatories contributing 10 cents per capita (totaling approximately $200,000) and by the federal and provincial governments providing the balance equally between them. This broad-based funding mechanism combined with a highly integrated institutional structure gives the full range of stakeholders in the basin some voice in the decision-making processes. The FBC's structure has provided a forum for discussions among the variety of government and NGOs, from local to national scales. The current set of initiatives operating through the FBC is extensive, ranging in scale from basin-wide to specific subregions of the basin. Each of these initiatives has input from the full breadth of stakeholders, from the federal to the local level, and from environmental to agricultural and economic interests. A number of successes have been achieved within these projects, both in building partnerships and in developing cooperative solutions.

The Murray-Darling has taken a flexible approach to apportioning costs in the basin. For example, a reduction of 80 EC in salinity is one of the stated goals of the MDB strategy, and the states are allocated prescribed "credits" or "debits" towards this target. Each state has the right to handle the credit or debit in its own territory from that point on, including the right to trade credits and debits with other States. The three main states involved, Victoria, NSW, and South Australia, have each taken a different approach in allocating the credits/debits. In Victoria and NSW, the state pays for the interception schemes that generate credits and then allocates these credits to land and water salinity management plans at the time of their approval. However, the beneficiaries of these credits are expected to bear some of the operation and maintenance costs of these large schemes. Further, the beneficiaries do not own the credits unless they also contributed to the capital costs of the interception schemes. In Victoria the allocation of salinity credits is based on established cost sharing arrangements and considerations of public and private goods. In NSW, the allocation is based on an assessment of whether a work or measure is in the public interest and has an acceptable benefit cost ratio. South Australia has taken a different approach altogether, allocating its share of the credits to improve the river salinity, and is committed to remain salinity neutral in its developments. To achieve this, South Australia rigorously imposes policies designed to improve the irrigation efficiency of existing developments and minimize the salinity impact of new developments. Since 1988 the strategy has achieved a 61 EC reduction against the 80 EC target through the construction of five new or enhanced salinity control schemes. These works cost $44 million in capital investment and require $1.8 million per year for operation and maintenance. The combined pre- and post-strategy schemes prevent 222,000 tons of salt per year from reaching the Murray.

Lesson 10: Watershed decision making at the lowest appropriate level is most effective.

Management of large-scale watersheds, such as the Mississippi or the Colorado, is notably lacking in the United States. One possible explanation for this lack is the concern at the smaller scales of a loss of decision-making authority. To allay such concerns and to encourage decision making by the most affected parties, decisions should be made at the lowest level appropriate for a given issue. For issues with basin-wide implications, policy decisions should be made at the highest watershed-wide level, while specific implementation issues and concerns impacting a limited part of the basin, should be made on a more local, sub-watershed basis.

The approach taken in the MDBI ensures that decisions that can be made at the local level remain in the hands of local decision makers. The initiative has taken a consistent approach in its decision making and implementation efforts; decisions with basin-wide impact are made by the ministerial council, but only in terms of the impact on or the apportionment to each state. The state then retains the right to implement the decision as it sees best without council influence. This approach has been successfully used in the salinity and drainage strategy, as well as the cap on total diversions from the Murray-Darling.

The challenge is to integrate the decisions made at the various levels so that all they are consistent with each other and a basin-wide plan. Such integration requires systematic ongoing communication among stakeholders and decision makers at all levels.

4.3 Conclusions

These 10 lessons have potential for broad application in the United States. While drawn from the five more detailed case studies, they are further supported by the examination of 16 additional watershed approaches reviewed in the Compendium (Appendix C) of this report. While some of the lessons would require the creation of new institutions and/or granting new regulatory authority, several can be implemented in existing frameworks. Certainly utilities can be a driving force for positive change in any of the institutional and regulatory frameworks currently in place in the United States.

APPENDIX A

INTERNATIONAL CASE STUDIES

APPENDIX A1

THE RHINE RIVER CASE STUDY

A1.1 Introduction

The Rhine River Basin offers an example of how a forum for discussions between divergent interests has lead to the successful management of the watershed as an integrated whole. These interests include not only the classic conflicting views of upstream and downstream users, but also the equally compelling conflict between human and ecosystem uses of the watershed. This case study gives the physical and historical context of the establishment of the effective management regime in the Rhine, followed by lessons that may support better watershed management in the United States. The parties on the Rhine have used win-win strategies to resolve these conflicts with a high degree of success. Involving stakeholders and the broader public has been a crucial factor in moving the relevant governments and institutions toward fully integrated watershed approaches.

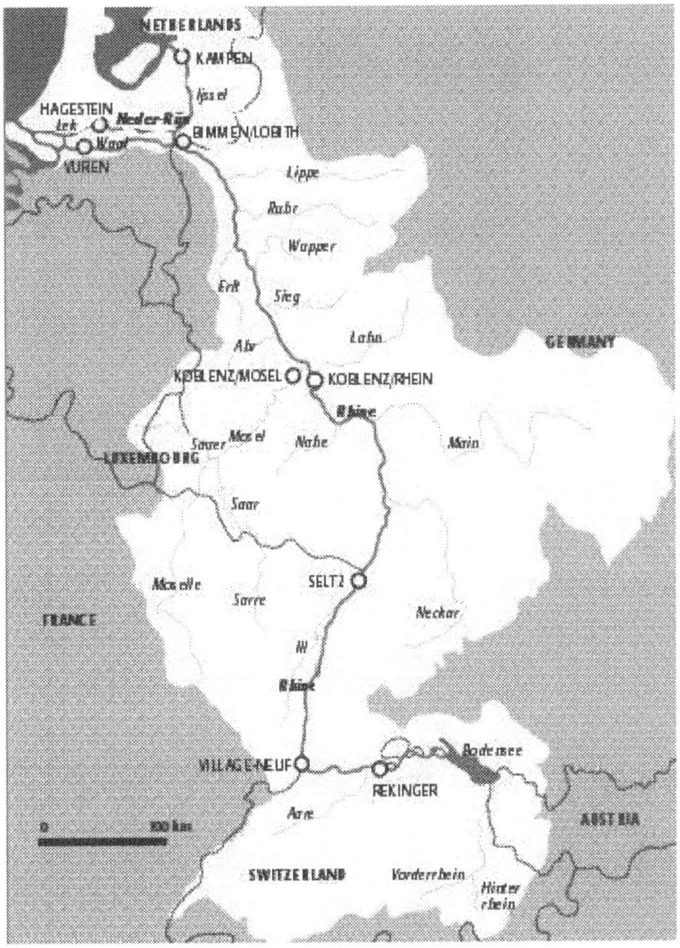

Figure A1-1. The Rhine River Basin
(http://www.ce.utexas.edu/prof/mckinney/ce397/Topics/Rhine.htm)

A1.2 Basin Description and Current Water Uses

The Rhine River originates in Switzerland, where small streams come down from a height of more than 3400 m, combine and meander over 1320 km before discharging into the North Sea. At the Dutch border the average annual flow is 2400 m^3/s, more than twice as much as at the Swiss border. The flow, length, and the size of the Rhine catchment area (200,000 km^2) make the Rhine the most important West European river.

There are nine riparian states on the Rhine, including Italy, Austria, Liechtenstein, Belgium, Luxembourg, Switzerland, France, Germany, and the downstream Netherlands, with a total population of 60 million. The latter four countries are the most dependent on the Rhine. While crossing the territories of these four countries, the Rhine fulfills different societal functions. Table A1-1 indicates that from a geographical point of view the downstream countries are the most dependent on the resource.

Table A1-1. Societal Functions in the Rhine Catchment Area.

	Switzerland	*France*	*Germany*	*Netherlands*
Drinking water			X	X
Process water	X	X	X	X
Irrigation				X
Hydro-power	X	X		X
Recreation	X		X	X
Fishing				X
Navigation	X	X	X	X
Receiving water for wastewater discharges	X	X	X	X

About 20 million people depend on Rhine as a source of drinking water, with the majority living in Germany and the Netherlands. The Dutch are the most dependent on the Rhine as a source of drinking water. In the western part of their country there are no alternatives because the groundwater is too brackish to be a source of drinking water.

Industrial water use occurs throughout the Rhine basin, with the highest concentration found in Switzerland near Basel, in Germany around Mannheim and in the Ruhr area, and in the Rotterdam port area of Holland. The Rhine catchment area is one of the most important industrial areas of the world. No other river has as many chemical plants on its banks as the Rhine (Ruchay, 1995).

Water for irrigation is particularly important for the Netherlands, not only for agricultural production, but to prevent saltwater intrusion in the half of the country that is below sea level. In contrast, although some hydropower is produced at the sluices in the Rhine in the Netherlands, the steeper parts of the catchment offer the best possibilities for hydropower development. Most hydropower stations are in France, where 10 stations operate in the canalized part of the Rhine (the Grand Canal d'Alsace). Similarly, the Swiss and German power stations are located in the steeper tributaries.

The Rhine catchment area also offers recreational opportunities in the upstream parts of the river, for instance Lake Constance and the Schaffhouse falls in the middle part, the German Romantic Rhine, and in the delta area in the Netherlands, where some wetlands with an international status (Ooypolder, Biesbosch, Waddensea) can be found. Several sites in the catchment area are also used for aquatic sports like swimming, sailing, and sport fishing. Professional fishing is mainly concentrated in the Netherlands.

Due to a favorable distribution of precipitation over the catchment area, the Rhine has a rather equal discharge, which historically has made the river useful for navigation. Of its 1300-km length, 880 km are navigable. To improve the natural situation, parts of the river have been canalized. Today, the Rhine is a very important inland waterway that connects Rotterdam Harbor, the largest seaport of the world, with Duisburg, the largest inland port of the world, as well as other industrial areas in Germany and Switzerland. Since 1992, it has been possible to go by boat from Rotterdam as far as the Black Sea. Over 180 million tons of goods a year are transported by boat on the Rhine (Huisman et al., 2000).

Finally, the Rhine is also used as a receiving water for wastewater discharges, both for industries and municipalities. Large companies like Sandoz and Ciba-Geigy in Switzerland; BASF, Hoechst, and Bayer in Germany; and MDAP, a French potassium mining company, all discharge wastewater into the river. Also, the 60 million people who reside and work in the catchment produce wastewater that also flows into the river.

A1.3 Impetus for Management on a Watershed Basis

Due to the multiple uses of the Rhine catchment, it is no surprise that conflicts have arisen throughout its history. Conflict around a particular use often proved to have a strong influence over the management of the Rhine as a watershed. As conflicts would arise, resolution was made to the specific issue, so that solutions initially tended to be fragmented. For example, there were separate agreements that developed around the issues of navigation in the 19th century, fishing in the early 20th century, and pollution in the mid to late 20th century. Although each of these agreements treated the Rhine as a watershed, they failed to integrate these issues within the function of the watershed as a whole, ultimately leading to more serious consequences for the Rhine's ecosystem as well as for human use. It was not until the late 20th century that a serious chemical spill forced the issue of integrating the watershed approaches for multiple functions of the Rhine.

The earliest motivating force for managing the Rhine as a watershed revolved around the use of the river for navigation. Although there were some physical obstacles, the major obstacles were due to the numerous rulers along the length of the river who sought to profit from the navigation through the imposition of tolls and duties. A peace conference in Vienna in 1815 led to the creation of the Central Commission for the Navigation of the Rhine, which ensured freedom to navigate the Rhine across country borders without tolls or duties.

A further impetus for looking at the Rhine as a shared watershed occurred in the late 1800s over the issue of salmon fishing. In the mid-1800s, fisheries were an important industry along the Rhine. Each town had its own fish market. At that time, more than half a million Atlantic salmon made their annual migration to the Rhine. Fifty other species of fish could also be found in the Rhine. In 1885 a quarter of a million salmon were caught in the Rhine. But the stocks were already declining. A treaty was agreed upon by the Rhine states in 1885 that prescribed that the riparian states would undertake all efforts to preserve the salmon stock at a high level. The treaty envisaged efforts to preserve the spawning grounds in the Rhine and its tributaries. Unfortunately, navigation and hydropower interests had priority over fishing interests. The steady construction of dams and weirs, as well as the canalization of the river continued, virtually destroying all fish spawning grounds in the Rhine by 1958 (Huisman et al., 2000).

A further watershed level issue arose with serious upstream-downstream conflicts in the early 20th century around the use of the river as a drinking water source. The Rhine had long had the nickname of the "sewer of Europe" because it had been used as a "sewer" by upstream countries. In addition, the Rhine Delta is very vulnerable to salinization by the North Sea. The Dutch government began protesting against the increasing pollution in 1932, but to no avail until the end of the World War II, when a new commission was formed in 1950 with the support of Germany, France, Luxembourg, the Netherlands, and Switzerland: the International Commission for the Protection of the Rhine (ICPR). The upstream riparian states publicly recognized that the pollution had an affect on the drinking water for the Dutch, but there remained a perception that it was unfair for the upstream countries to bear the financial burden of cleaning up their portion of the Rhine, while the Dutch could continue to pollute the Rhine estuaries and the North Sea (which was largely true at the time).

The pollution of the Rhine reached a critical point in 1971. A combination of heavy pollutant loadings and low flow contributed to oxygen levels insufficient to support aquatic life in large sections of the Rhine. The public was shocked at the state of the Rhine, and by 1972 the riparian governments began to take concrete steps to reduce the pollution of the Rhine. Two treaties were established in 1976 that required the step-by-step elimination of substances determined to be dangerous or on the "black-list" and reduction of substances on the less dangerous "gray-list." But due to mutual mistrust, the treaties were very detailed and complicated. Progress was made in terms of the instream water quality, but only incrementally.

A massive chemical spill occurred in 1986, killing more than 500,000 fish and thousands of livestock that used the river, as well as contaminating the downstream drinking water supplies. Some scientists at the time predicted that the river would be biologically dead for the next 20 years. This catastrophic spill catalyzed popular support and the political will to create a 15-year Rhine Action Plan (RAP) under the authority of ICPR with emphasis on ecological preservation and restoration. This plan was a turning point in management of the Rhine as a watershed. It integrated the diverse interests and uses of the river by covering water quantity, water quality, and the shipping trade and also including the exchange of data and joint research. Public outreach played a large role in the continuing public support of the plan. The first two phases of the plan have been completed: identifying and eliminating the main sources of pollution with a more active program of monitoring, including pollution patrols, and imposing large fines on polluters. By 2000, not only did the Rhine see the return of salmon, but also other species that had disappeared, including sea trout, flounder, river lamprey, bullhead, and perch.

A1.4 Institutional Arrangements

The institutional regime governing the use of the Rhine has changed, both in its role and structure as the perception of the issues to be addressed have changed. Some of the more interesting elements of the Rhine's institutional arrangements are how cross-jurisdictional issues are addressed, the extent of involvement of stakeholders, and the sources of funding to implement effective policies. Each of these elements is discussed in the following sections.

A1.4.1 Multi-jurisdictional Issues

The first watershed level institution for management of the Rhine, the Central Commission for the Navigation of the Rhine was established in the early 19th century to allow for free navigation of the river. The sovereign states at that time voluntarily agreed to the right of the commission to ensure free transport of goods along the Rhine. This commission was effective in achieving the goal of free navigation and remains the oldest active river commission in the world. However, the central commission did not have the authority to address non-navigational issues. Although the subsequent salmon treaty was ratified by the riparian states in 1885 to protect fishing interests, there was no institutional basis and the treaty therefore remained largely effectual due to the dominating interests of hydropower and navigation.

By the early 20th century, it was clear that there was no forum for a dialog between upstream and downstream interests. The Dutch had become increasingly dependent on the Rhine as a drinking water supply. Representatives of the French potassium mines in the Alsace had made a proposal prior to World War II to discharge waste salt water into the Rhine. Dutch drinking-water companies complained to the Dutch government that this would pose a serious threat to drinking water supplies. The war blocked any progress, but after World War II, when the mines had begun to discharge, studies showed that in addition to deteriorating drinking water quality, agriculture in the Netherlands would also suffer from the increased salinization. As a structure for debate was missing, the Dutch Ministry of Transport and Public Works decided to start consultations by making an appeal through diplomatic channels to other riparian states to discuss the Rhine problem informally. This appeal resulted in the establishment of the ICPR in 1950. A limited number of actors from the Rhine catchment area started to communicate in an informal and initially unstructured way on Rhine quality issues, with a small group of researchers meeting once a year to discuss national reports about the quality of the river (Dieperink, 2000).

Attempts by the Dutch in the 1950s to use the ICPR to discuss policies to prevent the already high levels of salt load from further increasing were barred by the French and Germans. They argued that the ICPR was exclusively a research organization and had no authority to discuss policy proposals. Several years of negotiations between the parties ensued that resulted in the first legally enforceable treaty in the Rhine region, the Treaty of Bern in 1963. This treaty gave a formal structure for discussions of issues, and further established that the ICPR was authorized to carry out research into the nature, extent, and origin of the pollution in the Rhine, and also had the authority to propose measures and prepare international regulations (agreements). However, the Germans and the Swiss successfully opposed Dutch proposals to extend the authority of the ICPR to the tributaries of the Rhine.

The 1963 Treaty of Bern also addressed a further institutional problem, inadequate feedback mechanisms from member states. The ICPR was put under the leadership of a president who was appointed for three years and was responsible for chairing an annual plenary conference. During these plenary meetings, delegates from the each of states along the Rhine would meet to discuss and/or debate the ICPR policies. In addition, it was formally agreed that information on different aspects of Rhine pollution could be exchanged in the context of working groups.

Although the Treaty of Bern successfully established a forum for discussion, the ICPR was said to operate in a vacuum, particularly from the Dutch perspective. As the Dutch saw it, the ICPR delegations lacked the political skill to get concessions from the various state governments. Given the continued increase in pollution, the Dutch government decided that the most effective strategy would be to convene a special ministerial conference on Rhine pollution. By making a number of diplomatic moves, they were able to convince the other riparian states that such a conference was important. A ministerial conference was held in 1972 (since 1972, ministerial conferences on Rhine pollution have continued to be held periodically, with the most recent held in January 2001) and charged the ICPR to develop two treaties for the Rhine basin, one on salt and one on chemicals. In 1976, both the Rhine Salt Treaty and Rhine Chemicals Treaty were ratified by the riparian states, but issues arose to delay implementation.

For the following 10 years, little progress was made. Problems arose with the Rhine Salt Treaty and Rhine Chemical Treaty. The negotiations on the salt treaty had to be restarted, with only marginal changes as a result. Efforts to move forward on setting chemical-specific emission norms progressed slowly. Few of the recommendations of the ICPR became binding.

The stalemate in progress was reversed by the chemical spill of 1986. The Rhine Action Program, established in 1987 as a result of the public outcry from the effects of the spill, has been very pro-active in working to reduce the pollution and restore ecological functions to the Rhine. The new Treaty on the Protection of the Rhine, which replaces the Treaty of Bern and the Rhine Chemicals Treaty, formally approved the involvement of the ICPR with ecological recovery and flood protection.

The current organizational framework of the ICPR is shown in Figure A1-2. The ICPR is supported through a small secretariat based in Germany. The plenary assembly has the authority to make decisions and meets once a year under the presidency of the ICPR. However, implementation of the ICPR decisions is the responsibility of the member states, and the decisions are not legally binding. The presidency is assumed by the delegations according to the order of signatures to the Bern Convention. The president supervises the technical scientific secretariat and presides over the meetings of the coordination group.

The coordination group has several tasks, including the drafting and directing the programs for the working groups, supervision of the implementation of the ICPR decisions, and initiation and supervision of public communication. There are three permanent and two temporary working groups comprised of senior national officials and technical experts, with the following responsibilities:

- Water Quality Group
  - Supervise the changes in water quality
  - Evaluate the results of research and develop new research programs
  - Assess unusual concentrations of substances
  - Evaluate monitoring results and prepare reports
  - Compare the state of the Rhine with target values on a periodic basis
- Ecology Group
  - Elaborate the Ecological Master Plan for the Rhine from an integrated ecological perspective
  - Draft guidelines for a network of biotopes in the Rhine corridor and propose any necessary measures
  - Support and complete the program for the return of long-distance migratory fish of the Rhine.
- Emissions Group
  - Register pollution sources and propose appropriate reduction measures, including, if necessary, economic incentives, for point and nonpoint sources
  - Harmonize the best available technology for significant industrial sectors along the Rhine and provide surveillance for its application
  - Prevent accidents and improve the safety of industrial plants
- Sustainability Group
  - Elaborate a program for sustainable development of the Rhine on the basis of the guidelines passed by the 12th Conference of Rhine Ministers
- Flood Protection Group
  - Provide international coordination with the implementation of the action plan on flood defense
  - Assess the efficacy of measures, organize public communications and cooperation of non-governmental organizations (NGOs) for the action plan
  - Improve flood warning systems

The most recent meeting of the ministers, held in January 2001, gave strong support for continuation of the work of the ICPR. Although stakeholder processes are not a formal requirement of the commission, the ministerial declaration noted, "that the exchange of information with nongovernmental organizations has become a permanent part of ICPR work and [we] welcome this development."

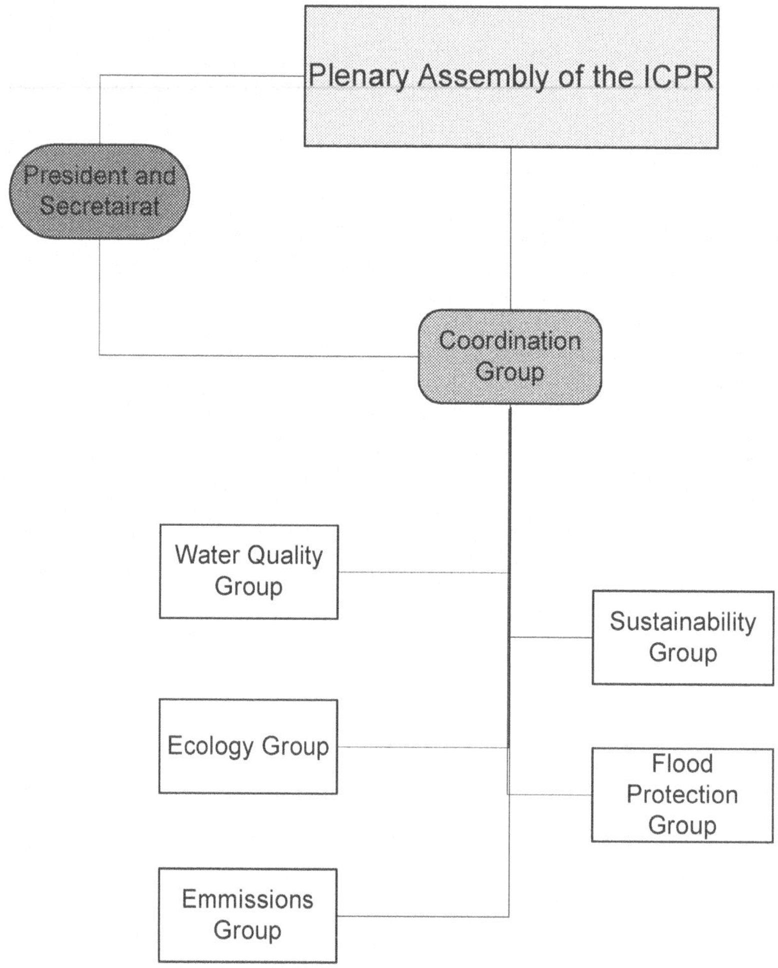

Figure A1-2. Current organizational structure of the ICPR (Based on figure from http://www.umweltbundesamt.org/wsektor/wasserdoku/english/kap48_e.pdf).

A1.4.2 Funding and Staffing

The Treaty of Bern established an official permanent and multi-lingual secretariat in Koblenz that would provide support for the commission and would, among other things, compile annual reports. The costs of the ICPR secretariat are apportioned according to a fixed factor to the four riparian states and the European Community, which joined the ICPR in 1976.

Funding for individual programs and policies is handled on a case-by-case basis, but typically costs are shared among the riparian governments. In this respect, the commission operates in a rather innovative way, which has proven useful. Environment ministers and officials from each member state gather at ICPR meetings on a regular basis. Decisions are taken jointly and each country adopts the ensuing measures. But neither the ICPR nor its secretariat has any executive or coercive power to speak of, such as the authority to order the construction of new facilities. Similarly, the commission draws up environmental projects for the Rhine, but if adopted they are funded and implemented by each member country. The secretariat simply monitors their application along the river's length.

For example, between 1970 and 1974, the river's oxygen content was between 2 mg/L and 4 mg/L, too low to support the growth of any organisms. In response, the countries along the river spent $38.5 billion to build a string of purification plants between 1970 and 1990. The oxygen level steadily rose and some of the river's biodiversity returned. Further examples of funding approaches are also given in the following section on implementation.

A1.4.3 Methods of Forming and Sustaining Partnerships with Stakeholders

There is a long history of stakeholder involvement in the Rhine, although not formally in any institutional structure. Some of the earliest influences came from the Dutch waterworks organization, Association of River Waterworks (RIWA) in the 1930s that formed in response to deteriorating water quality. Soon after its formation, RIWA contacted the Dutch government with proposals for action. Furthermore, to improve contacts with other governments this organization joined forces with fellow organizations in the Rhine catchment area to form the Internationale Arbeitsgemeinschaft der Wasserwerke im Rheineinzugsgebiet (IAWR). In its various Rhine memorandums the IAWR has proposed extensive demands on the governments of the riparian states along the Rhine. In addition, the waterworks were also able to influence policies because representatives of the drinking-water companies held administrative and political jobs in each of the riparian states.

The environmental NGOs in the region initiated fact-finding activities about pollution sources, which in turn influenced perceptions in the catchment area. While the IAWR provided the states with data concerning the pollution levels of the Rhine in their area, environmental organizations like the Dutch Clean Water Foundation (Stichting Reinwater or CLF) and Greenpeace started monitoring campaigns on the river to clearly identify specific point sources of pollution. Both succeeded in influencing perceptions concerning the quality of the Rhine.

NGO-initiatives proved effective in reducing some political barriers to change. For example, the environmental NGOs ensured that upstream governments and industry were kept under continual pressure, for instance, through legal proceedings against polluters. The legal proceedings were started in 1974. The Dutch Clean Water Foundation (CLF) and horticulturalists in the region whose cucumbers and other products were suffering from the high salt loads took action against the French potassium mines and were ultimately successful. After 14 years of proceedings an out-of-court settlement was achieved, with the affected horticulturalists receiving financial compensation for the damage they suffered from the mine discharges. These successes at court inspired Dutch drinking-water companies and other victims of pollution problems to start other successful proceedings in France in administrative and criminal courts.

Finally, negotiations concerning the ecological recovery of the catchment were also influenced by NGO activities. The successes in the civil proceedings of CLF improved the position of the Dutch government in its negotiations with Germany over problems of contamination from industrial discharges. The Dutch delegation threatened to support the municipality of Rotterdam, which wanted to start proceedings as well. Rotterdam Harbor was polluted with contaminated sludge, and upstream discharging industries were held responsible for the costs to Rotterdam for correct removal and storage of this sludge. This made the Germans more eager to accept the Dutch proposals.

The openness of the political system in the Rhine catchment area made governments receptive to the ideas of the NGOs. An open dialogue exists between NGOs and the governments. In the Netherlands this openness is probably the most far-reaching with Dutch government in some cases supporting their own opposition. Environmental organizations like CLF are partly funded by the government (Van Leeuwen, 1992). Public involvement is also stimulated by the French government, which started a foster parents plan for salmon. Recently, NGOs obtained an observer status at the ICPR. With the creation of the European Union (EU), the national political systems are more integrated. Apart from the contacts within the ICPR, many other contacts between the riparian states exist on political, business, and cultural levels.

NGOs have also had an impact on the issue of chemical pollution. Publicity campaigns influenced environmental awareness of industries, governments, and mobilized public opinion in the catchment area. Through buying shares in polluting companies, the Dutch CLF tried to influence the decision making of these companies. Major Rhine polluters were confronted with the results of monitoring campaigns in the so-called water tribunal, a non-official court in which they were condemned. CLF also published its arguments in newspapers, did grassroots work by educating kids at schools, and had contacts with ministers and political parties. These activities resulted in a growing willingness to adapt discharges to environmental standards. Industries intensified wastewater treatment. In response to these efforts concentrations of heavy metals on the Dutch border have decreased and an international agreement on specific standards could more easily be reached.

A1.5 Program Priorities and Goals

Priorities for the organizations involved with managing the Rhine evolved with the changing perceptions of the issues in the basin. At first, general knowledge concerning Rhine water quality was limited, and the focus remained on navigation, hydropower, flood control, and industrial/agricultural water use. Subsequently, a change in perceptions between the riparian states on the effects of chloride and other pollutant discharges occurred. The negative aspects of poor water quality were recognized, at least as they affected drinking water and agricultural uses. Most recently, the importance of the ecology of the catchment area was recognized. This historical change in priorities, starting with concern over water quality for drinking and agricultural purposes is briefly discussed below.

Initially, in the early 1950s, no data were available that could give an indication of the severity of the Rhine pollution. For this reason, the ICPR put emphasis on making an inventory of pollution problems in the widest sense. Their first move was to harmonize a number of measuring procedures and to set up an international monitoring network. In the four riparian states, nine monitoring stations were set up, at even increments across the catchment area. The measurement results were annually published by the ICPR, so insights became more widely distributed. Trend analysis of the data led to some concern, as it was established that the Rhine, particularly in its lower reaches, was under heavy pressure. Industrial centers such as the Ruhr Valley, the Moselle River, the potassium mines, and the German state of Baden-Württemberg were identified as the main sources of pollution.

In 1954, the Dutch ICPR delegation requested that there should be no further increase in the chloride load. This caused considerable disagreement. The French contested the definition of the problem developed by the Dutch, which identified the French potassium mines as being primarily responsible for the elevated salt load. The French contended that the potassium mines were not the only source of chloride discharge in the catchment area. In addition, the French suggested that the problems the Dutch ascribed to chloride discharge were because the Western Netherlands was below sea level. They claimed the main cause of salinization was intrusion of salty groundwater. The Dutch ICPR delegation reacted by elaborating their argument and by trying to underpin it more effectively. After lengthy discussion, the arguments presented by Dutch hydrologists to try and influence the French point of view led to a French proposal that ICPR working groups should be set up to study the problem. Research eventually showed that the potassium mines were the main source of pollution in the catchment area, but that soda works in France and Germany, coal mines in the Ruhr, and communal discharges originating in the catchment area as a whole also made significant contributions to the salt load measured at the Dutch border.

The ICPR had given a strong priority to water quality issues during this period, with an emphasis on water for human consumption and agriculture.

Following the 1986 chemical spill, and the subsequent development of the RAP, the ICPR had a more broad set of goals:

- Sustainable development of the Rhine ecosystem
- Guarantee the use of the Rhine for drinking water production
- Improvement of the sediment quality in order to enable the use or disposal of dredged material without causing environmental harm.
- Overall flood prevention and environmentally sound flood protection
- Improvement of the North Sea quality in accordance with other measures aimed at the protection of this marine area.

Still, multiple and differing perceptions remained concerning the ecology of the catchment. A special ICPR working group was established to bring together the fragmented knowledge on this topic at several research institutes in the Rhine catchment area. These efforts resulted in an ecological master plan, Salmon 2000, which not only gave an outline for a more ecologically sound catchment, but also outlined the obstacles to achieving this, which include water quality, hydrology, and morphology of the catchment area.

A1.6 Implementation Activities Over Time

The initial implementation efforts on the Rhine involved large-scale engineering projects to alter the course of the Rhine, both for navigational purposes and to reduce the impacts of floods. Although very effective in the short run, the riparian nations have found that this approach ultimately came with a high price, flooding and challenges to the survival of aquatic life (see Box A1-1).

A1.6.1 Instruments Used in Implementation

Both government and NGOs have used a variety of management instruments to implement policies and plans on the Rhine. For example, both types of organizations have used information to effect change through influencing public opinion and pressuring governments, as described in the previous section on institutional arrangements. Economic instruments have been highly effective in solving many of the upstream-downstream conflicts on the Rhine, and are described in the following section on innovative approaches. Regulatory and technological instruments have also been used at the recommendation of the Emissions Group of the ICPR to get industry to reduce the discharge of pollution into the river.

Box A1-1 Altering river hydrology

To enable the use of the alluvial areas for agriculture and other purposes, to improve shipping conditions, and to reduce the risk of the formation of ice walls in winter, the course of the Rhine river was drastically changed in the 19th and 20th centuries. As a result, the length of the river between Basle and the border of Hessen was reduced by almost 20%. Meandering sections and alluvial areas were cut off, causing great changes in the river ecosystem. The flow velocity was increased, leading to erosion of the riverbed and a consequent drop in the level of river water and groundwater.

These alterations led to a sharp increase in the risk of flooding. Only 15% of the Rhine's original flood plains remain, with an associated loss and/or elimination of many native plants, birds, mammals, and other species.

Over 450 dams have been constructed in the Rhine and its tributaries, making the migration of some species through natural means to spawning ground impossible.

The ICPR has relied heavily on the use of information and communication instruments to accomplish its goals. Establishing a common ground between parties with differing perspectives is essential to making progress in a watershed. One example of the use of information, as gained through research, is the approach taken towards resolving the reduction of salt in the Rhine. From the mid-1960s, research was done on how to reduce the discharge of chloride from the French potassium mines. Although a reduction of these discharges was considered the best solution to the problem, about 20 years of research had to follow concerning the methods of doing this. Multiple options were studied. First, research concentrated on assessing the possibility of storing salt in the Alsace. Subsequently, the feasibility of selling salt to a soda works and the possibility of solving the problem by transporting it to the North Sea were studied. The fourth option involved laying a pipeline to the channel. And the final option was to discharge the salt into the River Doubs in France. The Dutch actively participated in some of the subsidiary studies in an attempt to speed up the process. It took five years, however, before the message came from France that storing the salt was in principle the most favorable option, but that research had to be done concerning the exact location of the stacks. Disagreement between France and Germany about proposed locations, however, resulted in two new research projects. For another eight years injection of the waste salts in the sub-soil was reconsidered. Also, the feasibility of marketing the salts by a salt factory was studied. That the feasibility of all methods to resolve the problem was intensively studied by ICPR working groups contributed to the acceptance of a Dutch proposal to solve the problem. This proposal could finally be perceived as being the most efficient way of dealing with the chloride problem.

A1.6.2 Innovative Program Elements

The use of integrative negotiations between upstream and downstream parties has proved very effective in the Rhine basin. Historically, the nations have used both divisive and integrative tactics in negotiations. The Netherlands, the most downstream nation, has made the most far-reaching demands on its upstream neighbors, initially calling for the establishment of a commission with extensive powers in the early 20th century. In the latter part of the 20th century the Dutch promoted its ideas regarding the ecological recovery of the catchment area through the RAP. In both cases, other riparian nations were not initially amenable to the Dutch proposals. However, by offering incentives and employing more integrative tactics, the negotiations continued to move forward.

The incentive approach to negotiation gave the Rhine its first binding treaty, the Treaty of Bern in 1963. The Netherlands and Switzerland were able to induce Germany and France to agree to formalize the mandate of the ICPR. The Germans agreed to do so on condition that the offices of the ICPR would be established in Germany. This served to improve the international prestige of Germany, a reputation that had suffered from World War II. France agreed to sign because this compromise proved to be less expensive than earlier proposals.

With respect to cutting the amount of salt discharge from France, the Netherlands offered to help pay for the measures to be taken in France. Specifically, the co-payment would apply to an effort to explore the options to reduce the waste. Germany and Switzerland also contributed because their discharge of salts, which together was at least as substantial as that of France, did not have to be reduced. In exchange for their financial contribution, the Swiss were permitted to increase the volume of salt they could discharge. The Germans managed to have the French agree to close down some salt storage basins in the Alsace, from which salts had leaked into Germany's groundwater. The decision to devote the 1972 ministerial conference of the riparian states to a program for a full clean up of the river was partly based on the assumption that this would encourage France to take measures to deal with its salt. Nevertheless, it took many years before the Rhine Salt Treaty was finalized, and that only occurred when the Netherlands proposed a package deal that was not only less expensive, but also more flexible for France. It was agreed that the French would only have to shift to storing their waste salt when the water level was low. Furthermore, the riparian states would contribute to the measure that needed to be taken in the Netherlands. The Swiss got an extra reduction of their financial contribution.

Similar tactics were used in the Rhine Chemicals Treaty and the RAP. From the perspective of the Dutch, it was important for the various eutrophication-causing substances to be included in the chemicals treaty. In exchange, the Germans got the concession that the applicable area of the treaty was enlarged to include the Rotterdam port area. German agreement to the RAP was finally obtained when the Dutch dropped their intention to start a lawsuit against German industry. At the same time, the period over which the reductions were required was extended.

The successes of lawsuits indirectly offered the Dutch government the opportunity to change its strategy in international negotiations. The Dutch refused to accept French proposals for a temporary reduction of the chloride discharges. This stance was partly because the parties bearing the brunt of the chloride discharges, horticulturists and drinking-water companies, seemed likely to get compensation for damages. In addition, technical developments made horticulturists less dependent on the Rhine. So the Dutch government preferred to spend the money on the elaboration of the RAP. However, pressure from the drinking-water companies forced the Dutch government to revise its standpoint. If the French proposals were not to be carried out, the danger existed that provincial waterworks would be unable to use a water-softening installation that was under construction. In making this investment, it had been assumed that softening Rhine water by exchanging calcium ions for sodium ions would be possible without overstepping the norms for sodium in drinking water. However, if the reductions were not implemented, there would be no decrease in the chloride and sodium content of Rhine water, and the sodium content of the drinking water might exceed the legal maximum. Noord-Holland Waterworks introduced the idea of bringing the chloride problem to an end by a package deal of measures in France and the Netherlands. Introducing a policy mix costing significantly less than the original French proposal would eventually accommodate the drinking-water interests of the province of Noord-Holland.

A1.6.3 Monitoring Methods

The earliest investigations into the Rhine's surface water quality date back to the latter part of the 19th century. The number of water quality parameters monitored increased steadily, but with no joint management among the riparian nations. Prior to World War II, contacts between the Dutch drinking water industry and its German and French counterparts were established to investigate drinking water. But it was not until the creation of the ICPR in the 1950s that a formal international measurement network was set up with common analysis methods. Although initially tedious to organize, the common basis for objective assessment of the water quality proved invaluable in establishing trust among the nations involved.

The Netherlands and Germany (NordRhein-Westphalia) signed an agreement on August 28, 2001, allowing the German State Agency LUA and the Dutch RIZA (national water authority) to operate a joint international water quality monitoring station on the German-Dutch border, While RIZA and LUA have had their own stations on opposite sides of the Rhine for many years, these have now been merged, becoming one of the first truly internationally operated monitoring stations.

A1.6.4 Provisions for Public Education and Involvement

The institutional framework of the ICPR does not make explicit inclusion of a broad base of stakeholders other than the riparian governments. However, numerous NGOs are involved in the management of the watershed, including the umbrella organizations of the drinking water industry (IAWR, RIWA, Association of the Rhine Waterworks [ARW], and Association of Waterworks, Lake Constance/Rhine [AWBR]) as well as the association against water, soil, and air pollution, the Swiss water protection association, and the federation for European water protection. These organizations have played and continue to play an important role in disseminating information about water pollution to the public. The environmental movement and private sector are also involved. The environmental movement revealed the actual origin of certain pollution on the basis of a number of measuring campaigns. The private sector responded by providing concrete data on the discharges, which were subsequently published by the ICPR.

It is remarkable to note that even during the period of stalemate on the Rhine from 1976 to 1986, this community of organizations, including the ICPR, continued to build a knowledge base accessible to the public. The ICPR introduced new methods to monitor organic pollution, and developed a method to automate the communication of data. The environmental movement has continued its investigations.

A1.6.5 Level of Integration and Breadth

The management of the Rhine as a watershed began with a fragmented approach, handling each problem that arose in isolation. The basin-wide commission initially handled issues of navigation only, but by 2000 the Rhine was managed holistically, with ecological functioning considered in parallel with flood protection, drinking water, and agricultural water interests. The riparian parties are actively involved in the management, at the governmental level, but also through the basin's stakeholders, as represented by NGOs.

A1.6.6 Funding Sources and Levels

According to the Treaty of Bern, the costs of the ICPR secretariat would be apportioned according to a fixed factor to the four riparian states and the European Community, which joined the ICPR in 1976. Funding of projects and policies on the Rhine has primarily come from the riparian governments, with the share of the financial burden negotiated, as described in the section above on innovative approaches.

Examples of how project and policies obtain funding are the events that followed the catastrophic spill of chemicals in 1986, when a fire broke out at the Sandoz chemical plant, near Basel. The water that firefighters used to put out the blaze flushed huge amounts of insecticides and pesticides into the river, sparking an ecological catastrophe. Environmental awareness rose and the affected population and their representatives demanded that manufacturers take much tougher action against pollution.

Special chemical purification plants were built up and downstream from Schweizerhalle. "The 1987 program won credence for the idea that the Rhine was a total ecological system," says Schulte-Wülwer-Leidig. Under the program, the Rhine was to become a place where salmon, pike, perch, trout, and other fish could again thrive. But to achieve that goal, the water quality had to be further improved, an objective that was reached before the 2000 target date. Since 1991, more than a million young fish have been released into the Rhine and its branches (Weber, 2000).

But the troubles of the migrating fish were not over. In 1996, the first salmon caught in the river in many years was hooked just before the Iffezheim dam, near Baden-Baden. The fish came from the North Sea, but because of the dam could not reach the branch where it was born. Salmon only spawn in the places where they started life.

With the help of the environmental movement's focus on ecology, the reluctance of power station owners to build costly bridges or complex networks of passageways for the salmon is gradually breaking down. Such facilities are already under construction at Iffezheim, at the confluence with the river Ill (which has no dams) at a cost of about $6 million, and at Gambsheim in Alsace (Weber, 2000). Basel residents are being consulted about planned changes in the landscape that would help the salmon migrate. Egon Oberacker, a professional fisherman in the Swiss town of Nordbaden, is pleased that fish will be more plentiful now. But, he says, "We still can't live off fishing alone, as we once did."

A1.7 Degree of Success

Many of the goals set over the years have been attained. The agreed-upon international obligations have been converted into national legislation and regulations, and the actions of the relevant actors conform to these. Throughout the catchment area, purification measures have been taken. The time-series analyses that have been carried out on the parameters of water quality reveal a demonstrable improvement in quality. The concentrations of oxygen-binding and eutrophication-causing substances, chloride, heavy metals, and micro-pollutants have decreased significantly. The involvement and support of drinking water companies as well as the environment organizations underline the success of the Rhine management system. Perhaps even more remarkable is that this clean up took place during a period of substantial net economic expansion in the riparian states. The Rhine has changed from an open sewer into a river that supports aquatic life.

The following sections highlight the degree of success in the Rhine watershed in six areas: water quality, water quantity, public outreach, biological criteria, stakeholder processes, and cross-jurisdictional cooperation.

A1.7.1 Measures of Water Quality and Ecological Health

The ICPR did make steady improvements in water quality during the first decades of its work, primarily using end-of-pipe techniques that treated wastewater, rather than prevented pollution. Yet by 1975, only 29 of the more than 50 fish species remained, with no return of salmon. The Dutch water authority published a report in 1985 that shifted attention from end-of-pipe approaches to looking at the ecosystems of the receiving water bodies and what was necessary to support these ecosystems.

"The Basel disaster enabled the ICPR to set higher goals in its 1987 Rhine Action Program for Ecological Rehabilitation," says Koos Wieriks, of the ICPR secretariat. Commitments were made to cut the amounts of the most hazardous pollutants in half by 1995. Many experts thought this target could never be reached, but ICPR samples showed that from 1985 to 1992, mercury in the river at the German town of Bimmen-Lobith, near the Dutch border, fell from 6 tons to 3.2 tons, cadmium from 9 tons to 5.9 tons, zinc from 3600 tonnes to 1900 tonnes, atrazine from 10 tonnes to 3.7 tonnes and polychlorinated biphenyls (PCBs) from 390 kg to 90 kg (Weber, 2000).

In 1999, a third phase was agreed upon to broaden the scope of the plan. The result was the acceptance by the commission of a 40-year action plan in 2000. The plan includes measures on flood management and habitat protection and restoration in the alluvial zone around the river's banks. The flood management goal is to restore as far as possible the natural course of the river. The emphasis is on planning around the water system itself, rather than trying to control the water. This is a major shift, particularly for the Dutch, who have been building dikes for the past 1000 years.

The initial years of the RAP focused exclusively on water quality improvement. The Rhine states agreed on a target reduction of 50% of the pollution caused by priority compounds by 1995. In addition, water quality targets were set for about 50 priority compounds based on the requirements for drinking water supply, the protection of aquatic life forms, and on human tolerance levels for fish consumption (Schulte-Wulwer-Leidig, 1992). Rehabilitation efforts involved sewage and water treatment plants, decreases in point source discharges, reinforcing shorelines with large stones and planting of riparian vegetation, and use of seasonal chemical and biological concentration techniques to monitor water quality.

Recent monitoring studies indicate that the water quality targets for two-thirds of the priority compounds are within compliance levels. The total phosphorus and nitrogen, cadmium, and organics have been reduced by nearly 50%. However, these compounds periodically exceed compliance levels. These reductions are a result of reduced sewage and water treatment effluent, reduced toxic discharges, removal of contaminated sediments, and the increased use of phosphate-free detergents. The overall reduced toxicity of the Rhine River has allowed for rapid oxidation of ammonium, which is near compliance levels (De Ruyter van Steveninck et al., 1992). River regulation and shoreline stabilization have fixed processes of sedimentation and erosion in the floodplains and in the channel during low flow (Jongman, 1992). The targets for mercury, copper, zinc, lead, chloroform, hexachlorobenzene, and PCBs have not been reached (Van Dijk et al., 1995). Studies have indicated that contaminants such as zinc may inhibit sediment microbial breakdown of natural and artificial substrates (Van den Brink et al., 1994). This implies that otherwise biodegradable pollutants will continue to persist in the Rhine River sediments.

However, rehabilitation efforts have been successful in improving water quality enough to allow the biological communities to recover slightly. Concentration techniques have shown an increased relative growth rate of the natural phytoplankton community, insect community, submerged macrophytes, and macroinvertebrates. The numbers of anadromous and catadromous species, leeches, flatworms, mollusks, and some other crustaceans, and many exotic species such as *Corophium curvispinum* (Van Dijk et al., 1995), have also increased. The current success of exotic species is linked to abundant food availability (phytoplankton, suspended organic matter), resulting from increased eutrophication (Van den Brink et al., 1994).

Improvement in water quality allowed local rehabilitation projects to focus on restoring migratory fish. Several biotopes were identified that are essential for various fish species life history, including shorelines, islands edges, natural riverbed sections, harbors, mouths of tributaries, and stretches of side-water bodies (Lelek and Kohler, 1990). Migratory fish require the entire river as habitat for their life cycle. Rehabilitation efforts involved removal of concrete or metal walls in small areas in the channel, removal of small dams to provide some stretches of swiftly moving water for fish migration, installation of fish passes or ramps, restocking of fish, and reopening the Rhine River to its floodplains.

In the Rhine and its tributaries, 32 ha of spawning grounds and 250 ha of nursery grounds exist (Schulte et al., 1995). These areas were identified as sufficient sites for re-stocking efforts. In 1987 a Salmon *salar*-smolt release program for the Rhine River and its tributaries was started. In 1989, six Salmon *salar* were caught in the entire Rhine drainage basin. In 1992, 18 salmon *salar* were caught in the entire drainage basin (Schulte et al., 1995). Natural reproduction has been recorded since 1992 and in 1994 newly hatched yolk-sac alevins were found in natural spawning grounds in the Sieg River (a tributary of the Rhine River) and its tributary the Brol River (Marmulla and Ingendahl, 1996). The number of salmon *salar* caught continues to increase slowly. Success of stocking and reintroduction of fish indicates enough improvement of water quality to increase and sustain 40 indigenous and 15 introduced fish species (Van Dijk et al., 1995).

Terrestrial habitat improvements started later than water quality measures. Historically, the diverse flora of marshes, rivers, and dry, sunny grasslands characterized the Rhine River ecosystem. Almost 85% of the vegetation types consisted of grasslands (Jongman, 1992). The influence of the hydrologic regime was evident by the distribution of grassland types that showed distinct gradients from dry and sunny locations to wet and shady ones in short distances. By 1970, the tidal influence on the floodplains of all the states bordering the Rhine had disappeared because the estuaries were closed off from the main channel (Lelek and Kohler, 1990). The lack of tidal influence has reduced the number of dry, sunny vegetation types as well as wet, shady types. The focus of terrestrial habitat improvements has been the reconnection of floodplains with the mainstream by removing dams and other structures. Additional rehabilitation efforts include seed plantings of native grass and forb communities representative of the dry-to-wet gradients, weeding exotics and pests, and removal of contaminated soils and sediments. However, plantings of native grasses and forbs have failed on artificial soils in former clay pits filled with sand after excavation (Van Dijk et al., 1995).

Aquatic and terrestrial habitat rehabilitation has favored re-colonization of a few species. The rate of species recovery depends on maintaining water quality that can sustain aquatic populations. Species recovery also depends on the existence of donor communities within the Rhine River drainage basin. Other large rivers, floodplains, and dry-to-wet grasslands similar to the Rhine River drainage basin are hundreds of kilometers away. The degree of isolation of large rivers in Europe is significant enough to impede species re-colonization (Fuchs and Statzner, 1990). The aim of the RAP is to create an ecological network along the Rhine River and its tributaries, consisting of several large ecologically important reaches (1000 ha to 6000 ha each) with smaller connecting areas. At present, 7500 ha of floodplain have been identified as ecologically important areas, and 40% of this area has been reconnected with the main channel. The remaining 60% is expected to be successfully rehabilitated within the next 25 years along with an additional 5000 ha (Van Dijk et al., 1995).

The Rhine River Drainage Basin is influenced by pollution and regulation measures that have negatively affected the river system. The level of degradation is severe enough to hinder rehabilitation processes for decades. The lack of a reference system creates difficulty in determining appropriate biological criteria and measures of success. More detailed information is needed on the dynamics of fish communities such as predator-prey ratios, possible competition mechanisms between ubiquitous, exotic, and indigenous species, and fish movements between the mainstream and adjacent side-waters (Lelek and Kohler, 1990). Current publications do not provide adequate information on the organization, management, and administration of the RAP or on the progress in reaching the stated goals and objectives. Publications on rehabilitation techniques and successes are limited. An additional complication in researching the RAP is the lack of material written (or translated) in English. The lack of current detailed information creates difficulties in assessing whether the RAP is successfully rehabilitating the Rhine River and its tributaries. Improved water quality and ecological recovery of some species and their habitats in the Rhine River have been documented. However, pollution inputs, eutrophication, and human activities remain serious problems. In addition, the loss of species and isolation from donor communities impedes re-colonization. Recovery of the Rhine River ecosystem will take decades to recover rather than a 12-year rehabilitation program deadline.

A1.7.2 Extent of Public Outreach

A mix of strategies made NGOs successful actors in the Rhine catchment area (Van Noort, 1987). They persuaded other actors, provided them with information concerning discharges, started legal proceedings, and captured mass media attention.

NGOs could only be successful actors because the conditions in the Rhine catchment area were favorable. Characteristics of the political and legal system, characteristics of NGOs and industries, and finally the characteristics of the problems contributed to the positive results of NGO intervention.

A1.7.3 Stakeholder Processes and Involvement

The NGOs could only be successful because of their high levels of professionalism and internationalism. Both the waterworks and the environmental movement employed professionals. The waterworks have their own laboratories and the environmental movement had the possibility to start monitoring campaigns and succeeded in identifying point sources of pollution. At the international level, networking resulted in international agencies representing the waterworks, the IAWR, and the environmental movement, the International Rijngroep.

Another factor explaining the successes is that enterprises in the catchment area expressed a growing environmental awareness and had the financial resources to invest in more environmentally sound technologies.

Finally, the characteristics of the perceived problems allowed for progressive regime development. Disasters like the accidental discharges of the pesticide endosulphane and the accident at Sandoz, where large amounts of chemicals entered the water, raised public awareness of the risks of industrialization in the catchment area.

A1.7.4 Level of Cross-jurisdictional Cooperation

Transboundary cooperation is a time-consuming process, made by small increments. Cooperation requires trust. The decisions of individual countries must be left to each country. Common legal frameworks help the process of cooperation. Common monitoring infrastructure contributes to mutual trust.

The presence of the ICPR has proven essential for cooperation to take place across the various jurisdictional bodies in the Rhine watershed, by providing a forum that facilitates a good exchange of knowledge about the watershed, as well as a non-threatening environment for people to talk. By collecting and developing a knowledge base, the ICPR provides both government and NGOs a common basis or set of facts from which to build policies. The ICPR facilitated negotiations throughout its history, and adapted to the changing priorities recognized by society—what was initially a focus on concerns for drinking water ultimately became a focus on aquatic ecosystem health.

In the future, as diffuse sources of pollution form a growing part of the Rhine pollution issue, however, consultation at higher levels will become more important. In the context of the

EU, regulations for pesticide use must be introduced, and pollution originating from European-wide atmospheric deposition must be put on the other European agendas.

A1.8 Lessons and Applicability to the United States

The Rhine River basin may offer insights to management of U.S. water on several levels. In the broadest sense, the nine riparian countries of the Rhine share a similar set of political, social, and legal systems to the United States, with strong democracies, vocal citizens, and a court-based legal framework. The strong institutional framework that developed in the Rhine watershed, the ICPR, is not unlike the river basin institutions that have operated in the United States without explicit authority to build infrastructure or regulate, including the Ohio River Valley Water Sanitation Commission, the Interstate Commission on the Potomac River Basin, and the Great Lakes Commission. The environmental problems of the Rhine are also problems common to the United States: poor water quality caused by point and nonpoint source pollution, and deteriorating ecosystems caused by a combination of changing flow regimes, land use and water quality. Yet, through various watershed approaches, the Rhine has seen a reversal not only of deteriorating trends in water quality, but also in what had been the complete extinction of several fish species. Several lessons from the Rhine and their possible applicability to the United States are highlighted in the following sections.

Lesson 1: Integrative win-win methods support the resolution of upstream-downstream and human versus nature conflicts.

The Netherlands, the most downstream country on the Rhine, consistently adopted strategies that avoided open conflict with upstream stakeholders. The Dutch promoted the creation of the ICPR through diplomatic channels. Within the context of the ICPR, the Dutch government was persistent in its attempts to work with the country's upstream neighbors or stakeholders. Political maneuvers and the submission of memorandums and other policy proposals were among the most important strategies used by the Dutch to overcome barriers. In this way it was possible to create a formal consultative body with several working groups and to convene ministerial conferences. This made it easier to continue negotiations and to resolve substantive and political barriers.

The Dutch went one step further, however, with the consistent use of financial incentives to upstream parties to persuade them to modify policies and/or practices that degraded the Rhine. As early as the 1960s, the Dutch offered to help pay for measures to reduce the amount of salt discharged to the Rhine from French mines, and effectively won concessions and cooperation from France in other areas, such as the location of the ICPR secretariat. This strategy of downstream support for the introduction of upstream measures has been used frequently since then, and continues to be a part of the strategy taken by the Dutch in their efforts to restore the Rhine.

Is this lesson applicable to the United States? Certainly the upstream-downstream and human-ecosystem tensions exist in many U.S. watersheds. From an institutional and legal perspective, the complex of institutions revolving around water in the riparian nations of the Rhine are at least as complicated as river basins in the United States that cross county, state, and national borders. Yet, the parties in conflict have been able to find solutions that are satisfactory to all. The intervention of downstream parties in upstream areas is not unfamiliar in the United States. One historic example is the financial support New York City offered to farmers in the upstream Catskills region to adopt watershed protection measures, with the result of New York City avoiding the cost of building a multi-billion dollar water filtration plant. The Massachusetts Water Resources Authority, which primarily serves the Boston metropolitan area, used a similar approach to avoid the construction of a filtration plant, by purchasing and/or protecting land in upstream areas of the watershed for protection of the source. There is an important distinction between the case of the Rhine and the U.S. examples, however. The two U.S. cases were motivated by the prospect of constructing very expensive water filtration plants to meet the requirements of the Safe Drinking Water Act, while the motivation for parties in the Rhine was a recognition by private interests, such as horticulturalists and drinking water companies, that they were bearing the costs of upstream contamination.

Although this lesson is not unknown in the United States, it could be broadened to include the possibility of cost-sharing among upstream and downstream parties in cases where there is not a regulatory hammer to force the issue. One of the weaknesses identified in the United States is the relative paucity of large-scale watershed management efforts. Although the upstream-downstream conflicts are sometimes found in smaller watersheds, they are essentially unavoidable in efforts to manage watersheds on the larger scales, making this lesson extremely important for the U.S.

Lesson 2: Open communication and exchange of information builds trust among stakeholders.

The effective governance structure on the Rhine illustrates the importance of societal awareness. Both the ICPR and NGOs have placed great emphasis on creating broad public knowledge and understanding of the importance of preserving the Rhine's ecosystem. Data are made publicly available through the press and the Web, including access to real-time water quality data.

The history of watershed management on the Rhine demonstrates the need for balance among stakeholders, particularly with the lack of a voice for the fishing industry for much of the late 19th and early 20th centuries. The result was the complete elimination of the most valuable fish on the Rhine, the salmon, as well as the fishing industry. As drinking water supplies were contaminated, the development of NGOs such as the International Association of Waterworks in the Rhine began to voice concerns and slowly bring about solutions to the deteriorating water quality. But the interest in restoring the Rhine ecosystem was not voiced until the 1970 and 1986 crises brought media attention to the mortality of fish and aquatic life. Strong public support lent the ICPR the authority to develop and implement policies that fundamentally shift the approach to water management in the Rhine basin, from one of "working against the river to working with the river" as a vibrant ecological system. A prime example of this shift is the removal of dikes in the Netherlands to restore the natural flow regime to the Rhine.

The use of information and communication instruments is not new to the United States, with the U.S. EPA's Know Your Watershed program, which is accessible on the Web, as a prime example. Also, there are numerous NGOs from watershed, regional, and national levels that draw attention to water-related issues and that provide data to interested parties. What is unique about the Rhine is the relative balance that has been achieved among stakeholders in the basin.

This is a balance not only between upstream and downstream interests, but also among industrial, agricultural, hydropower, and ecological interests. Such balance is possible in the United States (and undoubtedly occurs in some watersheds), but the equal footing given to ecological protection and restoration remains the exception rather than the rule.

Lesson 3: An engaged civil society can provide authority that may be lacking in the watershed organization.

The ICPR does not have any executive or coercive power to speak of, such as the authority to order the construction of new facilities or the imposition of standards. Its role is to carry out research into the nature, extent, and origin of the pollution in the Rhine; to propose measures to improve conditions; and to prepare international agreements to help implement these measures. The commission identifies environmental projects for the Rhine but, if adopted, they are funded and implemented by each member country. The commission simply monitors their application along the river's length. Although this model is not one found in the United States, it could be a powerful integrative method for watershed management.

The non-regulatory approach in the Rhine is partly successful because of the role taken by NGOs, such as the waterworks lobby (including IAWR, RIWA, ARW, and AWBR) and the environmental movement, who put pressure on both governments and industries on a national and an international level (parallel to state and inter-state levels in the United States) to comply with the ICPR policies. Judicial proceedings have also been a powerful force in the Rhine. For example, the initiation of judicial proceedings against the potassium mines was the result of NGO actions and these proceedings had a positive effect on the willingness of the parties involved to take steps to improve water quality conditions, not only in this case, but also in other cases. The ability of NGO activities to result in changes in perceptions and behavior rested on the openness of the political and legal systems, the high degree of professionalism, and the willingness of business to respond to pressure as NGOs became stakeholders. The Rhine case emphasizes that NGO policies will only be effective if the conditions for an open dialogue are met, whereby decisions about water quality and other issues are made in a transparent multi-stakeholder process. These conditions are often, but not always, met in the U.S. context.

The cooperative relationship between the ICPR, its participating governments, and NGOs has allowed an organization with little formal authority to operate with a large degree of authority. Certainly there are cases of NGOs initiating lawsuits in the interest of protecting a watershed in the United States, but typically without the degree of collaboration with a multi-jurisdictional watershed organization, such as the ICPR. By working with the commission, lawsuits have been avoided and at the same time been a part of the negotiations between parties to find a commonly acceptable solution. Where watershed organizations in the United States do not have formal regulatory powers, the involvement of NGOs and the larger community may be a useful tool to create political support and achieve the goal of sustainable water management.

APPENDIX A2

THE MERSEY RIVER CASE STUDY

A2.1 Introduction

The case of the Mersey River Basin illustrates how significant investment, institution building, and the fostering of partnerships with the full spectrum of stakeholders can create remarkable gains in water quality. This case study describes the emergence and evolution of the Mersey Basin Campaign (MBC), a government funded public, private, and voluntary sector partnership, and the structure and programs that have made it an effective agent of change. The critical roles of the regional water and wastewater utility and government regulators are also discussed, as well as lessons that might be helpful to watershed managers in the United States.

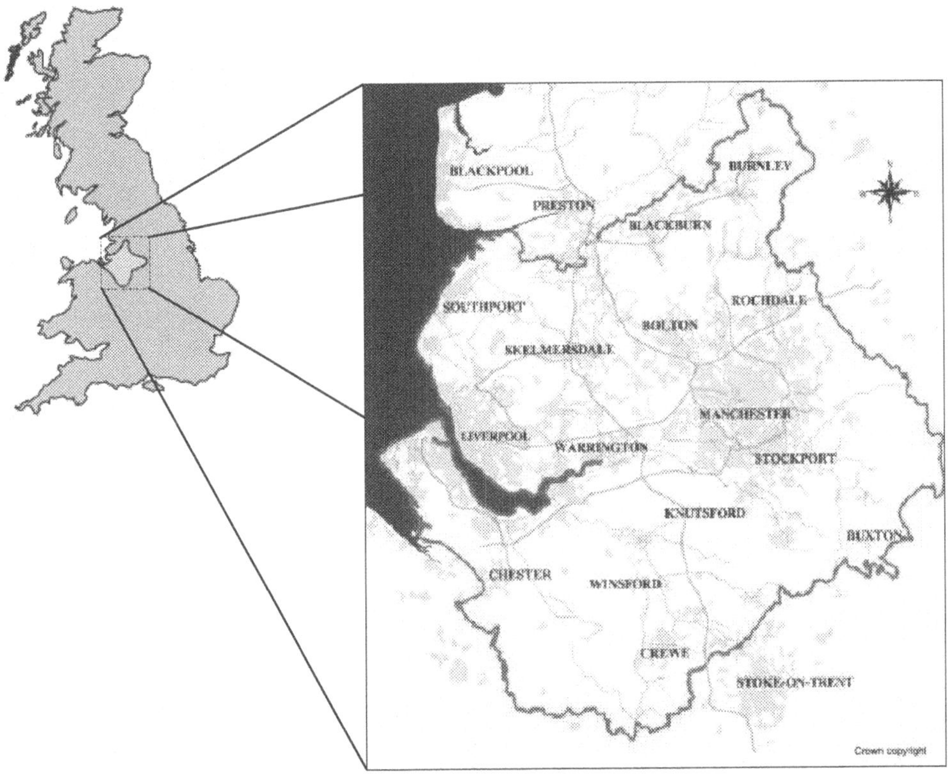

Figure A2-1. Mersey River Basin (Taken from http://www.merseybasin.org.uk/about-mersebasin/campaign-map/index.htm).

A2.2 Basin Description and Current Water Uses

The Mersey River is the largest river in northwest England and one of Britain's most famous (and infamous) rivers. Headwaters in the Pernine Hills of central England feed two tributaries, the Rivers Tame and Goyt, which join in the town of Stockport to form the Mersey. From there the river flows west through the southern suburbs of Manchester on to Liverpool and out to the Mersey Estuary and the Irish Sea.

The Mersey Basin covers an area 4680 km^2 (1789 mi^2) and encompasses approximately 200 km (124 mi) of rivers, streams, and canals. The network of canals, built to facilitate transport in the 19th century, is extensive.

Most of the Mersey Basin falls under the jurisdiction of the North West Development Agency and Regional Chamber, although a small area is within the West Midlands Region. There are two major cities within the Mersey Basin: Liverpool and Manchester. The basin includes five counties (Lancashire, Greater Manchester, Cheshire, Merseyside, and a small part of Derbyshire in the upper reaches) and 31 smaller municipalities. Approximately six million people live and work in the Mersey Basin.

Flowing waters in the Mersey Basin support the full spectrum of modern uses: drinking water, process water, irrigation, fishing, recreation, navigation, small-scale hydropower, and receiving waters for wastewater discharges.

A2.3 Impetus for Management on a Watershed Basis

Management of the Mersey Basin on a watershed basis grew out of the insight that the region's problems of industrial decline and decay and poor water quality were linked and required an approach that could accommodate both issues simultaneously.

In 1981, Michael Heseltine, then government minister for Merseyside, during a visit to Liverpool made a public statement describing the Mersey River as "the dirtiest river in Europe…and an affront to civilized society." The indictment brought high level and focused attention to the dismal state of the watershed and, through the continued efforts of Mr. Heseltine (first as government minister and later as secretary of state for the environment) and others, led directly to the establishment of the 25-year government-backed MBC.

The degradation of the Mersey had begun almost 200 years prior to Mr. Heseltine's statements. Until the late 18th century, the waters in the Mersey River Basin, like those in much of the world, ran clear and clean, and supported traditional uses, such as drinking water supply and irrigation. Fishing was an important industry. Degradation of the water quality and aquatic ecosystems in the basin began in the 19th century as the industrial revolution brought the rapid proliferation of coal mining, textiles, chemicals, and paper industries and the growth of densely populated cities and towns such as Liverpool and Manchester. By the mid-19th century, Liverpool had become one of the world's great seaports and the Mersey Basin the world's greatest manufacturing region.

As was the practice everywhere at this time, the many industries discharged their untreated effluent rich with biochemical oxygen demand (BOD), toxic heavy metals, and strong acids and bases, directly into the estuaries, rivers, and streams of the Mersey Basin. Cities and towns did likewise with their raw human sewage and the waters became increasingly polluted and less able to support human uses and aquatic life. These impacts were compounded by the construction of dams and levies to control flooding and improve navigation. Aquatic life all but disappeared from much of the Mersey Estuary.

The textile and manufacturing industries that had made the Mersey River Basin an industrial powerhouse began to decline in the 1920s and 1930s as manufacturers found themselves unable to compete with competitors in newly industrializing countries using cheaper labor. Inputs of some pollutants declined along with these industries, but other industries, notably the chemical industry, expanded during this period and into the 1940s, increasing the loading of toxic heavy metals such as mercury. During the 1970s chemical industry inputs still stood at 60 tonnes per year.

The condition of water quality and aquatic biodiversity remained extremely poor until the 1970s and 1980s when the impact of national laws regulating wastewater discharge enacted in the previous decades began to show their full effects.

In 1985 water quality in nearly 60% of sampled rivers in the basin was rated "poor" or "bad." Only 56% of the Mersey Basin's watercourses were clean enough to support fish, and much land along their banks was contaminated or abandoned.

Recognizing the negative environmental and economic consequences of the Mersey's condition, in 1981 the national government and water authorities launched a 15-year £170 million public works program aimed at cleaning up the Mersey Estuary. This smaller project generated momentum and interest among local communities for an expanded program. In 1985, the national government, in collaboration with the local governments, businesses, voluntary organizations, and others launched the 25-year MBC.

A2.4 Institutional Arrangements

"The best example by far of environmental co-operation between all partners who work so willingly and efficiently with the campaign."

- Judges for the Australian River Symposium's Thiess Environmental Services Riverprize (1999).

The environmental objectives of the MBC are driven by three main engines: the campaign itself and its subsidiary organizations and programs, infrastructure and related investments by United Utilities (the regional water and wastewater utility) and others, and government regulations and enforcement.

A2.4.1 The Mersey Basin Campaign

The MBC is, in its own words, "a government backed partnership, which brings together local authorities, businesses, voluntary organizations, and government agencies to deliver water quality improvements and waterside regeneration throughout the Mersey Basin River System." The organization provides the implied functions of forum, focus, and agent of change through a collection of institutions and programs that assist, coordinate, and guide the activities of its many partners.

Three organizations, each focusing on a particular community of stakeholders, evolved to form a coordinated effort under the MBC: the Mersey Basin Campaign Administration Ltd., the Mersey Basin Trust, and the Mersey Basin Business Foundation. Until recently, the organizations were entirely independent, each with an executive director and staff, separate operating budgets, and overseers, but operated in concert, through the monthly meetings of the Mersey Basin Campaign Council in which all three organizations were represented. The campaign administration and the business foundation are now in the process of merging. Staff members dedicated to each stakeholder constituency (public, private, and voluntary) are still represented on the Mersey Basin Campaign Council along with representatives from the Government Office for the North West (GONW), North West Development Agency (NWDA), who together guide development and implementation of a shared strategic plan.

A2.4.1.1 Mersey Basin Campaign Administration Limited

The campaign administration is responsible for general management, promotion, and marketing of the MBC and for maintaining contact with large institutional partners, such as the local and regional governments, United Utilities (the water and wastewater utility), the environment agency, and other national agencies. The campaign administration was formed in 1985 at the inception of the MBC as a nominally independent but regional government affiliated organization. It is now entirely independent, technically a subsidiary of the business foundation although it remains government funded through block grants from the central government.

A2.4.1.2 Mersey Basin Trust

Formed two years after the launch of the campaign, the Mersey Basin Trust is a registered charity whose mission is to educate, coordinate, and assist the voluntary, community, and educational sectors in support of the MBC's goals and to represent the interests of these sectors within the MBC. The trust has become an increasingly important player in the MBC, as development of a framework for informal intervention, local initiatives, and the fostering of community involvement has become a prime objective of the campaign (Wood and Handley, 1999). All types of organizations are welcome, with most falling into one of five categories: wildlife and open-space conservation groups, recreational organizations, community groups, heritage preservation organizations, and schools. Membership is free and the trust now has over 600 member organizations. The trust oversees the operation of campaign projects, including small grant programs, sub-watershed management groups, educational curricula for the public schools, and other educational and assistance programs. A 24-member board of trustees elected by member organizations oversees the trust.

Figure A2-2: Organizational Structure of the Mersey Basin Campaign.

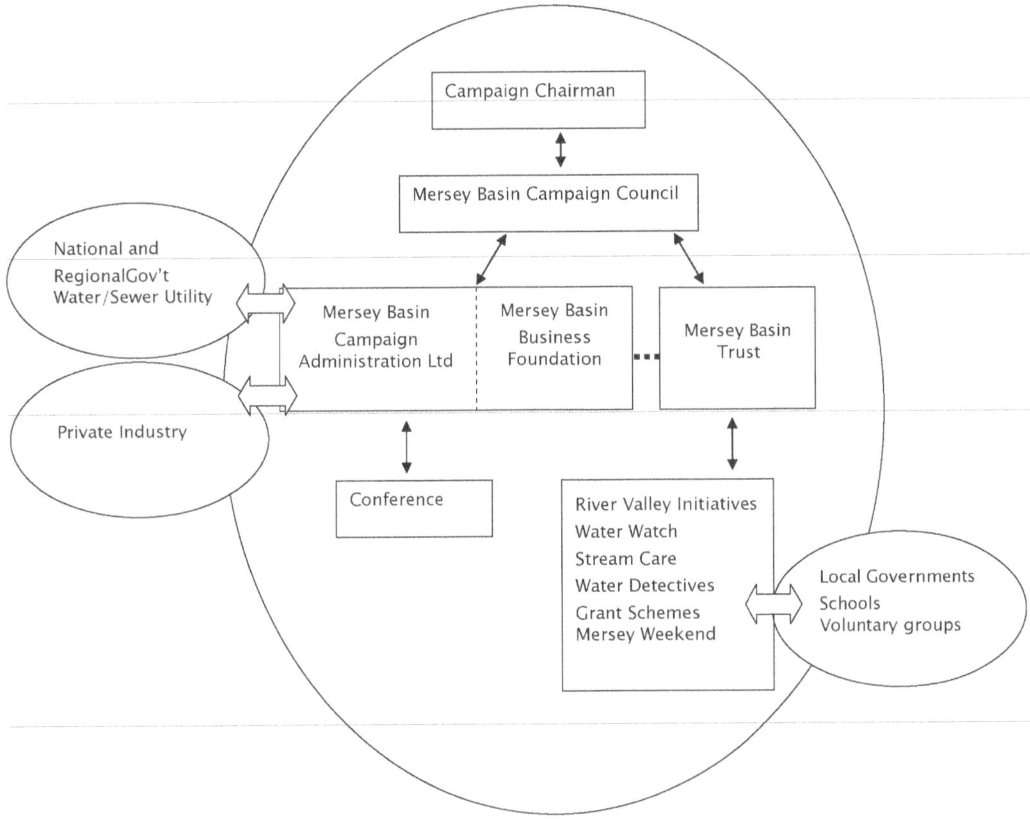

A2.4.1.3 Mersey Basin Business Foundation

Formed in 1992, the principal aims of the business foundation are to encourage business participation in campaign activities in collaboration with the public and voluntary sectors, to facilitate business input into strategic development of the campaign, to provide project support through sponsorship and/or in kind contributions, and to improve environmental performance. The foundation grew from the initial interest and commitment in 1987 of three large companies, ICI, Shell and Unilever, to support campaign-related projects. To join the foundation a business must make a minimum commitment of £30,000 over three years. Some corporations make more substantial contributions, such as Shell UK Ltd., who dedicated a senior manager to fill the position of executive director from 1992 to 1998. Although the foundation is technically open to businesses of all sizes, the minimum contribution requirement appears to select for relatively large, often multinational businesses.

A2.4.1.4 Funding and Staffing

The total 25-year cost of the campaign and related projects is estimated at £4 billion, with more than half of this funding (£2.5 million) contributed by United Utilities to clean up the rivers, and the remainder spent primarily to further investment in development of the region's watersides

Direct campaign operating expenses and grant programs represent only a fraction of the total investment in relevant projects in the Mersey Basin region. Other examples of major investments that contribute to campaign goals are the European Regional Development Fund's investment of £500 million in infrastructure regeneration and economic development projects in the region, and portions of the hundreds of millions of pounds for the upcoming Commonwealth Games in Manchester that will go to canal corridor revitalization.

The campaign has 18 staff located at the center, and 17 local coordinators based in individual sub-catchment areas. Eight local coordinators are employed directly by the campaign, while the others are employed by campaign partners. Of the 18 staff at the campaign center, six are employed by the trust. The remainder plus the local coordinators employed by the campaign fall under the employment of the combined business foundation and administration.

A2.4.2 Key Partners and Jurisdictional Issues

Although one of the defining features of the campaign is the array of partners from small, local NGOs to government agencies, some partners by dint of size, mission, and/or official role play a more significant part in the campaign.

A2.4.2.1 United Utilities

Direct release of raw sewage and inadequate wastewater treatment was responsible for much of the immediate ecological and water quality problems in the Mersey Basin. United Utilities, (formerly North West Water, Ltd.) the regional water and wastewater utility and a key campaign partner, has been a private company since Britain privatized water and wastewater utilities in 1989. United Utilities has contributed substantially to restoring water quality in the Mersey Basin through a massive ongoing program of infrastructure improvements approved by the Environment Agency. Begun in the 1980s and scheduled to continue through at least 2005, the program has included building new treatment plants, implementing treatment process improvements, and eliminating of sewer outfall pipes.

A2.4.2.2 Regional, National, and International Government Partners

The MBC was born of government initiative. Although it has become increasingly independent, the government remains a key stakeholder and financial backer. The organization receives ongoing financial support as well as technical assistance from multiple government agencies at all levels that share one or more of the campaign's goals. Core funding is provided by Department of Environment, Fisheries and Rural Affairs (DEFRA). The Environment Agency (the national environmental regulator) provides technical support and maintains strict regulatory enforcement. Department of Environment, Transportation and Regions (the transportation department) and the North-West Regional Government also provide support and coordinate planning. Much of the interaction with national government agencies occurs through the GONW, which represents these agencies at the local level. The campaign corporate plan and annual targets must be agreed upon by GONW, and GONW must be satisfied that the campaign has met the terms of its grant before allowing funding to be drawn down.

The devolution of power in Britain over the past decade has given the campaign far more influence in shaping development and environmental management at the regional level.

A2.4.2.3 Multi-jurisdictional Issues

To some extent the MBC is fortunate in that the Mersey Basin falls not only within the boundaries of a single country (Britain) but also almost entirely within the boundaries of a single political subunit (the North-West Region). The UK governmental structure further helps reduce potential multi-jurisdictional issues as the GONW represents all national government departments at the local level. As such, it is the MBC's main interface with the national government. To address the need to coordinate with many local governments (something common to organizations focusing on larger river basins), the campaign has developed a local operational model in the form of their River Valley Initiatives (discussed below) that deftly meets this challenge.

A2.4.3 Methods of Forming and Sustaining Partnerships with Stakeholders

The MBC's approach to forming and sustaining partnerships with stakeholders is sophisticated and multifaceted. A key method has been to establish institutions or initiatives designed to engage a particular group of stakeholders who offer valuable services and opportunities and provide a framework for independent initiative and action. Stakeholders in each group are collectively given input and control of these institutions and thus the campaign itself. These methods are strengthened and sustained by extensive and active communication with all partners, public recognition of success, and a strong commitment to transparency.

The method of developing institutions or initiatives dedicated to a particular subset of stakeholders is clearly exemplified in the specialization of the MBC Administration Ltd. in coordinating with government and utility stakeholders, the establishment of the Mersey Basin Business Foundation to recruit and involve businesses, and the formation of the Mersey Basin Trust to serve the voluntary sector, community organizations, and schools. Specialized focus does not, of course, require the establishment of entire institutions; initiatives and programs often work equally well. For example, the MBC recently appointed a scientific officer to interface with the region's universities. One strength of specialized institutions and initiatives is that they allow for the customization of messages and the development of an appropriate institutional culture.

Through the Mersey Basin Trust, the campaign has cleverly sought to reach the broader public by partnering with existing local organizations and institutions. The trust primarily seeks organizations, not individuals, as members. The trust maintains and strengthens these partnerships by providing valuable resources, such as improving access to financing, technical assistance, educational materials, and logistical guidance as exemplified in programs such as:

- *Water Watch*—targets the problems of litter in and around rivers and canals
- *Stream Care*—supports voluntary action to clean up and care for streams and rivers
- *Grants and Awards*—small grants to fund projects and awards to recognize achievement and dedication
- *Water Detectives*—river focused curricula and support for schools and teachers and
- *Mersey Basin Weekend*—a weekend offering over 100 conservation and recreational activities on and around watercourses in the Mersey Basin.

These services combined with clearly defined objectives and the campaign's regional focus create a framework for independent initiatives and actions that further the objectives of the campaign.

The campaign has also devoted considerable energy and resources to establishing new organizations that bring together the partners and stakeholders necessary for subwatershed planning. The organizations, known as River Valley Initiatives (RVIs) and catchment planning groups, are independent programs that bring together local authorities, businesses, and community groups along with important regional players, like the Environment Agency and United Utilities, to collaboratively tackle a broad range of environmental issues. RVIs have been encouraged to operate independently and develop according to the unique nature of the sub-watershed. These initiatives tap into local enthusiasm and foster a sense of ownership.

Stakeholders acquire input and/or control of the many constituent institutions and initiatives and the campaign itself through membership in executive committees and boards of directors and through participation in the annual conference. The conference brings all stakeholders together to see the breadth of the entire project, learn about activities and projects throughout the basin, publicize success, review progress, and contribute to the development of future strategies. Outstanding business and voluntary activity are recognized with public awards.

Partners and stakeholders are kept abreast of developments and activities within the MBC through several high quality periodicals including the bimonthly *Campaigner Newsletter,* which contains features and articles on activities throughout the Campaign area; the *Trust Newsletter,* which keeps members up to date on funding opportunities and voluntary activities; the *Water Detectives Newsletter,* which alerts schools about educational events; and an extensive website (www.merseybasin.org.uk). These sources are supplemented with occasional special publications, such as annual reports; the *Progress Report 1985-2000,* which details the campaign's progress in its first 15 years; and the *Manifesto for the New Millennium,* which plots how the campaign plans to move forward in the new millennium.

A2.5 Program Priorities and Goals

The MBC seeks to create and sustain a flexible partnership of organizations and individuals and to empower communities with the resources and expertise to act independently.

Reflecting the dual goals of environmental and economic improvement and the desire to engage the public in these endeavors, the overarching objectives of the campaign are:

- To improve water quality so that all rivers, streams, and canals within the MBC area are clean enough to support fish by 2010
- To stimulate the strategic development of attractive, sustainable waterside environments for businesses, residences, tourism, heritage, recreation, and wildlife
- To encourage people living and working in the Mersey Basin to value and cherish their watercourses and waterfront environments

In the *Manifesto for the New Millennium* the MBC expanded and refined the original three objectives into eight points to more accurately reflect the campaign's focus and priorities.

1. *"The campaign will set its three core objectives within the regional strategy, and commit itself to the sustainability of the region's environment for future generations."* This manifesto pledge formally introduces the concept of sustainability into the campaign's objectives and emphasizes the need to ensure that its goals are integrated into regional development strategies and plans.
2. *"The campaign will form agreements with partners to secure, support, and monitor commitment to waterways and waterway projects within their own business and development plans."*
3. *"The campaign will forge new partnerships and extend and deepen existing ones."* This manifesto pledge reflects recognition that not all important stakeholders have been drawn into partnership. In particular, property developers and others, such as local governments, are not participating fully, with only a subset of relevant departments involved.
4. *"The campaign will constantly urge action and investment necessary to achieve our water quality targets."*
   The importance of the extensive infrastructure investments of United Utilities and the Environment Agency in achieving water quality improvements is well recognized by the campaign, as is the need for continued investment at these levels to achieve a complete upgrade of the systems.

5. *"The campaign will unite all sectors of the community to make the most of the basin's watercourses and watersides."*
   The foul conditions of the waterways of the Mersey Basin until recently lead local cities and towns to make architectural and planning choices that minimized exposure to the water; the fifth manifesto pledge seeks to reverse these tendencies encouraging communities "to turn to face their rivers."
6. *"The campaign will encourage initiatives to enhance conditions for Mersey Basin wildlife, and to allow people to enjoy and respect the wildlife heritage."*
   As water quality and environmental conditions have improved, the Mersey Estuary has been designated a European Special Protection Area and a Ramsar convention wetland site. This manifesto pledge expresses the campaign's wish to build on these successes. Moreover, the campaign's original goals lacked a broad ecosystem perspective, which this manifesto pledge helps corrects.
7. *"The campaign will spread RVIs and related projects across the Basin, building on local knowledge, drive and enthusiasm to meet the wider objectives."*
   The campaign has identified as critical the need to establish institutions devoted to identifying needs and opportunities, setting priorities, and securing resources within each sub-watershed so as to foster ownership and draw on local enthusiasm and resources.
8. *"The campaign commits itself through education to securing the interest and active involvement of young people far into the future."*

A2.6 Degree of Success

The MBC and its partners in conjunction with massive monetary investment and strong regulatory standards and enforcement have been extremely successful. They have come a long way toward transforming a badly degraded river and canal system into a vibrant and healthy ecosystem. They have built an impressive array of institutions and networks and established broad and diverse stakeholder participation. Equally important, they have successfully cultivated public and institutional appreciation of the beauty and value of such a system. The MBC was awarded the first International River Prize for Best River Management Initiative at the Second International River Symposium in Brisbane, Australia.

A2.6.1 Measures of Water Quality and Ecological Health

Water quality in the Mersey Basin has improved dramatically during the last 15 years. More than 80% of the waterways in the basin can now support fish compared to less than 50% in the early 1980s. BOD has declined by more than 25% from over 200 tonnes/day in 1985 to just over 150 tonnes/day in 1996. Long absent species, such as salmon, ocean trout, and dolphin, have been sighted in the Mersey Estuary. Sightings of kingfishers in a triennial (loosely controlled) survey begun in 1992 were up by 28% in 1998 over 1995 levels. (However, levels of dragonflies and damselflies, other clean water indicator species, were down 9%).

Dissatisfied with the way the standard four-part water quality classification system reports outcomes, the campaign partners ICL, a software company, and the Environment Agency developed a more informative continuous 1 to 100 point scale system termed the

Mersey Measure, but which remains linked to the traditional Water Quality Index (WQI) classification system. The Mersey Measure is based on 8 parameters: dissolved oxygen, BOD, pH, nitrate, nitrogen, suspended solids, temperature, and chlorides.

Without the commitment of the Environment Agency to water quality improvements and the cooperation, support, and investment of United Utilities substantial water quality improvements in the Mersey Basin would not have occurred as swiftly as they have, if at all. United Utilities has invested over £1.2 billion in new sewers, treatment plants, and other infrastructure improvements approved by the Environment Agency. £700 million of these funds have paid for over 800 sewerage and wastewater treatment projects since 1989. The company will spend £300 million around the Mersey Estuary alone to ensure that dissolved oxygen never falls below 10% and is committed to spending at least an additional £1.3 billion on similar projects by 2010. These investments are driven by statutory water quality requirements established and defined in a series of EU directives implemented by the national government and policed by the Environment Agency. MBC has served as a consultant during development of these plans.

A2.6.2 Extent of Public Outreach, Stakeholder Processes, and Involvement

Public outreach is a strength of the campaign, as their name suggests. For example:

- The Mersey Basin Trust has attracted over 600 member organizations
- *The Campaigner Newsletter* boasts a circulation of over 30,000
- The campaign conducts programs with over 150 schools every year
- The 1998 Kingfisher and Dragonfly Survey reached over 140,000 people, of whom nearly 18,000 completed and returned surveys
- Over 2000 volunteers helped coordinate more than 120 events for the 1997 Mersey Basin Weekend

These data give some indication of the campaign's success in establishing stakeholder processes and involvement. The progressive establishment, growth, and increasing independence of Mersey programs and institutions designed to foster partnership and participation provides further evidence. Beginning with only a government-affiliated MBC Administration in 1985, the campaign spawned the Mersey Basin Trust in 1987, and the Mersey Basin Business Foundation in 1992. The campaign administration has become an independent entity, the trust has grown to encompass hundreds of organizations, and the business foundation has attracted more than 20 large businesses and generated over £310,000 in revenue to support campaign activities. The RVI program, begun in 1993, has spawned 17 RVIs that cover approximately half of the basin's major watercourses. Each of the many projects and programs the campaign has developed and/or supported, such as *Grants and Awards*, *Stream Care, Water Detectives, Water Watch,* and the annual Conference, has created additional opportunities for stakeholder participation and involvement.

As noted in the discussion of manifesto pledge (point six), the campaign has yet to involve all important stakeholders and is seeking better means of attracting property developers and branches of local government that lack an environmental protection related mandate.

A2.6.3 Level of Cross-jurisdictional Cooperation

The relative simplicity of coordinating jurisdictional issues at the regional scale, a single national government agency (GONW), a single regional development agency (NWDA), and a single water and wastewater utility (United Utilities), has allowed the campaign, together with its partners, to focus on multi-jurisdictional issues at the sub-regional level.

The MBC's RVIs are critical tools for fostering cross-jurisdictional cooperation. They operate at a geographic scale that encompass a modest number of adjacent municipalities enabling these governments to work together to pursue common cross-boundary projects. In addition, they provide a link between the campaign and the Environment Agency's local planning process with a member of the Environment Agency sitting on the steering committee of each RVI. Each RVI has a dedicated coordinator supported by the campaign or its partners to oversee projects; maintain continuity and momentum; and link to the MBC's strategy, operations, and networks. As noted above, the campaign has now established 17 RVIs covering approximately 50% of the Mersey Basin.

As an organization founded on the principal of partnership the campaign is extremely sensitive to the need for cross-jurisdictional cooperation. This is exemplified in the Mersey Estuary Management Plan, non-statutory management that provides the framework for the long-term sustainable development of the estuary zone, commissioned by the MBC. Implementation of such a plan necessarily depends on the cooperation and participation of a large number of different types of organizations. After development of the plan through a participatory process, the campaign obtained official pledges to assist in implementing the plan from over 30 entities including local authorities, NGOs, and government agencies.

A2.6.4 Economic Regeneration

The MBC was founded on the proposition that economic regeneration and environmental restoration were linked. While this may indeed be true, events have yet to bear this out unequivocally. The North West Region continues to struggle to establish itself as major economic force in the United Kingdom, even though the South East of England has witnessed an economic boom since the early 1980s. Merseyside has (for the second time) been awarded Objective 1 status by the EU due to low gross domestic product levels, high unemployment, and other economic and social issues. The area will receive high levels of EU funds to stimulate economic regeneration. Further, much of Greater Manchester received EU Objective 2 status.

The campaign, through its partners, has helped the region realize that its rivers and waterways are regional assets, and that the economic development agenda should embrace these assets. Large handsome waterside developments, such as Albert Dock and Salford Quays, have been constructed. Fifteen years ago, this would have barely been imagined.

The campaign has refined its view of how restoration of the regions water and watercourses will stimulate economic regeneration. Taking the lead from a study sponsored by the RK Mellon Foundation and Sustainable Pittsburgh, the campaign sees local environmental quality and access to water-based recreation as an important piece of the total "package" that draws knowledge-based companies and their employees. To help further economic regeneration following this view the campaign has become an important consultant in the development of regional economic development agendas and has contributed to the development of Regional Planning Guidance, the NWDA Regional Strategy to 2020, the NWDA Land Reclamation Programme, and the development of NWDA Sub-regional Action Plans. The land reclamation program provides a good example of how campaign environmental and economic goals are being combined. The multi-million pound strategic program of waterside reclamation seeks to provide opportunities for:

- Supporting the improvement of the Mersey Basin's watercourses and watersides, which are important regional environmental assets
- Investing in the creation of new waterside and riverine environments
- Projecting a positive image of clean and healthy waterways in the region
- Assisting in the delivery of urban renaissance
- Maximizing the use of waterways as public open space
- Improving access to waterside locations to allow residents and visitors to enjoy and respect the Mersey Basin's wildlife, culture, landscape, and heritage
- Uniting all sectors of the community in delivering waterside regeneration
- Identifying and tackling dereliction that exists along the regions waterside environment

A2.7 Lessons and Applicability to the United States

Lesson 1: Utilities play a critical role.

"The big improvements seen in the quality of rivers and canals have been due to a combination of investment by the water industry in better sewage treatment facilities and sewerage networks, and tight regulation enforcement and pollution prevention work by the Environment Agency."

–Jeremy Frost, Environment Agency's North West Environment Protection Manager

Utility involvement was necessary to effect the dramatic improvements in water quality in the Mersey, but alone it would not have been sufficient. The geographic scale of the utilities' territory and the coordinated, careful planning process under which improvements are being carried out have also been of great importance. United Utilities serves all of the Mersey River Basin and works cooperatively with the Office of Water Services (OFWAT), the UK water company's independent governing body, the Environment Agency, and the MBC in developing the long-term improvement plan and for each five-year asset management plan (AMP). MBC served as a consultant to United Utilities in the development of the current plan, AMP3 (2000 to 2005), and is already involved in shaping the next plan. In operational terms, MBC is developing mechanisms to ensure that United Utilities work very closely at a practical level with the local MBC coordinators in the implementation of AMP3.

While U.S. communities have made significant investments in wastewater infrastructure since the 1970s, often in response to regulatory requirements of the Clean

Water Act, (e.g., the Massachusetts Water Resources Authority's efforts to facilitate the clean up of Boston Harbor), it is rare to find a confluence of a similar set of factors found in the Mersey:

- A large watershed that falls within the jurisdiction of a single water/wastewater utility;
- The ability and willingness of that utility to make substantial monetary investments;
- Cooperative regulators and a regulatory process that encourages strategic long- and middle-range planning; and
- A well organized watershed initiative with deep voluntary and business support.

Nevertheless, the leading role played by United Utilities is something water/wastewater utilities and/or watershed managers in the United States might emulate. Adoption of this model is by no means limited only to large utilities with service areas encompassing medium- to large-scale watersheds. In regions where the utilities cover relatively small geographic areas, such as a single municipality or small region, groups of utilities might form partnerships defined by watershed boundaries as a platform for joint planning and management. Activities could then be coordinated with local and regional watershed management initiatives. As exemplified by United Utilities, water and wastewater utilities can serve as important sources of technical and financial resources for watershed management and frequently share many objectives with watershed initiatives.

Lesson 2: Large-scale watershed management initiatives can succeed by cultivating and integrating discrete sub-watershed and stakeholder initiatives.

A serious deficit in the U.S. watershed management movement is the absence of large-scale watershed management initiatives in most of the nation's river basins. Effectively involving the full spectrum of stakeholders, something the MBC does extremely well, is one of the obstacles to the establishment of such initiatives and a challenge to the few that do exist. The MBC approach of developing separate institutions/operational arms to serve subsets of stakeholders and to represent those stakeholders' interests within the MBC's planning and operations is a model that could be readily transferred to the United States. Enfranchising stakeholders by giving them a meaningful voice in setting goals and priorities is a hallmark of the U.S. watershed movement. What makes the MBC's approach distinct is that it reaches out to the three key stakeholder communities, government, NGOs, and businesses, by providing a dedicated institutional place, programmatic services and support, and funding avenues appropriate to each. The approach has been particularly effective in attracting partners within the business community.

The campaign's linking of economic and environmental improvements also contributes to its success, enabling it to attract businesses and local governments in an economically depressed region that might be hostile to an exclusively environmental message.

A second problem large watershed initiatives face is the absence of organizations dedicated to managing the nested smaller watersheds, organizations capable of generating local momentum and political will and able to help develop and implement management plans. The MBC's RVIs represents a possible model for developing and linking such organizations. This model is inherently flexible and emphasizes local goal setting and control, which is likely to have appeal in the United States.

Lesson 3: Sustained government financial support for core operational costs provides institutional stability and enables institutional growth.

Adequate and sustained funding especially to cover core operational costs, such as rent and administrative staff support, is one of the most commonly cited weaknesses in U.S. watershed management. Funding in general is extremely limited and is not growing quickly enough to meet burgeoning demand. State and local government funds that are available are often restricted to specific types of programmatic activities and rarely provide more than a year or two's worth of support for the designated activities, such as wetland and erosion protection.

The 25-year commitment of the British government to provide core-funding support for the MBC represents a radically different approach. The 25-year funding commitment, which only covers a small portion of administrative costs today, accomplished three things:

1. It created a sense of confidence in the long-term stability of the nascent institution.
2. It provided genuine financial stability during the first several years of operation.
3. It enabled the organization to concentrate on strategic and programmatic development.

The success of these programs and strategies allowed for robust institutional growth by broadening the partner base and improving the campaign's competitiveness in receiving grant awards.

An analogous grants program, even for shorter durations of 3, 5, and 10 years, would be of immense benefit to the watershed management movement in the United States. The opportunity is not limited only to federal government. For smaller watersheds and sub-watersheds the funding levels required are well within the means of state, county, or even city governments. In most cases, the cost for each watershed would probably be less than that of building and maintaining a couple of blocks of roadway for the same amount of time.

APPENDIX A3

THE FRASER RIVER CASE STUDY

A3.1 Introduction

The Fraser River Basin includes both large undeveloped areas and heavily urbanized centers. The conflict between environmental preservation and economic development has steadily increased over the past century, peaking in the latter part of the 20th century. The subsequent urbanization and industrialization have been associated with development of forestry, mining, agriculture, fisheries, energy, transportation, recreation, and tourism. This development has created a classic complex of water management challenges: water supply, pollution control, fisheries management, flood control, hydropower production, navigation and wetlands management.

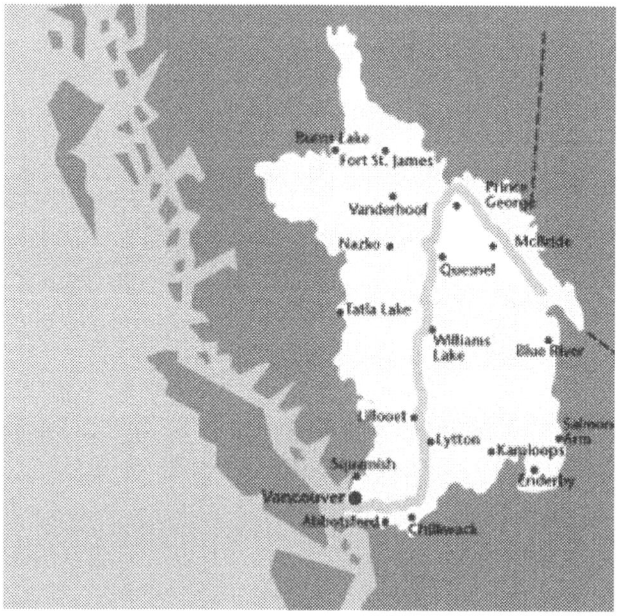

Figure A3-1. The Fraser River Basin (taken from http://www.fraserbasin.bc.ca/generalmap.html)

The perception of environmental deterioration led to an initiative by the Canadian government to restore and protect the Fraser, using a framework that was inclusive of the Basin's stakeholders. One of the products of the initiative was the creation of an institution to promote sustainability, but that was given no legal authority to regulate or implement. This governance model has been very effective in encouraging local action and projects in the best interests of a more sustainable Fraser River Basin.

A3.2 Basin Description and Current Water Uses

The Fraser River Basin is the fifth largest river basin in Canada, spanning an area of 238,000 km^2, an area comparable to Oregon. The Fraser is 1,325 km in length, flowing through diverse terrain, starting in the Rocky Mountains, traversing through vast forested areas, plateaus, canyons, rolling uplands, wetlands estuaries, and finally discharging to the Georgia Bay. While aboriginal peoples have lived in the Basin for more than 10,000 years and their populations likely never exceeded 50,000, settlement has expanded rapidly towards 2 million people since Simon Fraser's first voyage down the river in 1807. The current use of the Fraser varies along its length, with fishing and recreational uses dominating the sparsely populated upstream region, and industrial and municipal uses becoming more dominant in the more densely populated downstream region of Vancouver.

The Fraser Basin is one of the world's greatest salmon producers, and contains a wide diversity of fish and waterfowl species. As Canada's largest producer of salmon (Pomeroy, 1995), over 60% of the sockeye, 60% of the pink and 16% of the Chinook salmon catches are from the Fraser. An average annual return of $300 million is realized from commercial, recreational, and aboriginal fisheries (Marshall, 1998). The Fraser River estuary supports the highest density of wintering waterfowl, shorebirds, and birds of prey in Canada. In a typical year, about 1.5 million birds (150 species) use the estuary and its adjacent upland, especially between October and December (Butler and Canning 1989; Butler 1992) for birds migrating from eastern Siberia and western North America to South America.

The Fraser serves as the main artery for British Columbia's resource-based economy, which includes agriculture, forestry, and mining. The basin supports 48% of British Columbia's commercial forest area, 60% of metal mine production, and 44% of agricultural land. The Fraser and its tributaries are important shipping and transportation links for these resource-based industries. Over 60% of the province's population of 2+ million lives in the basin and depends on the river as a source of drinking water. The combination of industrial and municipal water use results in annual discharges to the river on the order of 900 million cubic meters. The basin is also a key attraction for the recreation and tourism industries, valued at about $300 million annually,

The basin has also been home to aboriginal peoples for thousands of years and is a key part of their culture, particularly with respect to fishing. The aboriginal community maintains a Web site on fishing issues in the Fraser Basin through the Fraser River Aboriginal Fisheries Secretariat.

A3.3 Impetus for Management on a Watershed Basis

Immediately following a devastating flood in 1948, the federal and provincial governments made a short-lived and unsuccessful attempt to develop a comprehensive management plan for the Fraser River Basin. But it was not until the early 1990s that a watershed effort resumed, with the emerging principles of sustainable development popularized by the Brundtland Report and the national, provincial, and local roundtables that became Canada's distinctive response to Brundtland's sustainable development imperatives.

The Fraser Basin Council (FBC) exists after more than a decade of struggles toward understanding on how to address integration of the social, economic, and environmental components of sustainability. In 1990, the two mayors at opposite ends of the Fraser River system, Vancouver and Prince George, challenged each other to "clean up their parts of the River." Until that time, the Fraser Basin was, as it had been since humans first became its companions, just "the sum of its parts." Suddenly, the reciprocal nature of river life became apparent to a critical mass of people. Since time immemorial, the river had been a contiguous, complete ecosystem. Many believed that it again had to become a single unit, *including* its human inhabitants, if it was to be sustained. The pressures of growth were apparent and the population was rising faster than ever. What was needed was a long-term sustainable growth strategy.

A3.4 Institutional Arrangements

A3.4.1 The Fraser Basin Management Board

In an attempt to address the deterioration of the Fraser River and its salmon stocks, the Canadian government initiated the Fraser River Action Plan (FRAP) in June 1991. The FRAP was conceived as a six-year, $100 million green plan initiative to be undertaken jointly by the Departments of Environment (DOE) and Fisheries and Oceans (DFO), each of whom was to administer half of the program's funding. In developing its green plan, the federal government concluded that building partnerships would be critical to its success. However, both agencies had for some time been developing their own plans for the FRAP, and severe restrictions had been placed on discussing these plans outside of their respective organizations. DOE and DFO intended to remedy this situation by initiating a collaborative partnership-building process.

An agreement was reached in 1992 to initiate the Fraser Basin Management Program (FBMP), to be led by a multi-governmental and non-governmental, multi-interest committee, called the Fraser Basin Management Board (FBMB). This board was required to make decisions by consensus and to encourage consensus-based decision making in all basin activities as well as to facilitate the development of locally based decision-making processes.

No formal basin-wide organization institution existed prior to the FBMB. The board was given five years to accomplish this goal. At the end of the five years, a new institution was established, the FBC. The institutional arrangements for the board are presented below, followed by those of the council. The institutional arrangement of the council greatly benefited from the experiences gained during the life of the board. The FBMP and its board set a path, with much public involvement, toward a very different kind of public-private partnership, one dedicated to balancing the components of a sustainable future for the Fraser Basin.

Public perception of the FBMP at the time generally equated it as "just another environmental organization." The task of incorporating and accepting inclusion of social and economic dimensions into the "sustainability agenda" was an on-going management board responsibility.

Establishing a multi-stakeholder consensus process proved quite difficult. The first issue was the process of selecting board members. Once selected, members then had to learn to recognize other points of view of both the basin and the agreement. The consensus process was also new to most board members and initially appeared slow and tedious. Finally, the board members represented larger constituencies, who also needed to believe that they were a part of the decision-making process. Communication with the larger public was essential. Each of these issues is discussed below, followed by the establishment of regional coordinators and a plan for sustainability.

A3.4.1.1 Selection of Board Members

The 19-member board consisted of an appointed "neutral" chair, three representatives of each of the four orders of government (federal, provincial, local, and First Nations [the aboriginal interests]) and six nongovernmental representatives from economic, environmental, and social interests in the basin. The process of selecting board members varied between each of the representative groups:

- The federal and provincial governments each selected three senior members of immediately relevant governmental agencies.
- The local towns grouped their communities into three areas of the Basin (upper, middle, and lower). Each group carried out a nomination and election process to fill the three local government seats on the board.
- The three First Nations positions were also filled by representatives from among the eight nations in the basin's upper, middle, and lower regions.
- The remaining six positions were filled through a process facilitated by the chair. Nominations were sought by advertising throughout the Basin. The criteria for selection were made explicit via public announcements (e.g., knowledgeable about sustainability and willing to commit to consensus processes). Over 100 nominations were received, reviewed, and forwarded to all signatory governments who eventually reached a consensus on the six appointees. This was not an easy task given a federal progressive conservative government in Ottawa, a provincial new democratic party government in Victoria, and a diversity of political interests among the local governments.

For the most part, the selection processes for the various board positions worked quite well and produced a diversity of membership that was generally acceptable to basin stakeholders. Nevertheless, two significant difficulties arose. First, the environmental interests represented by members of the British Columbia Environmental Network (BCEN) believed that they, just as the governments, should have three seats on the board and that they had obtained a commitment to this from the federal organizers of the program. Even though several of the non-governmental individuals appointed to the board did represent environmental interests, the BCEN considered this an inadequate alternative. It took more than a year and a great deal of effort to restore good working relations with members of the BCEN and to obtain their participation on the board's multi-stakeholder committees. The problem was not completely resolved until one of the BCEN's members became a board member when a vacancy arose in the third year of the agreement.

The second difficulty related to selection of the First Nations members of the board. Subsequent to drafting the agreement but before its signing, the federal and provincial governments decided that First Nations should also have three seats on the board. This was done to be consistent with these two governments' emerging recognition of First Nations as a fourth order of government. However, the First Nations had not been signatories to the agreement. Since there was no organization that represented the eight nations whose territories included parts of the Basin, there was no readily available mechanism for selecting their three board members. After extensive discussions with the chair throughout the summer of 1992, three members of First Nations agreed to join the board at its inaugural meeting and serve until they could complete their own internal selection processes. This agreement temporarily resolved the difficulties that had arisen from First Nations and ensured their participation from the outset of the board's activities.

The difficulties that arose with the representation of environmental interests and First Nations, while not unusual in such processes, illustrate how failure to ensure appropriate representation of stakeholders at the beginning of collaboration can cause substantial difficulties later on and, in the context of these two major stakeholders in the Fraser Basin, had the potential to seriously undermine the process.

A3.4.1.2 Diversity of Views of Board Members

A potential strength of the board was the diversity of knowledge and experience of the members about the basin, its governance processes, and management issues. At the outset, however, major challenges arose. While the civil servants were familiar with the administrative processes of government, this was not the case for many of the non-governmental members. Although individuals knew parts of the basin where they had lived and worked, no member was well acquainted with the basin in its entirety. Each of the non-aboriginal members had a great deal to learn about the First Nations in the basin. While some board members had expertise in environmental issues, the comparative advantage of others was in the economic or social issues. The elected local politicians and business members brought experience that the other members had never had. The lack of common knowledge and differing perspectives made communications very challenging in the early discussions, and the potential strength in the members' diversity only gradually began to be harnessed as people got to know each other, the board was informed by briefings, and board members traveled together to different basin communities. In facing these difficulties, the board was no different than most other multi-stakeholder processes, except in the remarkable diversity of its members and the breadth and complexity of the task it faced.

The board had its first monthly meeting in September 1992. It decided that the priority was to develop a strategic plan for implementing the agreement and to do this by involving stakeholders throughout the basin. During the autumn, the board developed its understanding of the complex agreement, prepared materials for informing and stimulating stakeholder involvement, and designed a consultation process that included a variety of formal and informal opportunities for discussion. Early in the year, the board held nine community open houses throughout the basin, followed by a basin-wide stakeholder meeting. Over 700 people participated, sharing their knowledge, experience, and concerns and offering advice on priorities. The concerns raised most frequently dealt with fish and fish habitats, water quality and quantity, water diversion and export, liquid and solid waste management, community and regional development, floodplain management, and government processes and programs. Based on input from these discussions, the board established basin-wide, multi-stakeholder committees with governmental and non-governmental participants to begin work on management strategies in four priority topic areas: water resources management (quantity and quality), waste management (solid and liquid), fisheries and aquatic habitat management, and community development.

A3.4.1.3 Understanding of and Commitment to the Agreement

From the inaugural meeting of the board, it soon became evident that there were varying interpretations about what the board was required and authorized to do. The agreement document was lengthy, contradictory on some important issues, and silent on others. The questions raised were wide ranging and fundamental, and included: Should the sustainability focus be on environmental issues and not the economic and social issues? Shouldn't the focus be on the water and not all the land issues? Is the board to provide leadership or only coordinate? What is the difference between the board and the program?

Through the early meetings, the board members struggled to reach agreement on the concepts and language that ultimately resulted in the statements of vision, mandate, and goals in the strategic plan in 1993. In the process, not only did the differing interests of the board members become clear, it also became evident that they had come to the initiative with varying ideas about what it was intended to do, how it would operate, and what it might deliver. For the federal government members, the board was the essential mechanism for building the partnerships, in particular for getting cooperation from the provincial and local governments and involving large numbers of stakeholders, that were going to be crucial in delivering on their ambitious goals of the FRAP. For the provincial government, it was an opportunity to influence where and how the $100 million in the federal FRAP would be spent. For the local governments, there were hopes of obtaining funds through the board to meet their own growing costs for building water supply and wastewater treatment systems. For the First Nations, the board was an opportunity to be recognized as one of four orders of government and bring forward their claims and, in particular, their immediate concerns about fishery allocations. Finally, in their varied interests, the non-governmental members were looking for new ways to influence the governments and hold them more accountable.

These were only some of the strong interests that became evident and led to questions about the nature of the parties' real commitments to the initiative. In developing the strategic plan, substantial progress was made during the first year in shaping a consensus among the board members. What remained, however, was the enormous and ongoing task of bringing all of the other governmental and non-governmental stakeholders in the basin into this consensus. What has never gone away, as is to be expected in such collaborative organizations, are the tensions created by the differing interests of the people sitting at the board table.

A3.4.1.4 Learning the Consensus Process

In the early meetings, the explicit use of consensus processes was a relatively new experience for most of the board members. For those who were accustomed to imposing decisions by virtue of their position and authority or by quickly calling for a vote, the process initially seemed to be time consuming and inefficient. As time went on, however, most members came to recognize the strengths of the consensus process and to use the process more efficiently. This acceptance was hastened by the experience of successfully dealing with a major challenge that arose at the board's second meeting. It took the better part of an intense day of discussions to reach agreement on the wording of one sentence and thereby avoid the First Nations members' threat to vacate the board: "The board recognizes four orders of government sitting at this table." This was wording that all board members could live with. This early success in averting a major setback by using the consensus process gave the new decision-making approach some credibility and the board a boost of confidence.

A second major step forward occurred in 1993 when the board had to make its first decision about taking a public position on an issue related to one of the participating governments' activities. The occasion was the opening of hearings to be conducted by the B.C. Utilities Commission into Alcan's Kemano Completion Project. The issue was politically very controversial and there were strong and differing views among the board members. A task force of the members was formed and a consensus on the board's position was gradually reached. This piloted a general approach for dealing with difficult consensus issues. It also set a precedent for how the board would handle issues when a governmental member of the board had potentially conflicting responsibilities as a member of the board and as an employee and officer of the government. From these early successes, the board's use of consensus processes continued to progress, and the board never found it necessary to develop and adopt formal consensus procedures, in contrast to many other multi-stakeholder processes (e.g., British Columbia Roundtable). While this can be seen as a result of the good working relations, it also can be viewed as a reflection of the relatively weak commitments made at the board table (i.e., even though there were differing interests, unless members are pressed to make substantial commitments for which the board will then hold them accountable, there are not going to be significant stresses and strains on the consensus process).

A3.4.1.5 The Use of Multi-stakeholder Processes

A key component of the board's strategy was using multi-stakeholder processes to fulfill its mandate. At an early stage, it became evident that there was much to be gained in moving toward sustainability by requiring that the diversity of stakeholders represented on the board also be reflected in appropriate ways in each of its collaborative initiatives. For example, diverse stakeholders were included in each of its steering committees, task forces, and demonstration projects. This provided the means of convening people to identify shared interests, create fresh perspectives on sustainability, and foster collaborative initiatives. In meetings throughout basin communities, the power of the approach was revealed as individuals and organizations discovered others who were working toward similar ends. At a time when resources of all kinds were coming under increasing demands, this lead to the identification of ways to collaborate in using limited resources more productively. Convening, facilitating, and spinning off these collaborative processes has been one of the board's most profound influences.

Ironically, as the board was beginning its multi-stakeholder processes it was the best and the worst of times to be doing this. By late 1992 and early 1993, the basin had an amazing number of new federal and provincial initiatives, all of which involved unprecedented levels of stakeholder involvement. The federal government delivered a variety of new projects and programs funded by green plan monies, including initiatives by not only Environment Canada and Fisheries and Oceans under the $100 million for FRAP, but also other federal agencies with their own green plan funds (including Forestry, Agriculture, and Health and Welfare). From the provincial government came consultation programs on new legislation and policies relating to water management, environmental protection, project assessment, protected areas, and land use. In addition to these initiatives and all of the existing activities of these governments came new planning exercises in various parts of the basin, including the regional land-use planning process of the Commission on Resources and Environment (CORE) in the Cariboo-Chilcotin, the sub-regional Land and Resource Management Planning (LRMP) process in the Kamloops area, and the Georgia Basin Initiative (GBI) for planning for growth management in the Canada-U.S. bioregion. At the same time these initiatives were coming from the top-down, stakeholders in the communities and watersheds were developing their own initiatives from the bottom-up, encouraged by the ideas for local roundtables being advanced by the British Columbia Round Table on the Environment and the Economy. Finally, the British Columbia Treaty Commission was getting its process underway with the expectation that First Nations would be coming into the negotiations with claims covering the entire basin.

Never before had there been such an urgent need for the kind of multi-stakeholder coordination and collaboration that the board intended to facilitate. From the perspective of opportunity, it was the best of times. But the realities of seizing this opportunity in practice were daunting. There was tremendous confusion about what was going on. Not only was their little effective coordination between the federal and provincial governments, within the governments there was often little collaboration.

All of this confusion made it extremely difficult for the new Board and its staff to comprehend what was going on, let alone to begin facilitating collaboration. To make matters worse, there was already the beginnings of a severe backlash from many stakeholders who were confused about what was happening and who felt they were being consulted to exhaustion with little evidence of tangible results. In its first stakeholder meetings in 1993, the board was told that it would be seen as just one more part of the problem unless it quickly demonstrated that it could help to resolve the problems of collaboration. In the Cariboo-Chilcotin, the CORE and related land-use planning processes were so contentious and time consuming that the board decided not to attempt to do much in this middle section of the basin until the CORE process stabilized. Headquarters and regional personnel within the federal and provincial agencies of government were often cautious in welcoming the board initiative and sometimes were strongly opposed. The reasons varied from simply feeling over-committed without any new demands, to fundamental opposition to the initiative because it was anticipated to be more of a hindrance than a help in achieving their own ends and interests.

Thus, at the outset, the board and its staff faced a huge challenge not only in developing its own understanding of its mandate but also in building support for its mandate among all the stakeholders. All things considered, the board and its staff were remarkably successful in avoiding a major disaster. They slowly began to build understanding and support. Little things gradually made a difference. For example, the board members went into the communities early, making clear that they were there to listen and learn. Then, they acted promptly to show that they had heard through reporting back promptly to stakeholders on what had been said and acting on demonstration projects. Progress has, however, been slow, and perhaps the greatest difficulties were encountered in winning the support of people in the government agencies. In the beginning, the board's staff had to spend inordinate amounts of time contending with active hostility to the initiative, allaying suspicions, correcting misinformation, clarifying the role of the board, and persuading people of its potential usefulness. What made the greatest difference and has been the key to the gradual progress on all fronts was the establishment and ongoing use of the multi-stakeholder processes, particularly as they increased regional and local involvement, facilitated collaboration at all levels, and began to show tangible results. Once again, the all too easy passage and acceptance by the governments can be seen to have been a dangerously flawed collaboration in that it produced an agreement that left out so many interests and key issues.

A3.4.1.6 Reinforcing Regional Input

During the third year, the board gave major attention to establishing a continuing presence in the basin regions. One of the key action items was to establish coordinators in each of four regions: the upper sub-basin down to and including Prince George; the middle sub-basin down to the Fraser Canyon and including Quesnel, Williams Lake, and Lillooet; the Thompson River sub-basin including Kamloops and Salmon Arm; and the lower valley sub-basin from Hope, through Greater Vancouver, to Georgia Strait. The establishment of full-time regional coordinators, paid by the board, was a critical step in moving toward fulfilling the agreement's commitment to foster decision-making processes that are based as much as possible in the communities and watersheds. Until this time, the board had relied on communications taking place through its regional board members, regional participants on basin-wide steering committees and task forces, the periodic community consultations, and continuing relations with the demonstration projects. The role of the community-based regional coordinators was to supplement and enhance these mechanisms by facilitating coordination among all government consultation and planning processes, making it easier for all stakeholders to keep abreast of regional programs and activities. The coordinators identified key regional stakeholders and established lines of communication to assist them in monitoring issues and collaborating on solutions. By the end of the year, coordinators were active in all four regions, and the board began to give more attention to the specifics of local situations, including the addition of two demonstration projects: Baker Creek Enhancement Society (Middle Fraser) and Alouette River Management Council (Lower Fraser).

A3.4.2 The Fraser Basin Council

The FBC was created in 1997 as a not-for-profit, NGO with a mandate of implementing the Charter for Sustainability, the strategic plan for the Fraser Basin. This strategic plan guides the actions of the FBC. The structure of the FBC is similar to the FBMB (see Figure A3-2 below), drawing on representatives from the four levels of government, as well as geographical and other interest groups.

With its unique financial contributions formula and equally unprecedented constitution and bylaws in place, the FBC was announced to the public beside its namesake's brown river waters in New Westminster on February 27, 1997. The council was hailed as a new kind of public/private partnership between "four orders of government" with the private and nongovernmental sectors mandated to work toward sustainability. On March 31, 1997, the FBMB was replaced by the FBC. It really was a new kind of 21st century organization, purposely designed to have no legal powers, but to be an impartial facilitator of sustainable decision making. It was a new mode of governance. The council is meant to be always infinitely flexible and constantly evolutionary. It is a trans-partisan, non-political, non-governmental, autonomous organization where governments serve as equals working together in ongoing management toward sustainability.

The council, with a unique constitution and bylaws, was legally incorporated as the Fraser Basin Council Society on June 3, 1997. Directors were selected, policies were developed and the first council meeting was held on June 17, 1997. The first board of directors established that this was never to be a "reports based, talking only" organization. This was to be an action-based organization. The three basin-based universities were to be accessed as required for research and support, but the watchword was action, to be facilitated by bringing together all those with an interest in finding solutions to unresolved or new sustainability challenges. The council's diverse and broadly representative directorate with a basin-based staff of regional coordinators, offices, and skills fosters solutions through large partnerships of interests, where no one entity has the power to act alone in seeking solutions. The council is dedicated to facilitating a sustainable Fraser Basin where, as the charter states, "social well-being is supported by a vibrant economy and sustained by a healthy environment."

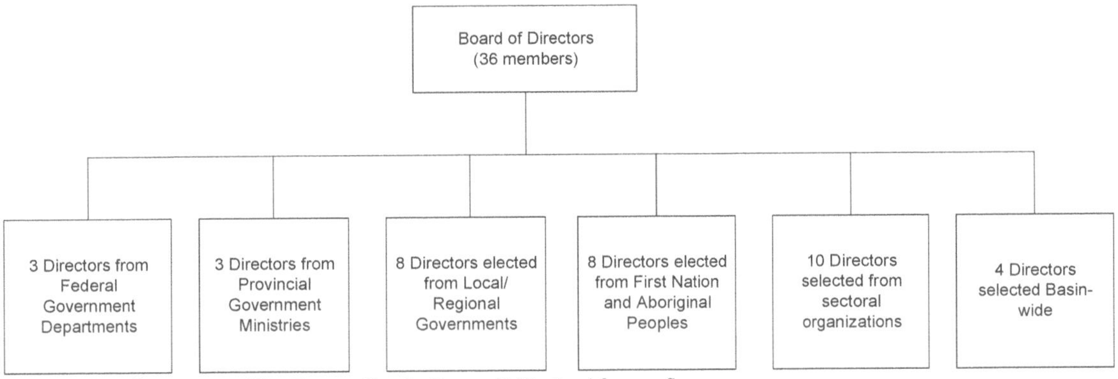

Figure A3-2. Structure of the Fraser Basin Council (Derived from a figure on http://www.fraserbasin.bc.ca/documents/FBC%20Org%20Chart.pdf).

Since 1997, the FBC has engaged in a number of initiatives with government, private, and nongovernmental partners to address complex issues that are threatening the sustainability of the basin. The role of the FBC is to facilitate dialogue, act as a jurisdictional and conflict resolution agent, seek comprehensive solutions, and move objectives to action. The council has encouraged development of new models of cooperative decision making, while striving to maintain impartiality. Initiatives have ranged from projects with basin-wide implications to those focused on sustainability challenges in specific regions. These are discussed in more detail in Section A3.6.

A3.5 Program Priorities and Goals

The FRAP recognized three dimensions of sustainability: environmental, economic and social. In its first phase (FBMB, 1993c, 19), the FBMB adopted a set of sustainability principles to guide implementation of its mandate, as follows:

- Influence the governments to ensure the fundamental needs and values of the people of the Fraser Basin, present and future, are met while recognizing the basin's ecological constraints.
- Incorporate and support First Nations in a manner that is consistent with the existence of their aboriginal rights.
- Ensure renewable and non-renewable resources are conserved and prudently managed.
- Foster equal and fair access to information and decision-making processes to reflect all citizens' needs and aspirations for ecological, community, and cultural diversity.
- Ensure that integrated and innovative approaches occur throughout the full range of planning levels in the basin by coordinating the collection, analysis, and sharing of data and information.
- Recognize the need for adaptive and precautionary decision making.

These principles were translated into a set of five goals for the FBMP. Five strategic programs, each with action plans, were designed to contribute to the production of a strategy for sustainability of the basin by the end of the first five-year agreement in 1997. The five strategic programs focused on: 1) basin management strategies for priority issues; 2) demonstration projects; 3) institutional development; 4) auditing sustainability; and 5) information, communications, and education. These programs and their action plans are discussed in the following section on implementation.

A3.6 Implementation Activities over Time

A3.6.1 Conditions in the Basin Prior to the Charter for Sustainability

The Fraser River had been degraded by wastewater and other contaminants at several locations between the upper reaches at Prince George and Quesnel and at the mouth (Dorcey, 1997). Over half of the industrial discharges in the basin came from interior pulp mills, whereas over 90% of municipal discharges occur in the lower Fraser Basin. In all, the average total discharge of all municipal, industrial, and agricultural wastewater into the Fraser River estuary was estimated to be 900 million cubic meters annually (British Columbia Ministry of Environment, Lands and Parks and Environment Canada, 1993). In 1985, more than half of the bioassays at the Annacis and Lulu wastewater treatment plants (WWTPs) showed toxicity to fish. Groundwater was not immune. The Abbotsford aquifer faced problems of nitrate and pesticide contamination, largely from agricultural sources.

Although there has been a substantial increase in the number of communities with tertiary sewage treatment, the largest producer of wastewater, the Vancouver metropolitan area, is not yet among them. Consequently, most of the wastewater entering the basin's receiving waters still receives only primary treatment. Further, discharges are increasing rapidly, with flows from the Annacis WWTP expected to double by 2036 as the eastern part of the Lower Mainland region grows (Dorcey, 1997).

A3.6.2 Implementation Activities of the Fraser Management Board

In the second year of its existence, the FBMB emphasized collaboration in the implementation of the action plans. During the spring and summer of 1994, multi-stakeholder steering committees were used to develop proposals for each of the strategic programs. Draft ideas were discussed during the autumn in a series of two-day workshops in five basin communities. Nearly 400 people participated in the workshop discussions and hundreds more attended evening open houses at which government and NGOs set up displays and made presentations about their activities. The workshop discussions focused on how to build on the strengths of existing management activities and how to remedy weaknesses, identifying who should do what, how, and when. All the discussions were summarized and distributed in a report and posted on a computer bulletin board system (Dorcey, 1997). Through this process, a new vision of watershed management emerged. The old, top-down processes of governance were replaced by new approaches that involved local decision making, the diversity of interests, and consensus processes.

A3.6.2.1 Selecting Demonstration Projects

The board invited proposals for demonstration projects in the basin and received more than 30 proposals in response. Using selection criteria based on the principles and goals of the program, a multi-stakeholder task force reviewed the proposals and made recommendations to the board. In November 1993, six projects from various watersheds were endorsed. The projects were the Salmon River Watershed Management Partnership (Lower Fraser); Nahatlatch Integrated Resource Management Plan (Fraser Canyon); Nicola Watershed Community Roundtable (Thompson sub-basin); Prince George Riverfront Trails Project (Upper Fraser); Salmon River Watershed Roundtable (Thompson sub-basin); and Williams Lake River Valley Corridor Project (Middle Fraser). The board's endorsement constituted a commitment to work with each of these multi-stakeholder initiatives (half had grassroots origins and all included government and NGO participants) to secure the technical and financial support needed to achieve their goals as long as these remained consistent with the program's sustainability principles.

A3.6.2.2 Facilitating Institutional Development

To be able to facilitate coordination and institutional development, the board began by identifying who does what in the basin. An overview was compiled and widely distributed. All of the stakeholders had complained about the difficulty of understanding an institutional system that included hundreds of departments, boards, commissions, programs, agencies, and organizations with different administrative boundaries each affecting some aspect of the basin's environmental, economic, and social health. In the spring, the first intergovernmental workshop on the basin, Integration of Government Program Delivery and Shared Responsibility in the Basin, was held. Eighty participants from federal, provincial, local, and First Nations governments discussed the newly emerging vision of watershed governance that flows from the bottom up as well as the top down. To guide assessments of how the institutional system affects sustainability and might be improved, a report was prepared on the criteria that should be employed.

A3.6.2.3 Auditing

The reports on institutional design (discussed in the paragraph above) and sustainability indicators were used to conduct the board's first audit, an assessment of the Fraser River Flood Control Program. Flood control had been identified during the consultations as an issue of great concern, heightened by the imminent prospect of the federal-provincial agreement for flood control coming to an end in 1995. A multi-stakeholder task force was established to assess the system that had been put in place as a result of the $300 million spent on the dyking program since it began in 1968. Its overall conclusion was that the program had done a good job of fulfilling its mandate under the 1968 terms of reference. However, it was now 1994 and flood protection needed to be improved, especially for those who lived in the floodplain and had developed a false sense of security. For example, most people were unaware that there was an estimated one in three chance that the greatest flood of record (1894) would be exceeded in the next 60 years. Many mistakenly believed they would be covered by insurance if it should happen. The report outlined recommendations on river management, dyke maintenance, new dyke control, flood-proofing, and integrated floodplain management.

A3.6.2.4 Communication and Education

Each of the board members faced different challenges in reporting back to their stakeholders. It took considerable time to build effective communication channels. The federal and provincial government members used interdepartmental committees to facilitate communications. The three local government members reported back to a steering committee consisting of a representative from each of the 60 communities involved. The three First Nations members and the six non-governmental members had no single mechanism for reporting back but used the various organizations and networks to which they had access. All of the board members were busy with other responsibilities and had limited time available for conveying the substance of discussions that had taken place at the board meeting. During the first two years of the board's activities, when there was little buy-in to the Fraser initiative (because of limited consultation during the drafting of the agreement and it was seen as one more of too many initiatives) the constraints on communications were very debilitating. It was not until the multi-stakeholder processes had begun to operate and a much wider diversity of governmental and non-governmental interests became directly involved that a two-way flow of information started to take place and a foundation of trust and collaboration began to build.

The initial communication problems were particularly great within the federal and provincial agencies, where members of the board's staff found that little clear information or instruction was filtering down from their respective government's board members. The governments instituted a coordinating committee consisting of senior representatives of their board members to facilitate intergovernmental and interdepartmental coordination. This became problematic as it was perceived by nongovernmental board members as the federal and provincial government bureaucracy beginning to predominate over the board's activities, and they raised questions about the independence of the board and its staff. These strains became particularly great when the committee got involved in the early negotiations over the board's budget and began to act as the representatives of the parties to the agreement. Once again, these difficulties slowly receded with the creation of the multi-stakeholder committees that involved a wider variety of governmental personnel from all four orders of government and began the process of building trust and commitment. Particularly significant was the involvement of the non-governmental board members.

The board responded to the immense demands for more and better information about sustainability and management initiatives in several ways. A multi-stakeholder steering committee was established to identify how the many programs and organizations developing and offering education programs relating to environmental, economic, and community sustainability might be better coordinated, particularly at the local level. In response to requests that information be relevant, concise, readable, and accessible, *The Source Book* and a computer bulletin board system were introduced. *The Source Book* was a binder designed to hold copies of the papers, newsletters, profiles, reports, and reviews produced by the board. The idea was that information would always be released in a brief format with references to the relevant fuller documents and designed for insertion in a tabbed sub-section of the binder. At the beginning of the year, the binder included a summary of the action plan. Over the next 12 months, items were distributed relating to each of its components, culminating in an end-of-year report summarizing achievements and laying out the action plan for the next year. All of this information was also posted on a computer bulletin board, which additionally provided a mechanism for feedback and exchange.

A3.6.2.5 Developing Management Strategies

The board's report at the end of the second year summarized the priority issues and more than 50 action items for 1993-1994 that had been identified by the four multi-stakeholder steering committees responsible for water resources management, fisheries and aquatic habitat management, pollution prevention and waste minimization, and community development. The action items identified the organizations involved in the program that had agreed to take the lead in forming and implementing appropriate partnerships. In most cases, this was one of the government agencies and only occasionally was it the board or an NGO. Some of the action items were already underway, but most were not yet initiated.

A3.7 State of the Basin Report and Board Report Cards

Drawing on the work of the various steering committees, the major and immensely time-consuming task in the third year was to produce the first *State of the Basin* report. More than 175 people with interest and knowledge in specific areas collaborated in the development of the report. Based on concerns expressed in the earlier discussions with stakeholders, the report focused on eight topics: population growth, water resource management, the basin's salmon fisheries, forests in the basin, the basin's economy, planning processes in the basin, building new relationships with First Nations' communities, and decision making. Each of the chapters has four main sections. An introduction establishes the importance of the topic and places it in the context of sustainability. The background section provides key indicator information on baselines, trends, and status. The bulk of each chapter is taken up by the issue definition and response sections, which define the component and interrelated issues to be considered and describe government and non-governmental responses. The conclusions identify progress toward sustainability and areas where work remains.

In conjunction with the *State of the Basin* report, the board prepared a report card "which indicates the progress that we, the two million-plus residents of the Fraser Basin, are making towards sustainability" with highlights of "good news," "bad news," and recommendations for next steps. High marks were given where measurable results had been accomplished. Middle grades were given where processes were in place to address sustainability issues but no results had been achieved or monitored. Low marks were given where there had been little or no progress towards sustainability and where development of new policies or legislation was a low priority. While certain areas of the basin may have made greater progress, the assigned grade was a "net grade" for progress in responding to the issue throughout the basin as a whole. Grades were only assigned to consensus issues.

Based on these grades, the board concluded that "[a] great deal of progress has been achieved in setting up processes that will bring us closer to sustainability...[however] at this point, there is little evidence of progress in some of the critical areas addressed in this report card." To get feedback on these conclusions, the report card also included a public report card, inviting responses from basin residents on key sustainability issues.

A3.7.1 Creating a Plan for Sustainability

The draft basin plan released at the end of 1995 was based on the work in the preceding four years as well as the results of the two-day workshop in which representatives of 80 of the partners in the program addressed what should happen when the first five-year agreement ended in 1997. The draft plan was presented as a workbook based on a vision for the basin, as illustrated in Figure A3-3. The vision proposed specific targets for each of the four "directions" to be taken to achieve sustainability, as well as strategies for implementation and who should take the lead. A five-year action plan was identified for addressing some of the more pressing sustainability issues. Recognizing that initiatives were already underway, the draft basin plan stressed the need to build on these initiatives. Some of the proposed strategies were recognized as already addressed in part by various organizations. Others required the assistance of new government institutional mechanisms, such as legislation, strategies, and programs. Many strategies, however, lacked a structured approach to facilitate implementation. Just as the board invited stakeholders to fill in the workbook and give comments and suggestions on each of its specifics, it also asked for input on overseeing the implementation of basin plan.

Five point challenge to implement the basin plan to:
- Every individual
- Governments
- Business and industry
- NGOs
- Educators

Four directions that must be taken in order to work toward the goal:
- Understanding sustainability
- Caring for ecosystems
- Strengthening communities
- Improving decision making

Three sustainability elements
- SOCIAL well-being
- Supported by a strong ECONOMY
- Sustained by a healthy ENVIRONMENT

Two principles
- Everything is connected
- We are all responsible and accountable

A single goal
- To achieve sustainability of the Fraser Basin

Figure A3-3. Our Vision: Sustainability Together

With the input of stakeholders, this basin plan or strategic plan for the Fraser River established a charter for sustainability. The charter for sustainability was FBMB's last and, in many respects, most pivotal "deliverable." It was released with the proviso that "an organization be found" to implement its contents. No such organization existed, so one was to be created. The board noted, if an organization or coalition of organizations oversees implementation of the basin plan, some of its important characteristics might include (Fraser Basin Management Board, 1996):

- Involving the federal, provincial, and local governments and First Nations with non-governmental interests
- Adding value through coordination and integration of efforts
- Not depending solely on governments for funding
- Structured to allow for contributions on a not-for-profit basis

This action proved a heroic task. Among many other critical considerations, directors determined through long dialogue that unless the future organization was fully inclusive and representative of all parts of society from the beginning, the movement toward sustainability would be fatally flawed. Through respectful, inclusive dialogue a way ahead was found so that First Nations and aboriginal peoples, who have been residents of the basin for at least

10,000 years, could confidently become equal partners in the decision-making process and founding members of the new council

It was ultimately agreed that without resolution of "aboriginal rights and title, now being defined, in a fair and just manner," genuine movement toward sustainability would be impossible. Directors accepted that Canada's constitution in Section 35 affirms the place of First Nations and aboriginal peoples. The resulting critical charter principle 11 on aboriginal rights and title was agreed, and weight was subsequently added to the principle with the Delgamuukw Directive of the Supreme Court stating the "duty to consult" First Nations and aboriginal peoples. Principle 11 re-assured basin First Nations and aboriginal peoples of a legitimate place at the council, and they chose to elect council representation from among the eight languages spoken in the basin. With parallel representation by the eight regional/local government elected directors, First Nations directors joined federal and provincial government directors selected from among departmental officials to ensure representation from Canada's defacto "four orders" of government on the new council.

The charter was conceived and confirmed as a "good-faith" document. It was a long-term plan to sustain the Fraser Basin. The charter for sustainability continues to be the guiding set of principles in the Basin, as implemented through the FBC. The principles are based on an agreed upon vision:

> **The Fraser Basin is a place where social well-being is supported by a vibrant economy and sustained by a healthy environment.**

There are four directions of understanding needed for realizing the vision, and a set of principles from which to operate (Fraser Basin Charter for Sustainability, 1997):

- **Understanding Sustainability**: where governments, community groups, and individuals recognize why and how they can contribute to building vibrant communities, developing strong and diverse economies, and maintaining the air, water, land, and living species that make up our ecosystems.
- **Caring for Ecosystems**: where we are all stewards of resources such as water, forests, fish, wildlife, and land. As stewards, we conserve and enhance our ecosystems to maintain strong and diverse economies and to support growing communities. In this way, we not only enjoy our natural environment, but also conserve it to support our high quality of life.
- **Strengthening Communities**: where communities benefit from local experience, skills, and values. Strong communities are built on a diverse economy, an educated workforce, safe neighborhoods, accessibility to basic commodities, shared goals, local action, and a sense of belonging.
- **Improving Decision Making**: where decision making is shared and we work together to reach creative agreements and achieve common goals that reflect the interests of a growing population mixed in gender, culture, religion, age, and interests; and where aboriginal rights and title now being defined are reconciled in a just and fair manner.

Principles

- **Mutual Dependence**
  Land, water, air, and all living organisms including humans are integral parts of the ecosystem. Biodiversity must be conserved.
- **Accountability**
  Each of us is responsible for the social, economic, and environmental consequences of our decisions and accountable for our actions.
- **Equity**
  All communities and regions must have equal opportunities to provide for the social, economic, and environmental needs of residents.
- **Integration**
  Consideration of social, economic, and environmental costs and benefits must be an integral part of all decision making.
- **Adaptive Approaches**
  Plans and activities must be adaptable and able to respond to external pressures and changing social values.
- **Coordinated and Cooperative Efforts**
  Coordinated and cooperative efforts are needed among all government and non-governmental interests.
- **Open and Informed Decision Making**
  Open decision making depends on the best available information.
- **Exercising Caution**
  Caution must be exercised when shaping decisions to avoid making irreversible mistakes.
- **Managing Uncertainty**
  A lack of certainty should not prevent decisive actions for sustainability.
- **Recognition**
  There must be recognition of existing rights, agreements, and obligations in all decision making.
- **Aboriginal Rights and Title**
  We recognize that aboriginal nations within the Fraser Basin assert aboriginal rights and title. These rights and title now defined must be acknowledged and reconciled in a just and fair manner.
- **Transition Takes Time**
  Sustainability is a journey that requires constant feedback, learning, and adjustment. In the short term, the elements of sustainability may not always be in balance.

The FBC and its five regional offices continue to pursue the goals of the charter for sustainability, and evaluate progress through rolling five-year plans, as described in the following section.

A3.7.2 Implementation activities of the Fraser Basin Council

The FBC engages in regional and basin-wide initiatives that are consistent with the charter for sustainability, and with the mandate of "educating on the need for sustainability." The other key mandate is to report on the progress towards sustainability. Some of the initiatives are described in the following section, while the work on sustainability reporting and indicators is described in a subsequent section on degrees of success.

A3.7.2.1 Regional and Basin-Wide Initiatives

Based on the priorities identified by the board and its committees, the FBC becomes involved as a conflict resolution agent, catalyst, and facilitator in specific issues of concern. Current initiatives, listed below, illustrate the breadth of issues and degree of cooperation taking place in the basin.

- Aboriginal/non-aboriginal partnerships document
- Rural development
- State of the Fraser Basin Conference
- Sustainability indicators
- Flood hazard management
- Britannia mine reclamation and remediation
- Debris management
- Nutrient management
- Shuswap lake water quality
- Fraser River Management Plan: hope to mission
- Economic networking directory and planning
- Cariboo/Chilcotin Fish Enhancement Society
- Nechako Watershed Council (NWC)
- Nechako Environmental Enhancement Fund (NEEF) consultation
- Mission of Streams Resource Centre
- Habitat stewardship in the Thompson

Further detail on some of these initiatives is described below, which provides insight into the role of the council and progress to date.

Nechako River Environmental Enhancement

The council has been working to help resolve over 50 years of conflict regarding the management of the Nechako River. The development and management of the Nechako has been the focus of social, economic, and environmental concerns, since Alcan constructed the Kenney Dam in the 1950s to generate hydroelectricity for the smelter in Kitimat.

Assisting the Nechako Environmental Enhancement Fund Management Committee (NEEF MC) with consultations on the downstream enhancement of the Nechako watershed area, the Fraser Basin Council facilitated workshops to obtain advice on enhancement options. Workshop participants included representatives from First Nations, federal, provincial, and local governments, environmental organizations, the private sector, and other interests.

In June 2001, the NEEF MC announced their decision that a cold water release facility was to be constructed on the Nechako River near the Kenney Dam. The announcement indicated a total estimated price tag of just under $100 million. This decision served as a key milestone toward the rehabilitation of fish that depend on the Nechako River and its tributaries, including sturgeon, sockeye salmon, chinook salmon, and many others.

The council was also instrumental in bringing together the interested parties to form the NWC, which includes representatives from all levels of government, First Nations, and business and community groups. The NWC supported the recommendation that a water release facility be built at Kenney Dam to better manage the timing, volume, and temperature of water released into the Nechako River and to allow the restoration of the Murray-Cheslatta system.

Nutrient Management in the Fraser Valley

To help avoid the sort of tragic circumstances that have become public issues in other parts of Canada, the council has facilitated the development of a nutrient management planning strategy focusing on managing agricultural manure and fertilizers "smarter" in the Fraser Valley, while still keeping farming viable.

The strategy, developed by a steering group which includes agricultural producer groups and government agencies, will also result in strong benefits to fish habitat, as runoff from manure and fertilizer into streams can affect water quality and fish health. Key actions and options include improving manure storage and exploring the use of on-farm plans, as well as research on alternative feed programs and enhanced soil testing. The British Columbia Ministry of Agriculture is taking a leadership role in the implementation of the plan, working together with provincial and federal agencies, local government and the British Columbia Agriculture Council.

Flood, Fish, and Gravel in the Fraser Valley

Gravel build-up in some areas of the lower Fraser River reduces the ability of local communities to protect themselves from floods and affects commercial navigation. At the same time, gravel removal can impact fish habitat and other river values, which has resulted in conflict over how this resource should be managed.

The council has facilitated a multi-interest process to develop a management plan for the Fraser River from Hope to Mission. This work has been assisted by funding from the DFO's Habitat Conservation and Stewardship Program and the provincial government, while the City of Chilliwack and other local governments have partnered with the Provincial Flood Protection Assistance Program to fund technical studies in support of the plan. Gravel removal for flood protection is one of the issues covered by the plan.

A testing phase took place in 2001, during which limited removals took place and limited progress was made toward addressing outstanding flood and erosion issues. At that time, participants recognized that a much more proactive approach was needed.

The strategy was to identify "where, when, and how much" gravel removal would be needed to address flood, erosion, and navigation hazards, while protecting and enhancing fish habitat. In a process facilitated by the council, a draft of this proactive strategy has been developed, recommending a balanced approach between gravel removals and dyking improvement. The British Columbia Ministry of Sustainable Resource Management agreed to take a leadership role in the implementation of this strategy.

Implementation of the strategy has been challenging, as disputes have arisen concerning federal and provincial government decisions allowing gravel removal and the Cheam First Nation's assertion of aboriginal rights and title. In the spring of 2003, a short-term agreement was reached that allows some gravel removal with a set aside of funds to support First Nations capacity building and flood management projects. A five-year gravel management plan is under development to define the timing, quantity, and location of gravel removal for flood protection, while protecting or enhancing traditional fish habitats.

Habitat Stewardship in the Thompson

Recent federal and provincial legislation is placing greater responsibility on municipalities and regional districts to manage growth and protect habitat. With the support of DFO's Habitat Conservation and Stewardship Program and the provincial government, the FBC has put two stewardship coordinators in the Thompson region. They are assisting the Columbia Shuswap and Thompson-Nicola Regional districts to gather information needed to adopt and implement new bylaws and zoning requirements.

The long-term goal is to prepare a template, which can assist other local governments with amendments to official community plans and zoning bylaws to reflect the legislation changes to the Local Government Act and other relevant federal and provincial legislation.

Britannia Mine Reclamation and Remediation

Situated approximately 50 km north of Vancouver, Britannia Mine operated from 1902 to 1974. In its prime the mine was one of the largest copper producers in the British Empire. Today, exposed ore in the open pits and underground mine shafts generate large quantities of effluent called acid rock drainage (ARD). ARD is very acidic, contains high levels of sulfate and metals, and is highly toxic to juvenile salmon stocks and other marine life. Environment Canada has characterized ARD at Britannia as the worst point source metals pollution problem in North America. In addition to the ARD problem at Britannia, there are other sources of contamination in soils, groundwater, and near-shore sediments.

Britannia residents hoped to see an early solution to the pollution problem and would like to have more certainty about the future. Economic development in Britannia depends on finding solutions to address the serious environmental problems.

Since 1998, the FBC has been instrumental in keeping the long-standing pollution problem at Britannia at the forefront. The council, acting as catalyst, facilitator, and sustainability facilitator, brought together government agencies, the private sector, Britannia residents, and members of the public to find a sustainable solution.

In 2001, an out-of-court settlement agreement was reached among federal and provincial governments and the present and former owners of the Britannia Mine, paving the way for a first phase of clean up. Remediation is underway, with construction of an acid rock drainage treatment plant scheduled in late 2004. A successful outcome would not have been possible without the extraordinary levels of collaboration between Britannia residents, the present landowner, past landowners, many departments of the government of British Columbia, and the federal Ministry of the Environment and Fisheries and Oceans Canada.

A3.7.2.2 Funding Sources and Levels

The FBMB was given a mandate in 1992 to provide leadership to the development of the FBMP and, through it, to facilitate a sustainability plan for the basin. To carry out its assignment, the board was neither given authority to control activities nor funds to allocate, but it was provided a budget (approximately $1.5 million per year) to carry out a five-point mandate focused on fostering collaboration, described below. The budget was funded by each of the local government signatories contributing 10 cents per capita (totaling approximately $200,000) and by the federal and provincial governments dividing the balance equally.

The FBMB was succeeded in 1997 by the FBC, which continues to be funded largely through the same mechanism. However, the council is working to become financially self-sustaining, aiming to have equal private and public support. The council has recently acquired charitable status, which has allowed it to develop a fund-raising strategy to obtain funds from sources other than government, such as corporate, individual, and foundation donors. Specific strategies and actions, such as the establishment of a sustainability fund (endowment fund), have been undertaken to capitalize on this opportunity.

A3.8 Degree of Success

One of the key initiatives taken by the FBC is the Sustainability Indicators Initiative (SII), which is unique in the breadth of factors considered in achieving sustainability in a watershed. Based on the lessons learned from the earlier FBMP report cards, the FBC worked for two years with government, nongovernmental and community organizations, and the private sector to develop sustainability indicators that track the trends of social, economic, and environmental sustainability.

The first full report on indicators of progress on sustainability was published in January 2003 as *A Snapshot of Sustainability: the State of the Fraser Basin* report. Ultimately, the sustainability indicators serve to increase public awareness of and facilitate action on sustainability issues in the basin, inform the development of policies and programs to address sustainability, and identify information gaps and research priorities.

The snapshot report provides a "medical check-up" for the Fraser Basin and its communities, measuring trends in 16 key issue areas including: population, air and water quality, fish and wildlife health, and forestry. The report results are mixed. In some aspects conditions appear to be normal and healthy, while in others improvement is necessary or additional information and monitoring is required to ascertain how the basin is doing. The second biennial indicators report, *State of the Fraser Basin Report 2004: Snapshot 2*, is expected to be released by the end of 2004. It will update the trends in the earlier report, identify new indicators and information sources, and provide regional profiles for the five major regions in the basin.

A3.8.1 Measures of Water Quality and Ecological Health

Although the water quality in the Fraser was good in comparison to other basins in Canada in the 1980s, some degradation has occurred due to the estimated 900 million cubic meters annually of wastewater discharge into the river. Most of this wastewater received only primary treatment. In 1985, more than half of the bioassays at two of the major municipal WWTPs showed toxicity to fish. Compared to a century ago, several of the Fraser's salmon species were less abundant and, in some cases, there had been a severe decline. A combination of factors was to blame, including lost habitat, pollution, overfishing, and natural causes. Fortunately, the spawning runs have much improved since the mid-1980s for several species, including sockeye, pink, and chum salmon.

Coho salmon, highly valued for social, environmental, and economic reasons, have continued to decline since the 1970s, despite efforts to protect and conserve them. Known populations of Fraser coho are now at or near historic lows. For Chinook Salmon, too, the spawning runs have been relatively low. Nevertheless, the Fraser remains the largest Chinook producer on the Pacific Coast.

Some of these problems were dramatically illustrated in 1994 when more than one million sockeye that normally would have migrated to spawning areas could not be accounted for. An independent investigative group, the Fraser River Salmon Public Review Board, set up by the federal government, identified problems with the methods used to estimate stocks, the enforcement of regulations, and management decisions. The board also pointed to aggressive overfishing and warned the consequences would be severe if the 1994 situation were to repeat itself. The board's recommendations are being implemented through a five-point action plan, which advocates:

- A more comprehensive approach to managing salmon
- Increased enforcement of regulations
- Better integration of priorities between fish science and fisheries management
- A tougher stance on the conditions of agreements under the Aboriginal Fisheries Strategy
- More measures to deal with the excessive capacity of the fishing fleets

According to the first snapshot of sustainability report, over the previous decade the number of salmon returning to spawn in the basin has increased in half of the streams and decreased in the other half. Moreover, the report found about 10% of the vertebrate species had experienced significant population declines during the same period. On the positive side, protected land areas have doubled since 1991 to comprise 13% of the basin, and substantial investments have been made to upgrade WWTPs in the region. There has also been a move by federal and provincial environmental agencies towards setting water quality objectives that are adapted to specific bodies of water or aquifers. While the objectives are not legally enforceable, they do form a reference point for judging water quality at a particular site. Water quality objectives have been set for several of the large river sections in the basin, including the lower Fraser River, the Thompson River, the Nechako River, and the upper Fraser River from Hope to the headwaters. However, the lack of legally enforceable standards remains a major obstacle in accelerating the improvement of water quality.

A3.8.2 Level of Cross-jurisdictional Cooperation and Stakeholder Involvement

The FBC's structure has provided a forum for discussions among the variety of government and NGOs from local to national scales. The current set of initiatives operating through the FBC is extensive, as described in a previous section, ranging in scale from basin-wide to specific sub-regions of the basin. These initiatives have input from the full breadth of stakeholders, from the federal to the local level, and from environmental to agricultural and economic interests. As described in the previous section, a number of successes have been achieved within these projects, in building partnerships and developing cooperative solutions.

Building this council will always remain a work in progress, if it is done well. Just as pursuit of sustainability is not a finite goal, neither should the council ever become finite. This council must meet its ever-shifting targets by adapting to change and remaining in an evolving state of "becoming." Other challenges will include the transition of regional committees to broadly representative regional councils as resources permit. Each to some degree has issues distinctive of their region and yet each is an integral part of an emerging, less competitive, more collaborative basin and/or provincial culture. The council is increasingly requested to embrace more operations than are circumscribed by the formal boundaries of the basin. It is expected that the council will increasingly be called on to operate at the provincial level, partnering with existing organizations and adapting in the process.

Over the past five years, there were seemingly insurmountable problems at least partially resolved through the council's encouragement of dialogue and facilitation. By acting as a catalyst and resolving conflicts and jurisdictional impasses and through sustainability education, long outstanding problems can give way to sustainable solutions. This council fills a critical need as an effective new governance model, where all opinions and sectors are deemed worthy of being respectfully included in the decision-making process.

A3.9 Lessons and Applicability to the United States

The emphasis in managing development in the Fraser Basin since the 1990s has consistently been "to create a place where social well-being is supported by a vibrant economy and sustained by a healthy environment," a goal that is consistent with the goals of most watershed efforts in the United States. The socio-political structure (and overlapping jurisdictions) of the water-related institutions in the Fraser Basin are similar to those in the United States, and could provide insights into a nongovernmental approach to watershed and ecosystem management,

Lesson 1: Establishing effective multi-stakeholder processes requires time and commitment.

The initiative in the Fraser Basin demonstrates not only the potential and essential role of collaborative multi-stakeholder processes in pursuing sustainability but also the challenges of putting them into practice in watershed management. Time must be allowed for this process to succeed. The FBC has been successful in raising awareness of the Fraser watershed because of its ability to provide a middle ground between government and NGOs and interests. The credibility of the council depends on its acceptance as an independent entity that operates efficiently and effectively and is accountable to stakeholders.

The Canadian government, at the federal, provincial, and local level, facilitated the establishment of an effective process by providing seed funding over a five-year period. Although there are cases where the United States has made similar commitments to establish a stakeholder process, notably the CALFED Bay Delta Program (CALFED) in Northern California, this is more the exception than the rule.

Lesson 2: Key stakeholders should be involved at the earliest possible stage.

The Fraser Basin experience illustrates the need for all of the parties to be present, to understand and to be committed to the intent and language of the consensus or agreement. The rapid drafting of the initial agreement in the early 1990s and its swift passage through bureaucratic and political approvals of three levels of government were misleadingly impressive. This could have been disastrous, and it took years to build trust among the stakeholders so that they could work jointly towards the shared goals.

One of the consistent obstacles to effective watershed management in the United States has been the difficulty in successfully engaging key stakeholders and gaining their support for new approaches. The Narragansett Bay Estuary Project provides an example of the high costs and lack of success that can occur when stakeholders are not engaged early in the process. The case of the Fraser demonstrates that important stakeholders can be involved at later stages but that delays are likely to result. Forward thinking about including all relevant stakeholders as early in the process as possible is advantageous for the goal of watershed sustainability.

Lesson 3: Multi-stakeholder processes provide a forum for addressing the joint management of land and water for sustainable watershed systems.

Nonpoint source pollution problems will only be solved with the ability to address decisions about changes in land use and application of fertilizers and other chemicals to agricultural and non-agricultural lands. Historically, this has proved difficult in Canada and the United States, where the governing institutions of land and water often do not share the same jurisdictional boundaries or common mandates regarding the preservation of the watershed. (Ontario is one exception. See the Grand River case study [Appendix A4] for additional information.) The Fraser Basin is breaking down some of these institutional barriers through the use of stakeholder processes that involve the full range of parties—from national to local on the government side, and from environmental to agricultural to economic interests on the non-governmental side.

APPENDIX A4

THE GRAND RIVER CASE STUDY

A4.1 Introduction

The Grand River watershed has been recognized around the world for its natural beauty, ecological health, and diversity—and for efforts to restore and protect it. This case study demonstrates how the formation of the Grand River Conservation Authority (GRCA) has led to outstanding success in watershed planning and management in the Grand River basin at both the local and regional levels and across an array of jurisdictional boundaries. The watershed comprises 34 municipalities and an array of local and provincial agencies. Of critical importance to successful watershed planning and management has been the institutional stability and established regulatory authority of the GRCA, along with the practice of integrating watershed planning with municipal land-use planning. U.S. watershed managers will find valuable lessons in these and other important elements of the Grand River watershed approach.

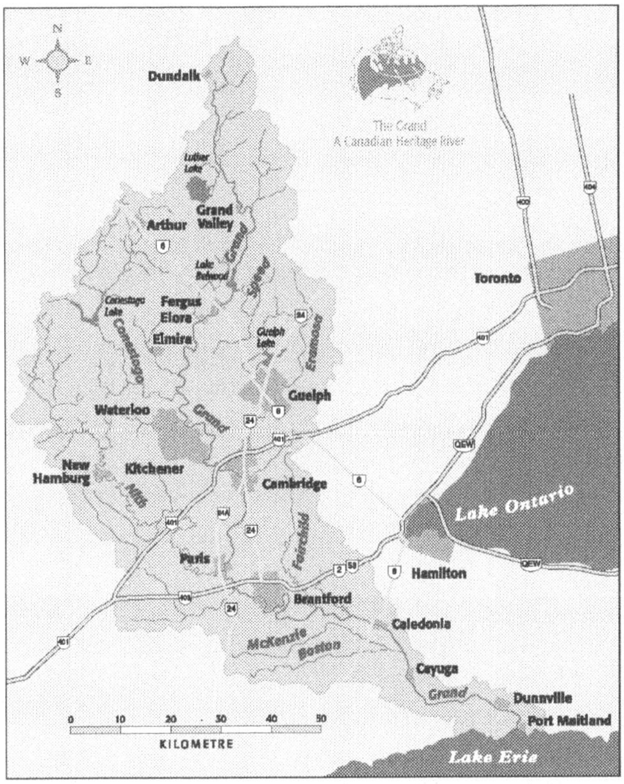

Figure A4-1. The Grand River Basin. ©Grand River Conservation Authority

A4.2 Basin Description and Current Water Uses

From its headwaters near Dundalk, the Grand River flows southward through the largest watershed in southern Ontario, draining an area of approximately 6800 km^2 (2600 mi^2) and meandering 298 km (185 mi) before reaching Lake Erie at Port Maitland. On its way, the Grand joins with its main tributaries, the Speed, Eramosa, Nith, and Conestogo rivers, as well as hundreds of smaller tributaries. The Grand is the largest Canadian river entering Lake Erie.

Over the years, the Grand has supported native cultures, pioneering settlements, and industrial enterprise, marks of which still can be seen in the mills, foundries, and factories on its banks. Remains of the old Grand River Canal demonstrate past resolve to make economic use of the ports of the Great Lakes, while old dams and breakwaters are evidence of efforts to regulate fluctuating river flows,

Today, the river and its tributaries serve a modern population of over 800,000 people living in 34 municipalities in nine regions and counties in southern Ontario. The river sustains these growing communities through a reliable supply of drinking water and process water, acceptance of wastewater discharges, transportation, hydro-electric power, and active and passive recreational opportunities, including boating and sport fishing. Waterloo Region, Guelph, Brantford, and Six Nations are among the municipalities that depend on the river for their public water supplies, while approximately 600,000 people throughout the watershed regularly discharge their treated wastewater into the river and its tributaries. At the same time, several communities, including Cambridge, Guelph, and Grand Valley, are focusing on the waterways for development of economically viable tourism industries.

A4.3 Impetus for Management on a Watershed Basis

The Grand and its resources have been utilized for hundreds of years for water supply, agriculture, fishing, navigation, and waste disposal. It has been diverted, dammed, and channeled. Yet, today the Grand represents a landscape of celebrated natural beauty and ecological health. This is the result of a coordinated watershed management approach that began in the 1930s to address growing problems of water supply, flood control, and sewage disposal associated with a rapidly industrializing society.

In the early 1900s, vast deforestation and drainage of wetlands for industrial development led to several devastating floods and sewage-laden summer low flows. “The river had become the enemy—at times a vengeful swollen entity in the spring, and a smelly sewer in the summer (Smith, 1997).” Public outcry over these events led to a government investigation of the Grand River in 1932. The Grand River Conservation Commission, a partnership of eight municipalities and concerned citizens with government support, was formed in 1938 to address these flood control and water quality issues on an ongoing basis.

Unfortunately, devastating land, water, and forestry practices continued in Ontario throughout the 1930s and 1940s. Agriculture interests, naturalists, and sportsperson’s groups began voicing concern that “all the renewable natural resources of the province were in an unhealthy state.” One report stated that “acres of sand, barely tied down with grass cover and scrub trees were typical of much of southern Ontario (Conservation Ontario, 2000).” These conditions, combined with alarming instances of flooding and pollution-choked waterways stimulated public and political recognition that natural resources needed permanent management and protection, not only in the Grand watershed, but throughout Ontario.

The scale of erosion, water quality and flooding problems required a new approach based on natural watershed boundaries and involving municipal councils as well as the provincial government. The government of Ontario studied the examples of the Tennessee Valley Authority (Tennessee, U.S.A.), Muskinghum Conservancy District (Ohio, U.S.A.), and the Grand River Conservation Commission before passing the Conservation Authorities Act of 1946 (Statutes of Ontario, C.11, 1946). This act empowered municipalities within a watershed to initiate joint land and water conservation measures with technical and financial help from the province of Ontario. The system outlined in the act was based on the general structure and function of the Grand River Conservation Commission. The Grand Valley Conservation Authority was formed in 1948 under the specific terms of the Act. The Grand Valley Conservation Authority's mandate applied to resource management issues other than those addressed by the Grand River Conservation Commission (e.g., reforestation, land acquisition, stewardship programs, and development of outdoor recreation areas). The conservation activities conducted by the Grand Valley Conservation Authority were guided by a 1954 watershed-wide management plan undertaken by the provincial Department of Planning and Development called *The Grand River Conservation Report—Hydraulics*. In 1966, in order to prevent duplication of efforts, the authority merged with the Grand River Conservation Commission to become the GRCA.

Because of the relatively early initiation of watershed management in the Grand River watershed and throughout Ontario, the Grand never experienced the disastrous level of pollution and deterioration that many other large river systems have endured. However, there is no doubt that the Grand has and is presently experiencing significant pressures from expanding industry, agriculture, and urbanization consistent with a rapidly growing population. One of the most important motivations for watershed management in the Grand River basin is the need to maintain adequate water supplies, water quality, and recreational resources for a rapidly growing population. The central part of the Grand River watershed is one of the fastest growing areas in Ontario, attracting a great number of new residents, businesses, and industries each year. The population in the watershed is expected to grow by 30% over the next 20 years, an increase of approximately 250,000 people. Continued comprehensive watershed planning has helped communities in the Grand begin to manage and accommodate this growth.

A4.4 Institutional Arrangements

The GRCA is the lead agency in most, if not all, of the watershed planning and management efforts in the Grand River watershed. The GRCA is one of 38 conservation authorities in the province of Ontario established under provisions of the Conservation Authorities Act. The institutional structure of conservation authorities in general, and of the GRCA in particular, has remained relatively stable through the years and incorporates important elements of multi-jurisdictional coordination and mechanisms for funding, staffing, and stakeholder involvement.

A4.4.1 Conservation Authorities Act of 1946

The Conservation Authorities Act of 1946 provides terms of reference and specific guidelines for the establishment of functioning conservation authorities with jurisdiction over natural resources within watershed boundaries in Ontario. Conservation authorities are considered corporate bodies with broad goals, as specified in Section 20 of the act:

> "The objects of the Authority are to establish and undertake, in the area over which it has jurisdiction, a program designed to further the conservation, restoration, development and management of natural resources other than gas, oil, coal and minerals." (Government of Ontario, R.S.O. 1990, c.27)

The act is based on three principles: local initiative, cost sharing, and watershed jurisdiction:

1. Local initiative is essential to the formation of conservation authorities. An authority can only be formed at the request of two or more municipalities within a watershed. In making the request, local citizens acknowledge the responsibility of working across municipal boundaries, recognizing and solving problems jointly, contributing financially to the works of the authority, and planning projects democratically, with economic and cultural sensitivity.
2. A system of cost sharing is also laid out by the act, specifying that the costs of authority operation and special projects are to be shared by the member municipalities and the provincial government. Specifically, the Ontario Ministry of Natural Resources (MNR) is to make "transfer payments" to the authorities to assist with regular operational costs as well as specific project costs. Funding from the municipalities comes through regular and special levies.
3. Conservation authorities are to have jurisdiction over all aspects of natural resource conservation in one or more watersheds, as approved by the MNR. The act gives authorities the power to undertake an array of conservation projects within their jurisdiction, as well as to establish and enforce watershed-wide regulations as necessary to protect life, property, water quality, and ecosystem health. Such regulations may include restrictions on floodplain development, filling of wetlands, and development on steep slopes.

With broad powers assigned by the provincial government, conservation authorities are responsible for a range of important programs, including flood control and floodplain planning, dam maintenance, water quality monitoring, erosion control, reforestation, acquisition and management of conservation areas, and public education. Some authorities may focus on implementation of erosion control regulations, flood warning systems, and urban stormwater management, while others may focus on outdoor recreation, forest management, and heritage conservation. Each authority's watershed management programs are geared to its needs and conditions.

A4.4.2 The Grand River Conservation Authority

Like all other conservation authorities, the GRCA reports to the Province of Ontario through the MNR, and has working relationships with other government ministries such as the Ministry of Environment and Energy and the Ministry of Municipal Affairs and Housing. Although there is frequent communication with and support from these provincial agencies, membership in the GRCA consists of representatives of the watershed municipalities. There are a total of 26 GRCA members appointed by the 34 watershed municipalities. These members are divided evenly between two chief committees: planning and operations; and administration, finance, and personnel. A GRCA chair and two vice-chairs are elected annually from the membership.

Each of the GRCA committees meets monthly to review respective watershed and authority-related issues and make recommendations for presentation to the full membership at its monthly meeting. The full membership then reviews committee recommendations and may endorse, modify, or defer them for additional research. This system of governance has worked extremely well and resulted in the passage of effective watershed-wide regulations, as well as the development and implementation of many valuable projects and programs.

A4.4.3 Elements of Multi-jurisdictional Coordination

Multi-jurisdictional coordination is a fundamental aspect of the organizational structure and functions of the GRCA. Elements of this multi-jurisdictional coordination occur in two realms: (1) among municipalities across municipal boundaries; and (2) between the GRCA and local and provincial agencies.

A4.4.3.1 Coordination Across Municipal Boundaries

Under the Conservation Authorities Act, the GRCA was created with a jurisdictional area defined by watershed boundaries. Municipalities that lie entirely within the watershed have full participatory membership in the GRCA. To the extent that a municipality lies only partially within the watershed, its membership and participation in the GRCA may be proportional to the proportion of its area within the watershed. While this situation makes the GRCA a truly watershed-based institution, it does not detract from the relative autonomy of its member municipalities. Municipalities appoint their own representatives to the authority. These representatives are expected to keep their local municipal councils apprised of the programs and projects considered and/or implemented by the authority. The representatives act as liaisons between the authority and the municipalities by expressing local needs and concerns. Municipal representatives also participate directly in the administration and operation of the authority and provide financial support through collection of regular and special project-related municipal levies. Moreover, through the GRCA, one or more municipalities can jointly undertake work in another municipality for the benefit of the watershed. The system of cost sharing in such situations is described in greater detail in Section A4.4.4.

A4.4.3.2 Coordination Between GRCA and Government Agencies

Through strong partnerships, the GRCA works closely with both provincial ministries and local municipal agencies. At the local level, the GRCA provides input to municipal land-use planning documents (official plans), zoning bylaws, subdivision, plans and municipal infrastructure planning. In addition, the GRCA often coordinates sub-watershed planning studies as part of a systematic watershed planning approach encouraged by the MNR. Sub-watershed plans reflect objectives of the greater watershed plan, but are tailored to tributary needs and local issues, providing detailed guidance to municipalities on site-specific water resource and land-use planning issues. Subwatershed plans may outline directives for stormwater management practices, sub-division drainage designs, and other site design details. Such information may then be utilized by the municipal agencies and incorporated into municipal land-use plans or specific site development plans. The MNR has produced guidance documents encouraging and detailing methods for the integration of watershed and sub-watershed plans with municipal land-use plans and site plans. This system is discussed in Section A4.7.

At the provincial level, the GRCA works very closely with the MNR. Indeed, MNR approval was required for the formation of the GRCA, and MNR approval is required for all the regulations that GRCA promulgates and all the projects it proposes to implement. Thus, while the provincial government has recognized the benefit of placing authority for natural resource management with watershed-based organizations tied to municipal representation, it is still able to ensure that local activities are consistent with the goals and objectives of Ontario.

Additionally, the close relationship between the GRCA, the MNR, and other provincial ministries facilitates communication and the organized collection and sharing of watershed information. Technical assistance and guidance on the development of monitoring programs is provided by the MNR, and useful stream flow, water quality, and other monitoring data is then transferred back to MNR from the GRCA. An example of this is GRCA's Real-time River Data Monitoring Network. Data is collected from a network of monitoring stations across the watershed, providing support for water management decisions at the local level and also contributing to the Provincial Flood Warning System. The province, through the MNR, provided support for the development of this real-time data collection network. Even Environment Canada plays a role in this program through the Federal-Provincial Cost Share Agreement. GRCA also works with provincial and federal agencies by providing input on regulatory processes and programs, such as environmental assessments (under the Environmental Assessment Act) and DFO programs.

At another level of multi-jurisdictional coordination, the GRCA is able to interact with other watersheds across the province through the umbrella organization for Ontario's 38 conservation authorities, Conservation Ontario (formerly Association of Conservation Authorities of Ontario). Conservation Ontario's membership consists of the 38 authorities, with two members from each authority serving on the governing council. Conservation Ontario provides a forum for discussion between watersheds, promotes the works of its members through media relations, and raises funds for the authorities collectively under the Conservation Ontario banner.

A4.4.4 Funding Mechanisms

The Conservation Authorities Act specifies that funding for the administration and operation of the authorities is to be collected through levies among the participating municipalities. The general levy, which tends to remain steady from year to year, is distributed among the municipalities on the basis of "market value assessment" in the portion of the municipality that is in the watershed. In this way, urban municipalities with a higher tax base pay more than rural municipalities with a lower tax base.

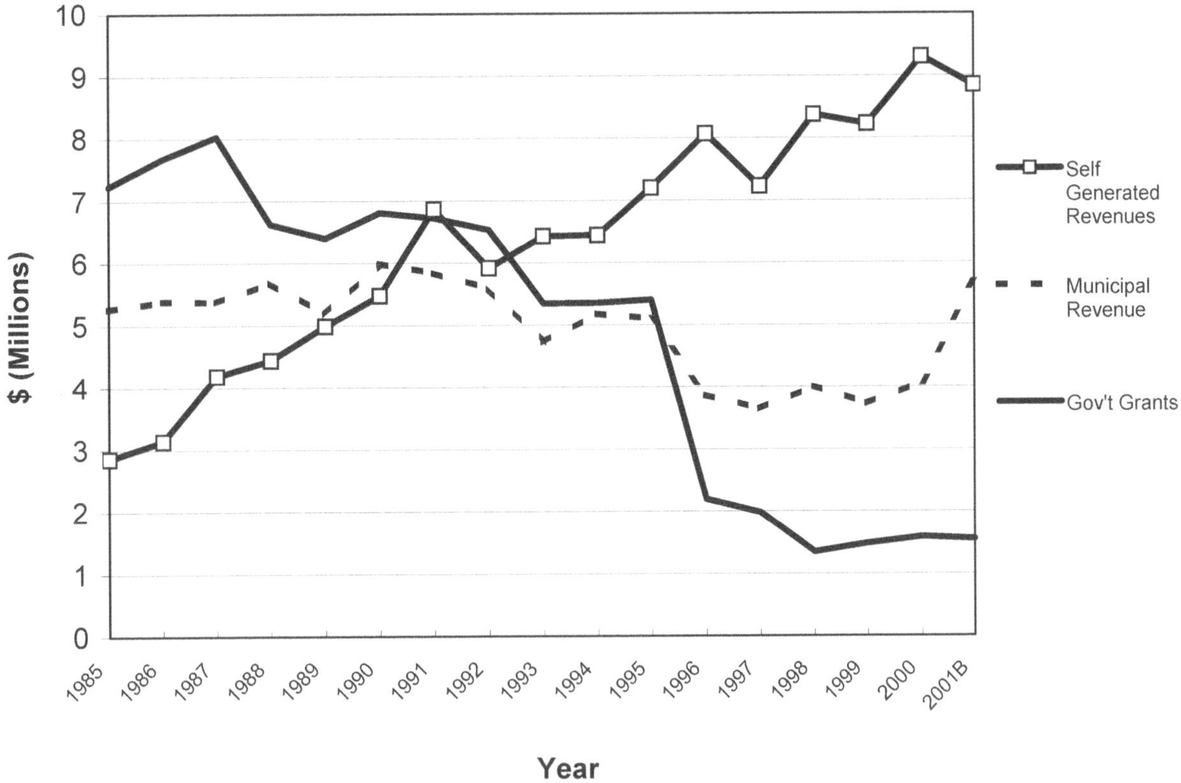

Figure A4-2. Grand River Conservation Authority 2001 Budget Revenue by Source, 1985 to 2001. ©Grand River Conservation Authority.

Costs for special projects are born by municipalities through levies, but according to the following system outlined in Section 27 (2) of the act:

> "...after determining the approximate maintenance costs for the succeeding year, the authority shall apportion the costs to the participating municipalities according to the benefit derived or to be derived by each municipality, and the amount apportioned to each such municipality shall be levied against the municipality."

This cost sharing system provides an equitable means of raising municipal funds for watershed improvement projects, where those benefiting most directly contribute a proportionately higher amount. Funding is also provided to the authority by the provincial government through transfer payments from the MNR. In the past, the province has provided 50% or more of the authority's project costs this way. Recently, however, it has limited its funding support to flood control and flood warning programs. Since 1997, provincial funding has no been provided for the development of watershed plans, capital works, recreation programs, and environmental improvement programs, such as soil and water conservation and stream rehabilitation. Figure A4-2 shows the changes in funding levels from different sources over the past 10 years.

As illustrated above, provincial grants to the GRCA have dropped in the last 10 years from 40% to 9% of the total budget. Meanwhile, self-generated funds, such as park user fees, land rentals, and private donations have increased to 61%. Municipal funding has remained steady at approximately 30% of the annual budget. Conservation program expenditures during the same time period have decreased from a high of $19 million in 1991 to $17.3 million in 2001 (Minshall, 2001). The following are current revenue sources for GRCA conservation programs:

- Watershed municipalities
- Provincial grants
- Income from conservation area user fees
- Rental income
- Hydro-electric production
- The Grand River Foundation
- Financial support from Grand Strategy partners

The Grand River Foundation began as a registered charity in 1965 and is comprised of volunteers from the corporate community with GRCA administrative support. The foundation raises funds from private interests primarily for visible, local projects in the watershed that provide tangible benefits to local communities, as well as other projects, as requested by the GRCA.

The GRCA funding strategy has evolved over the years, from a system supported mainly by provincial grants and municipal levies, to one that must seek significant support from its foundation, private partners, and stakeholders. In adopting this more business-like approach, the GRCA has been able to remain a viable and effective organization while building stronger partnerships within the watershed.

A4.4.5 Methods of Forming and Sustaining Partnerships with Stakeholders

The GRCA was formed out of local initiative and support, and one of its key objectives has been to provide opportunities for active involvement and education of the public, NGOs, special interest groups, and other stakeholders. The GRCA does this in a number of ways, including development of environmental education programs; encouragement of rural land stewardship; organization and training of student and volunteer groups; and public outreach through publications, forums, and the Grand River Foundation. Some examples are provided below.

- The GRCA conducts environmental education programs for over 50,000 students each year at four outdoor education facilities. The programs are run in partnership with local school boards,
- Rural stewardship takes place through outreach to local property owners and farming interests to educate and assist in the implementation of best management practices.
- GRCA has organized seminars, workshops, tours, and demonstrations for land-use planning and development professionals.

- Publications include the newsletter, *Grand Actions*, and the *State of the Watershed* annual report.

In the early 1990s, the GRCA began developing the Grand Strategy, in which stakeholder involvement and the programs listed above are key elements. Essentially, the Grand Strategy is GRCA's approach to more efficiently prioritizing and carrying out projects within the watershed through annual watershed assessment and action planning, and increased participation of individuals, community groups, and other stakeholders in GRCA programs. The Grand Strategy is discussed in more depth in Section A4.6.

A4.4.6 Staffing

The GRCA maintains a permanent staff of approximately 100 individuals. To some, this might seem a large number of permanent, paid staff for a watershed-based organization. But, the staff size is a reflection of the wide array of programs and services the authority is responsible for providing, including management of public parks and recreation areas, operation of dams and other flood management systems, and enforcement of land-use regulations, not to mention planning and implementation of specific watershed management projects throughout it's 34 member municipalities. The 100-member staff includes Geographic Information Systems (GIS) technicians, water quality specialists, and environmental educators, working as a multi-disciplinary team to carry out the authority's programs.

A4.5 Program Priorities and Goals

The GRCA has a broad mandate assigned by the Conservation Authorities Act "to further conservation, restoration, development and management of natural resources other than gas, oil, coal, and minerals." In addition, the GRCA has developed the following vision and mission statement.

> The vision of the GCRA is one of a healthy and sustaining relationship between the natural environment of the Grand River watershed and the demands on this environment by all forms of life.
>
> Its mission is "to conserve the natural processes and resources that support a safe and healthy environment for future generations in the Grand River watershed (GRCA, 1995-2004)."

These expansive goals are translated into resource planning and management programs tailored to the watershed's conditions and needs. Having identified its most important watershed management issues, the GRCA developed the following list of key program objectives:

- Keep the watershed healthy while accommodating growth
- Develop a viable tourism industry while protecting the resources on which it is based

- Improve water quality using a cost-effective balance between cleaning up urban wastewater and controlling rural sources of pollution
- Continue to manage water supply and water allocation
- Reduce flood and erosion damage using a mix of structural and non-structural approaches
- Conserve the natural environment and biodiversity
- Conserve heritage and a sense of place

To tackle the above objectives, the GRCA has developed successful programs that may be grouped into four core areas: (1) watershed management; (2) flood and water level management; (3) conservation land management; and (4) recreation and education.

1. **Watershed Management**
   In the broad area of watershed management, the GRCA manages a variety of programs designed to bring together its partner municipalities and stakeholders for watershed planning and decision making. The GRCA provides a forum for communication between partners, acts as a facilitator in decision making, and coordinates the design and implementation of projects. Such projects include watershed and sub-watershed planning, water quality monitoring, stream rehabilitation, technical assistance to landowners, and assistance to municipalities and provincial ministries in other environmental planning and permitting processes. Working with the municipalities and stakeholders, the GRCA has produced the Grand River Water Management Plan, several sub-watershed plans, a fisheries management plan, and a watershed forest plan. They have also delivered stewardship services to the public through tree planting programs and a rural water quality program designed to reduce rural soil erosion and nonpoint source pollution.

2. **Flood and Water Level Management**
   The GRCA undertakes numerous activities to protect watershed communities from flood damage and the adverse impacts of floodplain development and ensure adequate water supplies throughout the year. With responsibilities prescribed under the Conservation Authorities Act, the GRCA operates and maintains dams, reservoirs, and dykes; coordinates an advanced flood warning system; and implements floodplain management regulations. In support of these responsibilities, the GRCA uses state-of-the-art information tools, including GIS-based early flood prediction modeling and reservoir yield modeling. They collect data on the health of floodplain resources and undertake structural and bioengineering projects to stabilize stream banks and shorelines.

3. **Conservation Land Management**
   The GRCA owns and manages approximately 47,000 acres of conservation lands, including floodplains, wetlands, managed forests, and open space. With funding assistance from the Grand River Foundation, the GRCA has established four conservancy forests and purchased environmentally sensitive and wildlife habitat lands in the Grand's headwaters region. The GRCA also collects revenue through conservative use of the land, such as woodlot management and tree nurseries.

4. **Recreation and Education**
   Through maintenance of its extensive park system, the GRCA is able to provide passive recreational opportunities to the public through open space, river access, and trail networks. Thousands of residents and tourists experience the watershed's natural resources each year through use of these parks for camping, fishing, sailing, and other activities. The GRCA also operates four nature centers, providing environmental and outdoor education programs for children, community groups, and the public. Nominal fees help support the GRCA budget.

Implementation of this very comprehensive set of programs has enabled the GRCA to successfully move forward toward fulfillment of its mission and specific watershed goals.

A4.6 Implementation Efforts

In over 60 years of practicing watershed management, the GRCA has successfully implemented an impressive set of programs and watershed improvement projects. For much of this time, however, the GRCA has focused on solving flooding, low flow, and water quality problems. These issues were the impetus for creation of the GRCA and passage of the Conservation Authorities Act and received a great deal of support from the provincial government for many years. The GRCA's technically advanced floodplain management and water monitoring programs are the result of this early and enduring focus.

Gradually, however, the GRCA has developed a more balanced set of program areas, realizing that flood control, water supply, and water quality involve not only dams, reservoirs and riverbanks, but also forests, wetlands, natural stream channels, and developed areas. Today, implementation of watershed management projects is guided by the Grand River Basin Water Management Plan, Fisheries Management Plan, Forest Management Plan, and other action plans developed jointly by the GRCA and its partners over the past 25 years. Although a vast number watershed management projects are implemented directly by the GRCA, others are implemented by the provincial or local institutions that have more direct responsibility for the particular type of project. For example, implementation of the Grand River Basin Water Management Plan (which was developed by a multi-agency team in conjunction with local municipalities and public advisory groups and completed in 1982) has included the following players:

- GRCA: Responsible for building dykes and modifying reservoir operating procedures
- Municipalities: Responsible for upgrading wastewater treatment systems and developing municipal water supply master plans that are consistent with the water management plan
- Ministry of Environment: Responsible for issuing certificates of approval for wastewater treatment upgrades, ensuring that they are consistent with the water management plan.

For special projects for which there is no natural lead or established funding mechanism, federal and provincial program funds and foundation funds are sought. According to Lorrie Minshall of the GRCA, "For smaller projects, this is often successful. It has been particularly successful because there is a plan in place to put context and urgency around the proposal, and there is a group of government and non-government partners behind the project. For some larger projects, it has sometimes been necessary for the GRCA and the municipalities to collectively lobby the federal and/or provincial governments for assistance."

Unfortunately, provincial funding of watershed management projects has become increasingly scarce over the past 10 years. Since 1997, the MNR has not provided funding for conservation authority projects that are not directly related to flood control or flood warning programs. Conservation authorities have been left to devise new strategies for implementing their growing programs.

A4.6.1 The Grand Strategy

Where the GRCA's institutional arrangements and traditional funding system failed to keep up with the its expanding program areas and resource needs, the GRCA has sought out new partnerships and a new strategy for meeting its goals: the Grand Strategy. The Grand Strategy is a management plan that was initially used to gain support for the designation of the river as a Canadian Heritage River. The strategy was developed through a multidisciplinary participatory approach, facilitated by a coordinating committee co-chaired by the GRCA and the Ontario Ministry of Natural Resources, and was accepted in 1994. The Grand Strategy defines the GRCA's adaptive approach to watershed management. Although the strategy was originally developed to achieve the qualification requirements for designation of the Grand River as a Canadian Heritage River (e.g., management of heritage features and recreational opportunities), the plan's authors and contributors supported more comprehensive goals and approaches to resource management. The Grand Strategy has evolved from its original use into a collaborative process that provides a holistic approach to the watershed's resource management. The strategy entails rigorous data collection, annual assessment of progress, annual action planning, and significant efforts to involve more individuals, community groups, businesses, land-owners, and educational institutions in GRCA programs. The GRCA structure remains as it was created under the Conservation Authorities Act, and governance remains with the representatives of its 34 municipal members, but more opportunities for stakeholder participation and investment have been created. Under the Grand Strategy, the GRCA is providing opportunities for stakeholders to actively participate in solving watershed problems, identifying partners that can effectively carry out watershed activities, and recognizing contributions and achievements through publications, reports, and public forums.

For example, as part of the Grand Strategy, the GRCA produces an annual *State of the Watershed* report, publicizing the health of the watershed resources, the effectiveness of the year's activities and priorities for action. The GRCA also seeks to catalogue and recognize the accomplishments of its voluntary partners and stakeholders through a Registry of Accomplishments and Commitments, which is included in the *State of the Watershed* report. The monthly newsletter, *Grand Actions*, features stories about community-based projects and new opportunities for participation. Individuals and groups are encouraged to participate in GRCA working groups, provide input during preparation of the *State of the Watershed* report, and submit articles and information for publication.

In the words of the GRCA, the Grand Strategy is not a "sit on the shelf" report, but is "a living, dynamic process that identifies problems, develops creative solutions, pools resources, implements action, monitors results and evaluates projects on an ongoing basis. If the actions undertaken are not producing the anticipated results, the plan is adjusted so that actions that are not effective are not repeated."

A4.6.2 Monitoring

Cited throughout this case study, monitoring programs, whether for stream flow, water quality, or ecological health, represent a significant portion of GRCA operations. High quality data is recognized as the foundation of effective watershed management projects and is regularly collected, maintained, and made available by GRCA to its members, stakeholders, and the province of Ontario. The GRCA collects monitoring data in a variety of areas, including water quality and quantity, weather patterns, and aquatic and terrestrial ecology. Much of this data can be accessed from the GRCA Web site, including a variety of real-time river data and groundwater data.

A4.7 Innovative Program Elements

There is no doubt that the GRCA has evolved over the years into a highly effective, modern watershed management institution. But the GRCA's most innovative program elements are not the cutting-edge waste assimilation and GIS-based predictive flood models it has developed and employed, nor its successful involvement of watershed stakeholders as part of the Grand Strategy. The GRCA's most important innovations are certain elements of its institutional structure and function. Three main elements stand out as evidence of significant forethought and innovation on the part of the government of Ontario when it created the conservation authority model: 1) equity in costs and benefits of watershed management across municipalities; 2) statutory authority to promulgate and enforce watershed-wide regulations; and 3) integration of the watershed planning and municipal land-use planning processes.

A4.7.1 Equity Across Municipalities

In passing the Conservation Authorities Act in 1946, the government of Ontario recognized that active and continuous participation of the municipalities would be critical to the success of watershed management efforts. The conservation authorities, therefore, were to be created *by* municipalities with memberships consisting *of* municipalities. In this way, watershed management decisions would be made by and brought directly back to those in control of water works, roads, waste disposal, and land-use planning at the local level. Assuming municipal representatives could agree on priority projects for the benefit of the watershed, there would remain only the question of funding. At this point, implementation would have become problematic due to the fundamental autonomy and financial interests of the individual municipalities. Why would a municipality located at the headwaters of the Grand be willing to place a special levy on its citizens for a water supply protection project in a region near the mouth of the Grand? Certainly, it would not. However, if some indirect benefit to the headwaters town from the project could be shown, then perhaps the town council would be willing to contribute some proportional amount of funding. This was precisely the idea of the authors of the Conservation Authorities Act when they specified the mechanism for funding of special projects.

> "The portion of the money so required that each participating municipality shall raise shall be in the same proportion as the benefit derived by each such municipality bears to the total benefit derived by all participating municipalities."

In this way, the participating municipalities could be sure to receive some monetary assistance for projects directly benefiting them, without worrying that they would have to contribute disproportionately to projects directly benefiting others. In this equitable funding mechanism, the act created a truly cooperative arrangement between municipal partners for the benefit of the watershed as a whole.

A4.7.2 Regulatory Authority

Section 28 of the Conservation Authorities Act gives the GRCA and other authorities the power to establish regulations applicable to their area of jurisdiction. Such regulations may include restrictions on development or placing of fill in wetlands and floodplains, as well as restrictions on alteration of watercourses. Most importantly, the act gives the authorities power to appoint officers to enforce these regulations. This power to write and enforce regulations protecting watershed resources may be one of the most important aspects of the GRCA's long-term success. It has allowed them to create, administer, and enforce natural resource protection regulations without the limitation of municipal boundaries. A GRCA member from one municipality has equal vote on a project proposed to take place within another municipality. The GRCA staff provides guidance to the membership during this regulatory process based on GRCA approved policies and practices.

For example, building in a regulated wetland is prohibited, but building near one may be allowed provided that certain conditions are met with regard to distance from the wetland, drainage design, and vegetation removal. The GRCA staff includes engineers, biologists, and planners who can review the application and make recommendations to both the applicant and the GRCA membership. Often, an application will be approved with special conditions to ensure protection of the resource.

Certainly, floodplain and wetland regulations of this sort are not innovative in and of themselves, but regulation and enforcement of floodplain and wetland activities within a jurisdiction defined by natural watershed boundaries rather than political ones is as innovative today as it was in 1946.

A4.7.3 Integration of Watershed and Land-use Planning Processes

Because the GRCA and other conservation authorities in Ontario develop watershed plans and natural resource management plans for the areas within their jurisdiction, they often make recommendations about land-use and resource planning that should logically be incorporated into municipal planning documents and related development applications. The Ontario MNR recognized this in 1993 when it wrote:

> "Watershed planning is recognized by federal and provincial governments as being the most effective means of evaluating and developing water-related

resource management strategies and practices. Most decisions that are made on privately owned lands, however, are made in the context of the municipal land-use planning process on the basis of municipal boundaries or property ownership. It is very important, therefore, that there be adequate linkages established to incorporate water and related resource management directions into the municipal land use planning process."

For this reason, the Ontario MNR and Ministry of Municipal Affairs and Housing have entered in a memorandum of understanding (MOU) with the conservation authorities (through the umbrella group, Conservation Ontario) regarding the conservation authority's role in municipal planning. The MOU states, "The conservation authority is essentially set up as a technical advisor to municipalities." Therefore, the conservation authorities are responsible for providing advice on municipal land-use plans (called official plans) as well as participating in the review of development applications under the Planning Act (the provincial law governing the municipal land-use planning process.) In addition, the MNR emphasizes that the policies of official plans should recognize and incorporate relevant elements of watershed plans (where they exist), stating that, "all land-use planning decisions should be carried out in accordance with the recommendations of the watershed management plan (Ministry of Natural Resources, Ontario, Canada, n.d.)." Even when a watershed plan does not exist, the municipality is encouraged to consult the conservation authority or MNR to identify any special water management initiatives that may have relevance to municipal planning. The MNR further encourages the development of sub-watershed plans as a means of informing large-scale development proposals within a municipality. The sub-watershed plan is typically developed by the municipality and/or conservation authority and reflects the goals of the larger watershed plan while examining the sub-watershed in greater detail. A development proposal can then be considered by the municipality in light of more detailed water and natural resource management data and recommendations. The general process is depicted by the flow chart below.

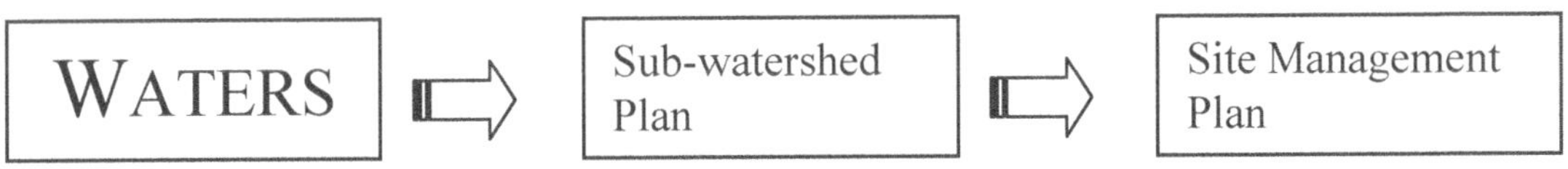

Once a sub-watershed plan has been developed, the municipality is encouraged to use it to amend its official plan and thereby instruct development proposals to incorporate specific water conservation, site design, and other recommended measures into their site management plans. This organized integration of watershed planning and local land-use planning has become an important means of achieving long-term watershed management goals.

A4.8 Degree of Success

Since the years of degradation that featured severely eroded landscapes, raging floods, and polluted sluggish river flows, the Grand River has not just recovered, but rebounded to become an example of outstanding natural beauty and abundance. In the 1930s, the Grand was almost abhorrent to its people. Today, it is a prized asset. As a result of the programs implemented by the GRCA and its partners over the last 60 years, the Grand River has seen dramatically reduced flooding, the return of sport fish, and increased recreational use and tourism. Implementation of floodplain regulations have resulted in green riparian buffers, while outreach to property owners has resulted in reductions in nonpoint source pollution.

A major achievement came in 1982 when the GRCA and its partners completed the Grand River Basin Water Management Plan. One result of this plan's implementation has been a 75% reduction in annual flood damages. In addition, the Grand River now does not dry up, but has significant sustained flow that allows for water supply and wastewater assimilation. All wastewater treatment systems have phosphorus removal and most remove nitrogen. As a result, water quality has improved significantly, and over 98 species of warm and cold water fish can be found within the river system (Minshall, 2001).

In 1995, several municipalities recognized the need to update the water management plan. Through the GRCA they established new basin planning models and tools that will soon result in integrated water quality, water supply, and forest and fisheries management plans. Through the Grand Strategy, new partners and stakeholders have become involved. And through the Grand River Foundation, new sources of funding have been accessed. Most importantly, the populations relying on the Grand for water supply now know that they have a clean source that will meet their needs into the future.

The Grand River has improved to such a degree over the years as a result of these activities that in 1994 it was declared a Canadian Heritage River "on the basis of its outstanding human heritage features and values and its abundance and diversity of outstanding recreational opportunities." More recently, in 2000, the Grand River received the prestigious Thiess RiverPrize at RiverFestival in Brisbane, Australia. The RiverPrize is acknowledges outstanding river management worldwide and is awarded based on criteria including on-the-ground performance and river improvements; cooperative participation by government, community, and industry; innovations in river basin management; and demonstrated levels of public accountability and recognition.

A4.9 Lessons and Applicability to the United States

The success achieved in the Grand River basin by the GRCA and its partners offers lessons that may help to guide watershed managers in the United States. The GRCA was formed out of enabling legislation with an effective and enduring institutional structure and clear regulatory authority. Its municipal membership has been able to equitably share the costs and benefits of watershed management across municipal boundaries. With provincial guidance, watershed planning and local land-use planning have been integrated for the purposes of protecting water supplies and water quality over the long term.

Lesson 1: Institutional stability and a clear mandate for watershed management can reduce fragmentation of authority and result in more efficient planning and implementation.

The institutional structure of the GRCA has not substantially changed since it was created in 1948 under provisions of the Conservation Authorities Act. This institutional stability, along with its consistency across the Province of Ontario in the 38 other authorities, has contributed to its success. The Conservation Authorities Act provides a clear mandate for comprehensive watershed management and specifies the structure of the watershed management institution's roles, responsibilities, jurisdiction, membership, and funding. Additionally, the GRCA's clear mandate supplies legitimacy for important actions, such as promulgation and enforcement of watershed-wide regulations and integration of watershed management with local land-use planning,

One weakness in watershed management in the United States is the lack of strong watershed-based institutions or agencies with authorities and expertise that are well understood and accepted by the public and municipalities. Too often, fragmentation of responsibilities among several different agencies involved in watershed management leads to duplication of efforts, unclear mandates, and conflicting missions. Strong enabling legislation that provides a clear mandate for watershed-based agencies or institutions may help solve some of these problems, as it has in Ontario. However, historical experience in the United States suggests that such agencies are unlikely to be created at the federal level and may be more likely to be established at the state level.

Lesson 2: Instilling regulatory authority in a watershed-based institution can facilitate effective watershed protection across political boundaries.

Section 28 of the Conservation Authorities Act gives the GRCA and other authorities in Ontario the power to establish regulations applicable to their area of jurisdiction, such as restrictions on development or placing of fill in wetlands and floodplains, as well as restrictions on alteration of watercourses. This has allowed the GRCA to monitor and regulate local development activities with watershed management principles in mind, unrestricted by municipal or other political boundaries when considering the natural resource impacts of a proposed activity. This regulatory authority has resulted in the creation of greenways along riparian corridors and the reduction of impervious surfaces and downstream flooding impacts.

Many U.S. states and regions have wetland protection regulations. The Federal Emergency Management Agency and local zoning bylaws restrict floodplain development, and the Clean Water Act and other state and federal laws regulate alteration of waterways. Further, municipalities frequently have local bylaws designed to protect their natural resources. But few, if any, of these regulations are created specifically for natural watersheds or implemented by watershed-based organizations. The result is that development is often permitted and/or poorly monitored in one municipality without noticeable adverse effects; meanwhile, such development may be contributing to severe flooding or water supply problems for downstream municipalities.

While regional authority has been achieved in specific U.S. locations (e.g., Oregon and the Lake Tahoe region of California and Nevada), in most parts of the United States there is strong political resistance by municipalities to sacrificing any measure of local control. Creating incentives for municipalities to accept and support a regional authority may be an important tool to overcoming this barrier. Such incentives might include priority access to state revolving funds or other grants for planning and improvement projects. Although politically difficult, the benefit of applying this lesson where possible in the United States could be significant, resulting in vastly improved administration and enforcement of resource protection regulations, where municipal boundaries no longer limit the view of costs and benefits.

Lesson 3: A system of apportioning costs and benefits equitably across the watershed can help secure consistent municipal participation and funding.

Because watershed management activities naturally take place at the local level, they frequently require the acceptance and involvement of municipal officials. Often, recommended watershed improvement projects, such as separation of combined stormwater–sewer systems, require substantial support from municipal agencies through long-term dedication of funds and disruptive, on-the-ground implementation efforts. For this reason, it is critical to involve municipal decision makers in watershed planning from the outset. One way to engage and secure municipal participation is to establish a system for equitable distribution of watershed management costs and benefits.

As we have learned, the Grand River example goes beyond just involving municipalities in the watershed management process and actually makes municipal membership the foundation of the watershed organization. Further, municipalities agree up front to share in the cost of the organization's operation and implementation of special projects. This has been made acceptable to the GRCA member municipalities because the costs and benefits of watershed management are distributed equitably. Participating municipalities receive monetary assistance for improvement projects of direct benefit to them, while they contribute to other projects only in the amount proportional to the indirect benefit they receive. This equitable system of apportioning costs and benefits throughout the watershed has made municipal funding of the GRCA and its programs more palatable. In addition, it has provided stability to the GRCA budget.

The sharing of costs and benefits among divergent municipal upstream and downstream interests is not new to U.S. watersheds. A common example is of downstream water users offering assistance to upstream communities in the adoption of water conservation and water quality protection measures. Cambridge, Massachusetts, for example, works closely with communities upstream from its water supply to implement stormwater best management practices (BMPs) and protect land around important tributaries and wetlands. Cambridge's investment of money and resources in the upstream communities is paid back through a clean and protected water supply. Many similar examples exist in the United States and elsewhere, but they differ from the GRCA in that in the Grand municipalities work together on a more consistent basis for more collective benefits that are common to all. In addition, apportioning costs and benefits equitably among municipalities requires a strong watershed or regional organization, such as the GRCA, with a system of tracking municipal projects and contributions. Where they exist in the United States, regional utilities could provide this system and could be instrumental in the coordination and balancing of municipal contributions to watershed management projects. Such a situation could promote increased municipal involvement by helping to ensure that costs and benefits are distributed equitably,

Lesson 4: Explicit policies and guidance documents can be used to promote the integration of watershed and land-use planning at the local level.

Perhaps the most important lesson that the Grand River case study can offer U.S. watershed managers involves the integration of watershed planning and local land-use planning. In the Grand watershed and throughout Ontario, municipal land-use planning and watershed planning processes are seen as inextricably linked. Guidance from provincial agencies in the form of legislation and MOU has led to establishment of a process by which 1) conservation authorities produce watershed plans; 2) watershed plan recommendations are incorporated into municipal official plans; 3) sub-watershed plans are developed when necessary to guide impending land-use changes at the local level; 4) official plans are then amended to incorporate the more detailed recommendations of the sub-watershed plans; and 5) specific site development plans are tailored to meet the requirements of the newly amended official plans. This straight-forward process, outlined by the provincial government and implemented locally, fosters cooperation between planners and watershed managers and leads to development that meets the needs of the watershed. Without this integration, some of the most important watershed management recommendations might not be implemented at the local level.

In the United States, most municipalities have town or city plans, zoning ordinances, and may undertake redevelopment plans or other planning initiatives. Many of these plans focus on transportation issues, historical preservation, open space and recreation, and the characteristics of special zoning districts. Zoning language typically specifies acceptable uses, building heights, floor area ratios, parking limitations, and aquifer or water supply protection districts. Such planning documents and associated local permitting processes are often the only guidance a municipality uses to manage growth and review development proposals. Even where a detailed watershed plan has been developed, essential pieces may not be referenced in the local land-use plan. Yet issues of water supply, waste assimilation, groundwater recharge, and ecology are critical to the sustainability of the municipality and its population.

The process outlined by the government of Ontario for integration of watershed plans with municipal official plans could be applied to the United States. Where watershed plans have been developed and approved by state or local agencies, local land-use planners, planning boards, and boards of zoning appeal could be directed or encouraged to incorporate relevant elements into local planning documents. However, one reason why the process works so well in Ontario is that municipal representatives, through the conservation authority, are directly involved in development of the watershed plans. In the United States, this is less often the case. Municipalities may be unwilling to adopt land-use recommendations from a watershed plan that they had no part in writing. Because local approval of and commitment to the watershed plan and its recommendations are essential to its integration with land-use plans, it is important that local officials are directly involved in the watershed planning process. Similar to the approach in Ontario, clear guidance for the integration of watershed planning with local land-use planning could be provided by the state through policy statements, guidance documents, or MOUs, along with incentives for local follow-through. Incentives could include state grants and/or assistance for projects in municipalities that have incorporated elements of approved watershed plans into municipal policies.

APPENDIX A5

THE MURRAY-DARLING CASE STUDY

Figure A5-1. The Murray Darling River Basin (Taken from http://rubens.anu.edu.au/student.projects/wine/stats.html).

A5.1 Introduction

Management of the resources of the Murray-Darling Basin (MDB) typifies many of the problems faced in the task of unifying contrasting biophysical regions and socio-economic circumstances in a river basin. Among the world's largest river systems in terms of its catchment area (Figure A5-1), the Murray-Darling's surface runoff is among the smallest and most variable (Crabb, 2001). One result has been the accumulation of salts over geologic time, making the basin effectively a natural salt trap (Blackmore et al., 1999). This problem has been exacerbated by recent developments of water and land use, causing significant negative economic consequences for agricultural, domestic, and industrial water users, as well as serious effects on the basin's flora and fauna, which have evolved around the unique conditions found in the basin. Formerly widespread species are now restricted to a few localities, and entire ecosystems now have a very limited extent. In the early 1980s, the condition of the Murray-Darling River system was described as, "…the major agricultural, environmental question in Australia today (Pigram, 2000)."

These pressures lead to the creation of a multi-jurisdictional organization through the MDB Initiative (MDBI) with three decision-making bodies: a ministerial council, a commission, and community advisory committee. This structure has allowed technical bodies to operate in conjunction with political bodies, which has been an effective approach to decision making. Working together toward the goal of "equitable, efficient, and sustainable use of the land, water, and environmental resources of the MDB," difficult decisions have been made to move the basin towards sustainability, including a cap on the total amount of water used, encouraging a market in salt reductions, and examining ways to address the lack of sustainability of current farming systems.

A5.2 Basin Description and Current Water Uses

The Murray-Darling River Basin is Australia's largest, encompassing more than one million square kilometers in the southeast of the continent. The major rivers flowing through the region rise in the eastern highlands draining subtropical climate zones in the north and alpine areas further south. In spite of its size, the total runoff is relatively low, and the quantity is unevenly distributed, both seasonally and spatially, with annual rainfall varying from 1200 mm in the mountains to 150 mm in the arid west, and is considered to be one of the most variable in the world (Finlayson and McMahon, 1988). The flora and fauna of the basin have adapted to these circumstances. For example, the mulga, an arid-land acacia tree, can survive up to 50 years with little rain but requires regular soakings in summer and winter of the same year to reproduce. The climate phenomenon known as El Niño produces such a climate for Australia. However, the human approach to cope with this variability has been to make the Murray-Darling into a highly regulated system, with storage almost 1.4 times the total annual flow. These changes to the hydrology have a significant impact on the flows of almost all of the basin's rivers, in particular reducing the mean annual flow at the mouth of the Murray by 80%.

The population living in and supplied with water by the basin is nearly two million. In addition, water is diverted out of the basin to supply the city of Adelaide for domestic and industrial purposes. However, it is agricultural water use that dominates in the basin, accounting for 95 percent of the diverted flow (Crabb, 2001). As the nation's most important agricultural region, the Murray-Darling supports 25% of the nation's cattle and dairy farms, 50 percent of the sheep and cropland, and 75% of the irrigated land (Chenoweth and Malano, 2001).

One of the dominant features of the MDB is its relatively high naturally occurring salinity. Salt has accumulated over geologic time in the basin due to its exceptionally flat terrain, low rainfall, and high evaporation that combine to concentrate salt in the soil profile and groundwater, and preclude thorough flushing of the accumulated salts. During periods of low flow, the salinity of the streams and rivers rises dramatically. The aquatic ecosystems have become accustomed to these significant variations in salinity.

However, a combination of land- and water-use changes over the past century have caused the salt to rise to the ground surface while simultaneously reducing the flows to the rivers, thus reducing the already limited flushing of salts from the land and increasing instream salt concentrations. The increasing regulation of river flows and the abstraction of water for consumptive uses has further compounded the lack of natural flushing. Australian rural production systems were built by drastically changing the nature and seasonal patterns in the hydrological and nutrient cycles. The use of water for irrigation not only reduces dilution flows but also speeds up the rise in groundwater levels, bringing salt to the surface and increasing the discharge to the rivers. While drainage provides short-term relief, it also adds to river salinity. The result has been a serious decline in fish and other aquatic species, as well as a loss of productive farmland.

A5.3 Impetus for Management on a Watershed Basis

The increasing awareness of the escalating problems, primarily around the issue of salinity, emerged in the basin through the 1960s to the 1980s and resulted in a series of inquiries, many of which called for urgent action (see the MDBI Web site, "A Brief History of the Murray-Darling Basin Agreement"). In the mid-1980s studies showed that 96,000 ha of irrigated land in the basin had visible signs of salinization, and it was estimated that the area affected by high water table could increase from 559,000 ha in 1985 to 869,000 ha in 2015. It was also estimated that the average salinity at Morgan, a town on the River Murray upstream of the water uptake for the capital city of Adelaide, would dramatically increase, possibly exceeding levels appropriate for drinking water.

These mounting pressures led to a meeting of ministers responsible for land, water, and other environmental resources from the governments of New South Wales, Victoria, South Australia, and the Commonwealth in 1985 in Adelaide. The meeting was called to discuss the resource and environmental problems of the MDB and, in particular, salinity and land degradation. The meeting was followed by two years of intensive meetings and negotiations by politicians and bureaucrats from the four governments. At the inter- and intra-state levels, the people involved came to know each other in a way that had never occurred before. The outcome was the MDB Agreement in 1987. This agreement provided the foundation for the MDBI by putting in place a process for the management of the water, land, and other environmental resources on a basin-wide basis.

A5.4 Institutional Arrangements

Under the Australian constitution, primary responsibility for water and environmental management rests with the Australian state governments rather than the commonwealth government of Australia. Specific powers are conferred by the constitution on the commonwealth government, most of which are concurrent in that they are not specifically withdrawn from the states. The constitution states that all powers that are not allocated exclusively to the commonwealth government remain state powers. The net effect is that management of the Murray-Darling is shared by the states of Queensland, New South Wales (NSW), Victoria, South Australia and the whole of the Australian Capital Territory (ACT).

A5.4.1 Multi-jurisdictional Issues

Managing the water and land resources of this inter-jurisdictional basin has long presented difficulties, from before the time of federation in 1901. In spite of many debates, it was not until 1915 that the River Murray Waters Agreement (RMWA) was signed. Though at the time pioneering legislation, it was a limited mandate, essentially confined to the sharing of River Murray water between NSW, Victoria and South Australia. It took another two years to establish the River Murray Commission (RMC), which had the task of putting the agreement into effect. The main limitations of the RMWA were that it was confined primarily to the main stem of the Murray, which is why Queensland was not a signatory, and that it was only concerned with quantitative issues (although a 1982 amendment did give the RMC limited powers to deal with water quality issues).

A5.4.2 The Creation of the Murray-Darling Basin Initiative

By the mid-1980s, the RMWA and the RMC were unable to tackle the emerging environmental and resource management issues. The result was a heads of government meeting that gave rise to the MDB Agreement, signed by the governments of Australia, NSW, Victoria, and South Australia in 1987. In its initial form, it was as an amendment to the RMWA, becoming fully legal in 1988. The agreement finally put into effect a recommendation of a 1902 Royal Commission on the Murray that "the river and its tributaries must be looked on as one." A revised and totally separate MDB Agreement, replacing the RMWA, was signed in 1992. Queensland was also a signatory of this revised version. The ACT joined in 1998 and participates in a limited way by means of a MOU.

The agreement created three decision-making bodies: a ministerial council, a commission, and a community advisory committee (Figure A5-2). The responsibilities of each body is described below:

- **Murray-Darling Basin Ministerial Council**: Ministers from each of the signatory governments make up the MDB Ministerial Council. This council is the decision-making forum for basin-wide policies and programs.
- **Murray-Darling Basin Commission (MDBC)**: The MDBC is comprised of senior executives from each partner government and has an independent president. It gives advice to the ministerial council and carries out the decisions made by the council.
- **Community Advisory Committee**: The Community Advisory Committee provides a two-way communication channel between the ministerial council and the community. The committee provides the council with advice from the basin-wide community. It comprises representatives from the basin's catchments as well as from four key special interest organizations and an indigenous representative.

There are many other inter-jurisdictional agreements in the MDB that continue to operate. Continuity is provided because government ministers and officers involved with them are also involved with the MDBI, although there is no formal link between the two.

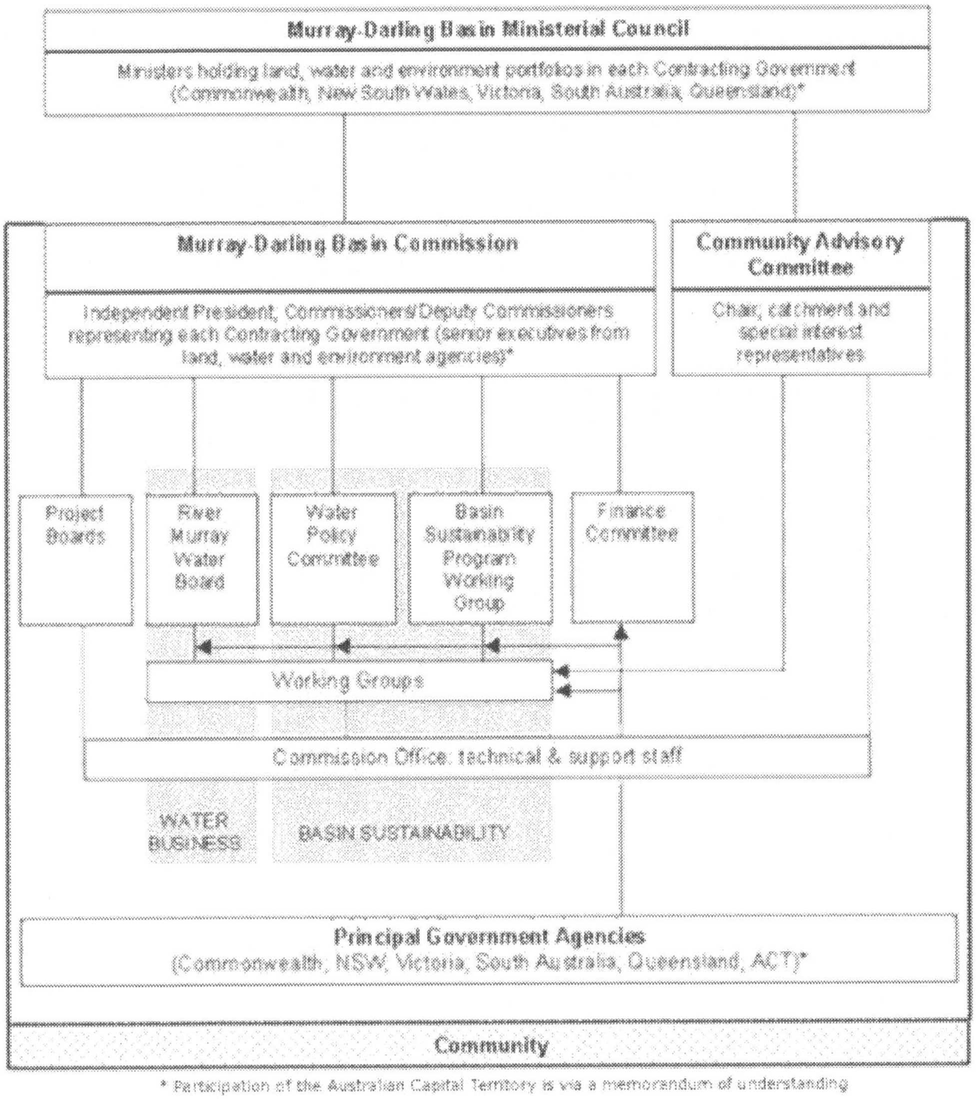

Figure A5-2. Governance of the Murray-Darling Basin Initiative (Taken from http://www.mdbc.gov.au/about/governance/overview.htm).

A5.4.3 Funding

Funding for the initiative comes from the state and federal levels, with substantial community contributions. Cost sharing has achieved growing acceptance in the MDB (Hooper et al., 1999) as an efficient allocation method of distributing costs between many resource users. There are ongoing efforts to better establish property rights in the basin, both in the context of allocation of costs and for the support of water markets. Funding for infrastructure, for example large salt interception schemes, is shared equally by the commonwealth, NSW, Victorian and South Australian governments. However, the cost of operation and monitoring of the schemes is shared by the three states.

A5.5 Program Priorities and Goals

The purpose of the MDB Agreement is:

"To promote and co-ordinate effective planning and management for the equitable efficient and sustainable use of the water, land, and other environmental resources of the Murray-Darling Basin."

The strategic and philosophical framework for achieving the purpose of the agreement is the Natural Resources Management Strategy, which was approved by the ministerial council in 1990. The strategy provides the broad charter for a community-government partnership to develop plans for the integrated management of the basin's water, land, and other environmental resources on a catchment basis. The Natural Resources Management Strategy established two fundamental pillars for handling natural resource management in the basin. The first pillar was the philosophy of integrated catchment management (ICM), recognizing the links between various biophysical processes, which affect or are affected by water, its movement, and its uses. The second pillar was the community/government partnership, recognizing that neither party working in isolation can protect the basin's natural resources.

In order to improve the performance and accountability of work undertaken under the Natural Resources Management Strategy, the ministerial council put in place the Basin Sustainability Plan in 1996. Following consultation with key groups, the objectives of the Basin Sustainability Plan were reviewed and updated in 1999 to ensure their relevance to priority natural resource management issues across the basin. The plan provides the framework for the co-ordination of planning, monitoring, evaluation, and reporting for the Natural Resources Management Strategy, and covers all government and community investment for sustainable natural resources management in the basin. The Natural Resources Management Strategy and Basin Sustainability Plan together are the foundation of the commission's basin-wide planning processes for natural resource management.

The Basin Sustainability Plan contains long-term productivity and resource condition objectives for sustainable agriculture, water quality, nature conservation, and cultural heritage. For each of these priority areas, specific objectives apply to irrigated and dry land basin regions and its river environments. Reporting against these objectives is designed to show short-term achievements (empowerment), medium-term achievements (implementation), and long-term outcomes (resource condition).

The plan also contains direction setting and management implementation objectives, which focus on people and management arrangements. These objectives are designed to help ensure that the arrangements for natural resource management enhance the partnership between community and government, and help the managers of land and water protect the basin's catchments.

As part of the ICM Framework for the Basin for the decade 2001 to 2010, the commission developed a framework for monitoring progress in achieving the desired natural resource outcomes in the basin. The ICM policy employs a system for setting targets for catchment health designed to protect social, economic, and environmental assets threatened by problems such as dry land salinity. Basin-wide targets are set for water quality, water sharing, river ecosystem health, and terrestrial biodiversity.

A5.6 Implementation Activities Over Time

The initiative has given rise to a range of measures, dealing with such issues as algal management and nutrient pollution, native fish species, floodplain wetlands, biodiversity, and irrigation and dry land agricultural systems. The focus is on implementation activities revolving around three crosscutting problem areas for the MDB: salinity, conserving sensitive ecosystems, and diversions of water for environmental purposes (discussed below). These activities taken under the MDBI illustrate the range of problems faced; the approaches taken, including the decision-making process; and the remaining challenges.

A5.6.1 Salinity and Drainage

The management of salinity problems in the MDB has been one of the major focuses of the initiative throughout its operation. Irrigation and salinity was considered by the ministerial council at its inaugural meeting, and it was agreed then to establish a number of working groups to study various salinity and related issues in the basin, in addition to the salinity investigations already carried out by the MDBC's predecessor, the RMC. A framework for cooperative action by the participating governments to manage salinity problems in the basin was subsequently developed, called the Salinity and Drainage Strategy. The council formally adopted the strategy in 1988. This inter-governmental agreement imposes rights and responsibilities on each government member of the commission, and holds them accountable for all future actions impacting river salinity.

The strategy's objective is to effectively manage the pressing problems of water logging and land salinization in the irrigation districts of the Murray Valley in NSW and Victoria, and the river salinity in the lower Murray River. The essential features of the Salinity and Drainage Strategy are:

- River salinity levels from 1975 to 1985 are adopted as the benchmark for attributing impacts of all future actions that affect river salinity.
- The rules for river operation are changed to provide for improved dilution of salinity at critical times.
- Each state is responsible for its actions after 1988 that affect river salinity.
- No actions that increase salinity are allowed unless they are offset by works to ameliorate them.
- These works are authorized with limited salt disposal rights to the river.
- These rights were negotiated on the understanding that the upper states would provide financial support for salinity reduction works further downstream.
- The rights have become known as salinity credits; they are a tradable "pollution right."

In the words of Sen. Peter Cook, chair of the MDB Ministerial Council at the time the agreement was reached on the strategy:

> "For the first time in our history, the states have agreed to tackle a major environmental problem through a common effort across their borders. The strategy paves the way for further interstate cooperation in promoting the sustained use of our land, water, and environmental resources."

A5.6.1.1 Trading in Salinity Credits

A reduction of 80 electrical conductivity (EC) in salinity is one of the stated goals of the strategy, and the states are allocated prescribed "credits" or "debits" towards this target. Each state has the right to handle the credit or debit in its own territory from that point on, including the right to trade credits and debits with other states. The three main states involved, Victoria, NSW, and South Australia have each taken a different approach in allocating the credits/debits. In Victoria and NSW, the state pays for the interception schemes that generate credits and then allocates these credits to land and water salinity management plans at the time of their approval. However, the beneficiaries of these credits are expected to bear some of the operation and maintenance costs of these large schemes. Further, the beneficiaries do not own the credits unless they also contributed to the capital costs of the interception schemes. In Victoria the allocation of salinity credits is based on established cost sharing arrangements and considerations of public and private goods. In NSW, the allocation is based on an assessment of whether a work or measure is in the public interest and has an acceptable benefit cost ratio. South Australia has taken a different approach altogether, allocating its share of the credits to improve the river salinity, and is committed to remain salinity neutral in its developments. To achieve this, South Australia rigorously imposes policies designed to improve the irrigation efficiency of existing developments and minimize the salinity impact of new developments.

One of the toughest issues the initiative had with respect to the strategy was the establishment of benchmark salinities. Some states had licensed diversions that had not yet been implemented, and felt that those allocations should be counted as if implemented in the benchmarks. The issue was consistently raised in council and commission meetings, and several studies were conducted by technical committees to better understand the implications of these new developments. In the interim, the states would not be held accountable for these new, previously approved diversions, and the implementation of the strategy was not delayed. Results from the modeling of possible development scenarios together with the resulting salinity impacts were presented to the responsible committee in 1992. It was noted that there could be a significant impact on salinity levels, depending on the nature of the developments but the two main states concerned with these results differed on this issue. The committee agreed that these two states would prepare information on how they would develop the pre-1988 commitments for consideration at the next meeting. After several subsequent meetings, in 1993 the committee made a recommendation to the commission that states should be accountable for all salinity impacts resulting from new irrigation developments regardless of when the commitments for such developments were made. The commission decided to further study the matter, which continued from 1993 to 1995.

By 1996 the ministerial council had another major decision to make regarding a cap on total diversions to protect the environment. No decision was made on the issue of benchmarks, other than to review the strategy and conduct a salinity audit, primarily because new research showed that the magnitude of the salinity problem had been greatly underestimated. The studies, completed in 1999, had two main conclusions. First, the studies concluded that the strategy had successfully reduced the salinity in the river measured at Monroe. Since 1988, the strategy has achieved a 61 EC reduction against the 80 EC target through the construction of five new or enhanced salinity control schemes. These works cost $44 million in capital investment and require $1.8 million per year for operation and maintenance. The combined pre- and post-strategy schemes prevent 222,000 tons of salt per year from reaching the Murray.

But the second conclusion was that in spite of these measurable successes, the scale of the salinity problem was so large that it would not be feasible to manage it in all areas at risk. According to Blackmore et al. (1999), the only way for farming systems to operate such that dry land salinity is prevented is to have these systems mimic natural vegetation in their use of rainfall and reduce leakage to groundwater to a very low rate (1mm to 2 mm per year or less than 1% of rainfall). In a recent review of field studies an important Australian research institution, the Commonwealth Scientific and Industrial Research Organization, concluded that current leakage rates under farming systems, even when practiced at a "best practice" level, generally far exceed native vegetation. In other words, the management interventions in dry land catchments over the past 20 years have yet to demonstrate a level of salinity control such that future costs and impacts will be reduced below those predicted in the audit. Current farming systems for large areas of the agricultural zone that are capable of reducing salinization are not, at the same time, capable of providing adequate farm income. Clearly, this is a major challenge. In response the new Basin Salinity Management Strategy (2001 to 2015) was prepared with substantial input from the six governments and other stakeholders within the basin. It was approved by the ministerial council in September 2001. The new strategy provides guidance to communities and governments for controlling salinity and protecting key natural resources. It also establishes salinity targets for each tributary and the entire Murray-Darling system, reflecting a shared sense of responsibility.

According to the MDBC's *Salinity Update 2003, Basin Salinity Management Strategy*, the strategy provides the framework for salinity management in the basin, with the following objectives: (1) maintain water quality of the shared water resources of the Murray-Darling for all beneficial uses, and keep river salinity (at Morgan, SA) less than 800 EC 95% of the time; (2) control the rise in salt loads in all tributaries to meet the end-of-valley salinity targets; (3) control land degradation and protect important terrestrial ecosystems, productive farm land, cultural heritage, and infrastructure; and (4) maximize net benefits of salinity control across the basin.

A5.6.1.2 The Decision-making Process Around Salinity

The decision making around the issue of salinity in the basin reveals a process that was largely dominated by MDBC sub-committees and working groups. It was from this more technical level that the issue of establishing benchmarks was pushed. While these expert groups provided technically sound advice, it was unacceptable to some of the parties for political and economic reasons. Although it could be argued that this delayed decision making, the accumulated evidence of the need for action forced and continues to force the commission and council to address these pressing issues.

A more subtle principle has been advanced due to the extensive studies on the hydro-geological processes and the degree of sustainability of farming systems. That is, there is a greater willingness to undertake joint actions to address natural resource management issues, because the studies have increased awareness concerning the downstream impacts due to actions upstream. As stated in the 10-year review of the strategy, "There are clear signs that this has led to behavioral changes among water users in the basin." This is not to say that all upstream-downstream issues have been resolved, but there is a better appreciation of the geo-spatial nature of impacts in a river basin.

A5.6.2 Protecting Sensitive Ecosystems

One of the outcomes of the MDB Agreement was the 1990 Natural Resources Management Strategy (NRMS), a framework for a community-government partnership to develop plans for the integrated management of the basin's natural resources on a catchment basis. The first application of the strategy regarding ecosystems was for the Barmah-Millewa Forest. This is one of the most important wetlands in Australia and is listed in the Ramsar Convention as a wetland of significance (Crabb, 1997). Located between NSW and Victoria, development has significantly altered the flow and flooding pattern of the river, thus changing the forest ecosystem. The council early in its inception asked the commission to develop water management strategies for the Barmah-Millewa Forest. The MDBC appointed a team of consultants who prepared a final report early in 1992. A community group with representation from all major local interests was set up to review the report and give comments. These were compiled and presented to the council in 1993. The council approved in principle the recommendation to allocate 100 million cubic meters of water per year for use in the forest, with the water coming equally from Victoria and NSW. However, the council also recommended that the MDBC continue to refine the plan and to further develop community support to gain ownership of the plan.

Following the ministerial council meeting, the MDBC held discussions with the community reference group and relevant state government agencies in NSW and Victoria. The largest concern at that time was voiced by the community group, who stated that the state agencies were not interested in cross-border water management, which would be essential as the forest needed to be managed as a single system. The group also thought that although an annual environmental allocation of water for the forest was likely to be acceptable to the community, this forest should not be considered in isolation from the river basin as a whole, and therefore the entire Murray River should have an environmental plan. In response, the MDBC agreed to establish a working group to conduct such a study. In the interim, the implementation of the water management plan for the forest began, with community input to continue in the form of a formal organization, the Barmah-Millewa Forum. The larger issue of environmental flows for the river would not be addressed in this forum, but would be in the subsequent decisions taken on establishing a cap on diversions, as discussed below. The stated vision of the forum is:

- To maintain and where possible improve the ecological and productive sustainability of the Barmah-Millewa Forest; and

- To establish a planning and operational framework to better meet the flooding and drying requirements of the riparian forests and wetlands.

The process of establishing a plan for the Barmah-Millewa Forest demonstrated the decision-making ability at the mid-level, namely the commission. However, the community was heavily involved in the process of developing the plan for managing the forest, and continues to play the managing role through the forum, which has representation from communities, tourism, and water user groups, as well as state government agencies responsible for land and water resource management.

A5.6.2.1 Other Ecosystem Activities in the Basin

The MDBI also has established activities to address ecosystem issues elsewhere in the basin, including the following four strategies:

- **Riverine Environment Sub-program Strategic Plan (2000)**: This plan outlines the commission's strategic activities for policy development, knowledge generation, and on-the-ground action relevant to the riverine environment.
- **Floodplain Wetlands Management Strategy (1998)**: This strategy aims to guide the identification of significant wetlands in the MDB and the preparation of appropriate management plans. It also provides a framework for interstate coordination of policy and management and for monitoring and reporting on the success of floodplain wetland management activity.
- **Algal Management Strategy (1994)**: This was developed to help reduce the frequency and intensity of algal blooms and other water quality problems associated with nutrient pollution of the MDB. It aims to reduce nutrient concentrations in streams and improve in-stream flows.
- **Fish Management Plan (1991)**: This provides a framework for remedial action designed to reverse the decline in the native fish populations of the Murray-Darling river system. An updated fish management strategy was recently prepared to help increase native fish populations in the basin, reduce the impacts of carp, and improve ways for fish to move through dams, locks, and weirs.

A5.6.2.2 Proposed Ecosystem Activities in the Basin

Recently, the basin ministerial council initiated the Sustainable Rivers Audit (SRA) to measure river health in a consistent way across the basin. According to the basin commission, the SRA aims to:

- Determine the ecological condition and health of river valleys in the basin;
- Provide insight into the variability of river health indicators throughout the basin and over time; and
- Promote improved natural resource management practices.

Groups of indicators are organized into themes, with an initial focus on fish, macroinvertebrates, and hydrology. Further measures will include an index of water quality, a measure of flow regime, and an assessment of habitat. These three elements are drivers of the biological outcomes and serve as useful diagnostics in interpreting the reasons why the biological measures may have changed. The proposal is that the biological measures, along with the water quality and flow index would be reported annually for each river valley in the MDB. A more comprehensive sustainability assessment would be carried out initially and thereafter every five years.

A5.6.3 Cap on Water Diversions to Protect the Environment

Possibly the most significant management decision taken by the initiative has been to cap water diversions, in part to preserve in-stream environmental demands. The decision-making process for the cap is closely related to the salinity management issues, particularly benchmarks and the amount of water committed in yet-to-be-used licenses for water diversions. In the context of the Salinity and Drainage Strategy, the effect of growth in diversions was explored and it was determined that the over commitment of surface water in the basin was a serious problem that required political intervention. This was raised at a meeting of the ministerial council in 1993. At this meeting it was decided that the MDBC should prepare a report on the scope for continued growth in diversions in the basin. The MDBC prepared a report, *Limits to Surface Water Diversion in the Murray-Darling* in 1994 that concluded that the then current licensing and allocation system served only to limit water diversion during droughts. During periods of more plentiful supply, various practices in the basin encouraged additional diversions. Without a change to the system, diversions would increase over time. At a subsequent meeting of the council in 1995, it was agreed that there was a need for an appropriate balance between water diversions and in-stream water use. It was decided that water diversions must be capped and an immediate moratorium was placed on further diversions of water from the rivers of the basin (the cap on diversions). A number of working groups were then established to examine the various issues around the cap, including determining the appropriate level for the cap.

The cap was a significant decision since it represented a distinct modification in how water was allocated among the states. Water use by the different states was to be determined based on the level of development in place that existed in the basin in 1994 rather than solely by the formula agreed upon by the states several decades before and that had been accepted by the MDB Agreement.

Since the introduction of the cap, an annual water audit has monitored its implementation in order to ascertain whether the states are meeting their commitments. Each state has been required to develop reliable monitoring and reporting systems.

The cap has had the effect of explicit recognition of the importance of striking a balance between consumptive and in-stream (environmental) demands. The cap has also had the effect of generating interest in a market for water trading. In response, the MDBC established a water market reform working group to fast track advancing water markets and the trading environment. The commission recently piloted permanent interstate water trading in the southern part of the basin.

A5.6.3.1 The Decision-making Process Around the Cap

The decision-making process for the cap differs from the Barmah-Millewa Forest strategy in that it had almost no community input. Community input was taken at the level of how to implement the cap, but was not taken in the establishment of the cap.

A5.7 Degree of Success

In virtually all of its activities the MDBI has established clear metrics for its goals and has had successful progress towards those goals. Two examples discussed above are the Salinity and Drainage Strategy, which set a goal of an 80 EC reduction in salinity at one location in the river, 61 EC of which has been met, and the cap in diversions, which has been successfully imposed since 1995. The MDBI has also seen success in restoring ecosystems. One of the goals of the program in the Barmah-Millewa Forest is to meet the flooding and drying requirements of the riparian forests and wetlands. Evidence of this success was seen in the recent flooding of the forest using the ministerial council's environmental water allocation. The 2000-2001 flood event in the Barmah-Millewa Forest was appropriately timed for the successful breeding of many frog species. (Diversity and numbers of frogs are good indicators of the health of the forest environment.) Nine species of frog inhabit the forest. This included at least two species not previously recorded in the Barmah-Millewa Forest.

However, these metrics only partially address the broader goal of the initiative of "sustainable use of the water, land, and other environmental resources." The MDBI is trying to address the biological integrity of the basin (beyond the Barmah-Millewa Forest) through a proposed program called the sustainable rivers audit, which uses appropriate metrics for the biological health of various ecosystems in the basin. The initiative also struggles with agricultural practices in the basin that are not sustainable with respect to salt. The issues surrounding the problems of biological integrity and agriculture and salinity, are problems that the initiative continues to address. Perhaps the most significant success has been a gradual change in the public perception of the problems.

A5.8 Lessons and Applicability to the United States

The MDBI is an example that covers many issues plaguing the United States. The MDBI is a well-organized and documented example of a large-scale approach to watershed management that crosses over many jurisdictions. The MDBI addresses the integration of authority from local issues to ministerial councils and strategies for increasing awareness and participation among stakeholders. The initiative also addresses point and non-point source pollution, as well as ecosystem protection. These lessons are summarized below.

Lesson 1: The political nature of decision-making around water issues should be directly addressed in any management process.

Decision making that occurs in multi-jurisdictional rivers basins is different than many other forms of organizational decision making. In particular, politics not only plays a role, it can dominate the decision-making process. In the case of the Murray-Darling, its management authority was set up with a political body at its head. As such, there was open acknowledgement that politics is absolutely central to management decision making, with processes adapted accordingly. By being up front about the political nature, it was possible to improve the decision-making processes by making external political factors more transparent, allowing other forms of decision making to play their role within the context of politics. Technical considerations are thus permitted to become more influential.

This blending of technical and political factors occurred in the decision making leading to the cap on water diversions in the MDB. Although there is no doubt that the different ministers at the ministerial council would have reiterated arguments favorable to their individual states during discussions, and that the issue was pursued by some states more than others, it was technical information that ultimately lead to a decision by this political body.

This political–technical model may be applicable in the United States for problems occurring in large river basins that cross state borders. Allocation issues, whether in terms of quantities of water or pollutants that can be discharged into a river system, are inherently political in nature, as well as technical. An institutional structure that openly contains these elements will be more likely to succeed in implementation of watershed-wide policies.

Lesson 2: Nonpoint source pollutant trading can be facilitated with well defined property rights/responsibilities.

The trading in salinity that takes place under the Salinity and Drainage Strategy was established through a negotiated agreement that imposes rights and responsibilities on each government member of the agreement and holds them accountable for all future actions in their territory that impact river salinity. Each government has discretion in handling the salinity issue within its territory. The costs associated with addressing salinity at larger scales (i.e., costs that cross state boundaries) are allocated among the states and the federal government with respect to capital costs and among the states only for operation and maintenance costs. Explicit in the strategy is the statement that "upper states would provide financial support for salinity reduction works further downstream," thus establishing the principle that upstream users are responsible for the impacts downstream.

Nonpoint source pollution continues to be a difficult problem in the United States, primarily because land-use decisions are made at the municipal level and are subject to pressures from powerful development interests. These interests often do not participate in watershed planning efforts, and are not held accountable for the impact of the developments on water quality and watershed ecosystem health. Using the Murray-Darling model in this context, the municipalities would be given some allocation of a pollutant (e.g., x kg of nitrogen per year) and could not exceed that allocation unless they purchase an additional allocation from an adjacent municipality. This approach would not require a change in who makes decisions in land-use zoning, which would be very difficult politically, but would rather increase the responsibility of the decision-making body to include impacts on affected watersheds.

Lesson 3: A strategy of decision-making at the lowest appropriate level is effective

The MDBI has taken a consistent approach in its decision making and implementation efforts. Decisions with basin-wide impact are made at the highest level, namely the ministerial council, but only in terms of the impact on or the apportionment to each state. The state then retains the right to implement the decision as it sees best without council influence. This approach has been successfully used in the Salinity and Drainage

Strategy, as well as the cap on total diversions from the Murray-Darling. Also, in the case of the Barmah-Millewa Forest, the council made a decision about the total water to be allocated and the states that would supply the water, but all of the implementation decisions were then passed on to the affected states and a newly created body with community input, the Barmah-Millewa Forum. This forum has been very successful in restoring parts of the ecosystem in the forest.

Management of large-scale watersheds, such as the Mississippi or the Colorado, is notably lacking in the United States One possible explanation for this lack is the concern at the smaller scales of a loss of decision-making authority. The approach taken in the Murray-Darling would allay those fears by ensuring that decisions that can be made at the local level remain in the hands of local decision makers.

APPENDIX B

The Watershed Management Approach in the United States

B.1 Introduction

B.1.1 What is a Watershed?

In the simplest terms, a watershed is a catchment or drainage basin, an area of land that drains to a common river, stream, or lake. The boundaries of a watershed are natural. They are determined by the interaction of water and the topography of the land. As interest in using watersheds as an organizing structure for managing water and water-related resources has grown, the definition of a watershed has broadened to include not only the hydrologic system within a common drainage area, but all resources and activities within those boundaries that affect or are affected by the water, including human structures and activities as well as ecosystems.

B.1.2 Scale and Terminology

Watersheds vary greatly in scale. The Missouri-Mississippi rivers watershed covers approximately one third of the contiguous United States. However, a very small creek may also define a watershed (or sub-watershed) covering only a fraction of a square mile. Although they exist along a continuum of scales, watersheds have a nested structure that is inherently hierarchical. For example, water in the catchment area of a small creek drains into a larger brook or river with a watershed that encompasses all the creeks that drain into it. In turn, this may be part of a larger watershed comprising many sub-watersheds.

The United States Geological Survey (USGS) has developed a system for classifying watersheds at different scales (Table B-1). The largest USGS hydrologic units are the 21 water resource regions that have been given two-digit identification numbers. Some of these define the drainage area of a major river, such as the Missouri or Columbia, while others combine the drainage areas of several rivers and consist of multiple watersheds. The USGS has divided the 21 water resources regions into 222 sub-regions, and given these four-digit identifiers. The smallest USGS unit is the "cataloging unit," with an average area of 1700 mi^2. There are 2150 watersheds at this scale, each identified by an eight-digit number. Most watershed management activity in the United States is occurring at scales at the cataloging unit or smaller.

While the USGS classification system can be useful for distinguishing watershed scales, the nomenclature is awkward and has not been adopted by planners, policymakers, water professionals, academics, or others. Indeed, the literature has not adopted a consistent nomenclature of any kind. Thus, this report employs the terms watershed, river basin, and basin interchangeably.

Table B-1 USGS Watershed Hydrologic Units.

USGS	Identifier	Total Number	Average Size* (mi^2)
Water resource regions	2 digit	21	170,000
Planning regions	4 digit	222	17,000
Cataloging units	8 digit	2150	1,700
Sub-cataloging unit		>20,000	

*Average sizes of USGS watershed categories derived from land area of United States, USGS.

B.1.3 The Watershed Management Approach

Increasingly, nongovernmental organizations (NGOs), states, and the federal government are embracing an approach to managing human activities and natural resources within watershed units. Simply using watershed boundaries as the organizing unit for management is, narrowly speaking, watershed-based management. However, over the last two decades, watershed management has come to mean far more than this. It is an approach characterized by science-based planning and decision making on a watershed scale, integration of a variety of competing water resource priorities and goals, cooperation from an array of stakeholders and governmental agencies, and increased levels of public participation.

Using watershed boundaries as the organizing unit for management has been promoted as an effective method for managing water and water-related resources for several reasons:

- The watershed is a natural geographic boundary and therefore the most logical organizing unit for managing water resources. Integrated multi-objective management of water resources is only possible where all of the elements that influence water quality, quantity, and related resource values are considered and managed together within watershed boundaries.
- Using the watershed as the management unit transcends problems created by political boundaries, which rarely reflect hydrologic units. Political entities, such as cities, counties, and states, have not succeeded in managing water and water-related resources effectively, in large part because their boundaries do not reflect natural watershed boundaries. Rivers and streams often serve as the boundary between cities or states, perfectly bisecting the watershed, a situation that frequently leads to conflict.
- The nested structure of watersheds enables issues to be managed at the appropriate scale.
- The impacts of land use, such as agriculture and development, on water resources are most clear when the system is analyzed within a watershed context.

This more comprehensive definition of watershed management emerged in the late 1980s and 1990s. It includes the following elements. However, in practice few initiatives include all features and most emphasize one or two key elements.

- **Multi-objective integrated resource management.** Watershed management seeks to pursue multiple objectives simultaneously, such as water quality protection and flood control, while balancing competing uses, such as in-stream flows and irrigation.
- **Inter-governmental coordination.** Watershed management coordinates and balances federal, state, and local government agency objectives.
- **Multi-stakeholder participation.** All relevant stakeholders are included in the management process.
- **Collaborative participatory processes.** Decision making uses an open, participatory process often striving for consensus-based or super-majority decisions. The management process involves creation of a framework of formal and informal agreements between government and non-government partners to implement plans.
- **Science-based decision making.** Identification, prioritization, analysis, planning, project design, and project implementation are based on scientific and technical information and tools.
- **Action orientation.** Development and implementation of tangible projects are a primary objective.
- **Adaptive management.** Recognizing that scientific information and analytical tools are imperfect and actual impacts of projects may differ from expectations, the management process is iterative and adaptive, comprising a continuous cycle of data gathering and monitoring, analysis, planning, implementation, and assessment.

This broadly defined watershed approach is seen by many as the best way to effectively and efficiently manage water and water-related resources.

B.2 Evolution of the Watershed Management Approach

The history of water resource planning and management in the United States can usefully be divided into two broad periods, 1800 to 1965 and 1965 to the present. The period through 1965 was characterized by large, federally sponsored water development projects and successive attempts to establish a unified national system of institutions for multi-purpose river basin management. These efforts were thwarted by the many federal agencies with some authority over water resource management.

B.2.1 Water Development and River Basin Management (1800 to 1965)

Historically, water resources have been managed for many purposes, including navigation, flood control, irrigation, drinking water, hydropower, industrial cooling, recreation, and habitat protection. In order to oversee these efforts, a proliferation of federal, state, and local agencies were created. These agencies often had different and often competing missions.

The federal government was the primary force behind most water management projects between 1800 and 1965, largely because it was the only entity with sufficient financial and technical resources. Projects constructed in the 19th century were generally for a single purpose, most often navigation or flood control, and were developed and implemented by the Army Corps of Engineers (Corps). In the early 1900s new water resource needs emerged along with new agencies to mange the resources for those ends. Most notable was the need for water supply and distribution for the development of agriculture in the arid western states, managed by the Bureau of Reclamation (BOR) within the Department of the Interior, and the development of hydropower, overseen by the Federal Power Commission.

The concept of managing water resources within hydrologic boundaries, specifically within large river basins, was developed at the end of the 19th century and first articulated by John Wesley Powell in his 1879 report to the U.S. Congress on developing the arid western states and territories. However, the idea was not adopted by the federal government until the beginning of the 20th century, during the administration of Theodore Roosevelt. Successive federal commissions and numerous congressional bills over the next 60 years developed and promoted the idea as the best way to manage the increasingly numerous factors that influence water resources. The proposals generally focused on creating a system of federal institutions to manage water resources on the large river basin scale. Federal water resource agencies, such as the BOR and the Corps, afraid that a new integrated system would diminish their authority, resources, and autonomy, joined local water development interests in opposing such proposals. As a result, virtually all of these bills were either blocked in committee or given an unfavorable recommendation.

The idea that water development projects could be designed to serve multiple purposes was also championed during Theodore Roosevelt's administration and linked to the concept of basin management. Proponents of this idea convinced the U.S. Reclamation Service (renamed the BOR in 1923) to include hydroelectric turbines in several of its initial projects. Interest in multi-purpose water development and river basin management expanded with the success of these projects.

These experiments also highlighted the vast untapped potential of hydroelectric energy leading to the creation of the Federal Power Commission in 1920. Power, navigation, flood control, and irrigation were integrated for the first time in the comprehensive plans developed by the commission in collaboration with the Corps. The plans, however, were never implemented because the commission lacked an institutional vehicle to effect the proposed integration.

Interstate compacts that emerged at this time might have served as such a vehicle, but they were developed with the much more narrow goals of resolving conflicts over water allocations and enabling particular water development projects.

Beginning in 1933, a number of experiments with different institutional structures for managing water resources within large river basins were established in selected basins around the United States, including the Tennessee Valley Authority (TVA) and the Delaware River Basin Commission. While many of these programs are considered successful, each one is essentially unique and, because of opposition from federal agencies, there was never successfully replication.

In the 1930s and 1940s, the establishment of the U.S. Soil Conservation Service and similar state and local programs resulted in the successful creation of programs that managed water resources within a watershed unit. As they worked in only small-scale watersheds and limited their focus to erosion and flood control until the 1990s, the programs had little impact on the development of more comprehensive federal water resource management proposals until relatively recently.

Twice, once in the 1940s and 1950s and again in the 1960s and 1970s, a system of institutions to manage large river basins or water resource regions was established. However, because they lacked authority to implement their policy decisions, participating federal agencies continued to pursue their individual objectives. As a result, neither attempt succeeded in effecting integrated management of water resources at the basin level.

The first organizations created, known as Interagency River Basin Committees, were designed to coordinate the activities of federal agencies within a basin and provide for some state involvement in planning. This marked the first time the federal government attempted to involve the states in basin management. However, the committees were given no authority to make or implement decisions. Rather, decision-making power remained with U.S. Congress, rendering the committees ineffective.

Similarly, the Tier II River Basin Commissions established by the Water Resources Planning Act of 1965 were not granted the power to make enforceable decisions. Like their predecessors, they failed to elicit coordinated action from the various federal and state agencies. They did, however, provide greater state participation and served as effective forums for communication and debate. Moreover, these commissions were essentially water development agencies and, by the 1980s, the era of large water development projects was coming to an end.

B.2.2 Emergence of the Environmental Movement and Evolution of the Modern Watershed Approach (1965 to present)

The emergence of the environmental movement in the 1960s, combined with the creation of new federal and state programs and agencies designed to protect the environment (most notably the U.S. Environmental Protection Agency [EPA] in 1970), set the stage for the transformation of water resource management in the United States. The introduction of ecological concerns added a critical new dimension to integrated resource management and remains at the core of the modern watershed management approach. Initially, these programs validated public concern by including natural resource degradation, water quality, and habitat preservation, including in-stream flows, as competing uses and providing some regulatory tools to protect these uses. Later, the ecological perspective led to recognition of the need for a more holistic ecosystem and watershed approach to manage this inherently integrated system.

The problems of toxic pollution and environmental degradation entered public awareness in the 1960s with the publication of Rachel Carson's *Silent Spring* and a number of high profile environmental disasters. Public concern translated into federal concern, and led to the passage of a series of federal pollution control laws including the Water Quality Act of 1965 and the Clean Water Act of 1972. Although federal legislation regulating water quality is generally thought to begin with the Water Quality Act of 1965 and the Federal Water Pollution Control Act of 1972 (Clean Water Act), these mark the beginning of federal concern with toxic pollutants. Concern about pathogenic pollution, however, arose several decades earlier with the Water Pollution Control Act of 1948 and its amendments in 1956. These acts authorized federal subsidies to local municipalities for the construction of municipal wastewater treatment plants (WWTPs).

The Water Quality Act mandated the establishment of federal ambient water quality standards and the development and implementation of plans to control pollution from individual point sources to meet those standards. It also created the Federal Water Pollution Control Administration to set general guidelines for state standards, but gave states primary responsibility for issuing discharge permits to individual sources of pollution, taking enforcement action, and imposing fines.

The U.S. EPA was established in 1970, following the Santa Barbara oil spill, over strong opposition from the traditional federal water and resource agencies. Although one of the principal motivations for its creation was the recognition of the need for a single environmental regulatory agency that could manage the environment as a unified interrelated system, water issues outside of water quality, such as hydropower and flood control, remained split among different federal agencies.

The Clean Water Act of 1972 was the first major piece of legislation to provide new power and financial resources to the U.S. EPA. It set a goal that all U.S. waters should be fishable and swimable by 1985. States soon created parallel environmental agencies with similar missions and programs.

While the objectives of the programs were broad, they were first pursued using an approach that focused on large point sources of pollution and considered each medium (air, water, land) independently. These efforts were essentially successful at controlling pollution from the targeted sources. However, success in addressing point sources of pollution only revealed the much more difficult problem of nonpoint source pollution, such as oil contaminated runoff from streets, erosion from construction projects, and nutrient runoff from farms. It also highlighted the intimate relationship between nonpoint source pollution and land use. Concurrently, the tendency of the media-specific approach to create cross-media transfer of pollutants, rather than eliminate pollution, was leading U.S. EPA and others to reassess the wisdom of this approach. Similar conflicts emerged at the state level. This resulted in increased recognition of the environment as an integrated system, an ecosystem, that requires integrated management. The watershed approach provided an obvious framework.

In 1991, U.S. EPA began advocating a watershed approach to water quality and more general environmental protection. In 1993, the agency articulated a detailed watershed protection approach as a solution to the shortcomings in the agency's existing programs. The new approach was intended to be a framework for integrating existing programs. It encompassed all waters (surface, ground, inland, and coastal) and heavily emphasized elements that affect the entire ecosystem, such as toxic and conventional pollutants, temperature, flow, habitat quality, and biodiversity.

The U.S. EPA's watershed approach integrated many of the elements included in the modern definition of watershed management: integrated resource planning within a watershed framework, science-based decision making, multi-stakeholder processes involving all levels of government, conservation district representatives, public interest groups, industry, land owners, etc. A dramatic departure from its earlier top-down structure, the watershed approach attempted to deal with environmental management at a community level because it recognized that critical land use and other issues are managed primarily at this level. The U.S. EPA watershed approach acknowledged the issue of appropriate scale, but did not present a method for its determination. More importantly, it did not offer procedures for identifying trade-offs among different options, and implied that mutually acceptable solutions can always be reached. Some critics suggest this assumption leads U.S. EPA to understate the need for compensation tools to balance unequal costs that certain stakeholder interests may bear. Nevertheless, its adoption by U.S. EPA raised the profile of the watershed approach in general and marked a significant milestone in its evolution.

The first home of a watershed management approach, the Soil Conservation Service within the U.S. Department of Agriculture (USDA), also underwent important transformations over the last decade. In the mid-1990s the agency was renamed the Natural Resources Conservation Service (NRCS) to reflect its expanded role in natural resource conservation, specifically in protecting wetlands and water quality on private agricultural, range, and forest lands. Perhaps the most valuable contribution of the Soil Conservation Service to modern watershed management was the development of the coordinated resource management process. In the 1950s and 1960s agency staff developed a collaborative non-adversarial decision-making process for resolving conflicts between different stakeholders and developing action plans that were mutually acceptable. The coordinated resource management approach is part of the NRCS's ecosystem-based approach that focuses on scientific management of natural systems and processes. The NRCS approach is often applied within hydrologic boundaries, but will adjust its framework in recognition of socioeconomic, political, and legal constraints. The process encourages active stakeholder involvement including all levels of government, special interest groups and the public. Like the U.S. EPA approach, the NRCS process also assumes that stakeholders can arrive at a mutually acceptable compromise that does not overburden any one person or group and thus does not include a method for developing compensation measures.

The Soil Conservation Service and later the NRCS played an important role in the emergence of the modern watershed approach at the community level. Their ubiquitous local presence and focus has made the NRCS frequent federal partners to local watershed initiatives, and many such initiatives have adopted the service's principals of coordinated resource management.

In 1998, the two federal agencies that had most fully developed a watershed management approach, the USDA (home of the NRCS) and U.S. EPA, joined forces to forge "The Clean Water Action Plan," a coordinated watershed-based approach for protecting and restoring the nation's water resources. Development of the action plan represented a new cooperative approach to watershed protection at the federal level. The plan attempted to integrate the various programs of each agency into a coordinated framework, without significant reorganization or consolidation. The plan outlined a process in which federal agencies, states, and tribes develop assessments of all the USGS cataloging unit-scale watersheds, prioritize them based on greatest need for protection and restoration, and then develop and implement restoration strategies.

After its submission in February 1998, the plan became a quasi-independent initiative with the aim of coordinating and integrating water resource management across federal agencies. The plan has succeeded, at least to a limited degree, in generating agreements for integrated management between virtually all of the federal agencies with water resource management authority. In 2000, the Corps, the Department of the Interior, Department of Energy, Department of Commerce, Department of Defense, and the TVA joined U.S. EPA and USDA in the development of a Unified Federal Policy for a Watershed Approach to Federal Land and Resource Management. These agencies also signed a federal Multi-Agency Source Water Agreement promoting the integration of state, tribal and local drinking water source assessment and protection initiatives. Whether these policies will result in actual cross-agency coordination of activities remains to be seen. However, they have the potential to herald a new era of interagency cooperation and collaboration.

While the Clean Water Action Plan may appear to represent a return to the top-down regulatory approach, it is attempting to be something quite different—to define a new role for the federal agencies as conveners, facilitators, and purveyors of funding and technical assistance who act in a collaborative, coordinated manner. The extent to which these new roles can be successfully carried out without modifying their enabling legislation is an important question. So far, the signs are promising, but it is still too early to assess definitively.

B.2.3 State and Local Activity

While federal policy and programs have played and will continue to play an important role in water resource management and the evolution of a watershed management approach, the greatest innovations in recent years have been at the state and local levels.

As federal powers have devolved to the states over the past 20 years, states have developed the institutional and technical capacity to carry out their growing responsibilities. This devolution of power has also prompted state-level experimentation and innovation in watershed management. In virtually all watershed initiatives, local and regional governments and other stakeholders are recognized as closest to watershed management, and they play significant roles.

Massachusetts is generally regarded as one of the states with the most highly developed watershed management approach. The Massachusetts Watershed Initiative, launched in 1993, significantly reorganized the state's environmental agencies in an attempt to more fully integrate watershed protection and planning concepts into core activities. State-sponsored "basin teams" convene and institutionalize a process involving federal and state agency representatives, local officials, and citizens' organizations. The objective of the process is to develop an action plan for each of the state's 27 basins in a coordinated and collaborative manner. Basins follow a five-year planning cycle that synchronizes wastewater and water withdrawal permitting and various grant programs, including state revolving fund (SRF) loans. Washington, Oregon, and Florida have also developed sophisticated programs to promote watershed management. Many other states, including Idaho, Maryland, Minnesota, New Jersey, Pennsylvania, and Wisconsin, are also implementing watershed approaches and programs.

The past 20 years witnessed the birth of a broad-based watershed movement. Hundreds, if not thousands, of watershed initiatives have sprung up at the regional and local level, with the most dramatic growth occurring during the last decade. The definition of the modern watershed approach is in large part derived from empirical assessments of these efforts. Most appear to have developed independently. While each incorporates many of the elements of the modern watershed approach, they exhibit a breathtaking level of diversity. In some sense each represents an independent experiment with a variation of the watershed approach.

B.2.4 Summary

Water resource management in the United States has evolved from single purpose water development projects to integrated resource management efforts within hydrologic units. Historically, the federal government was the principal agent of water resource development activities and created a large number of agencies to oversee different types of development projects. The idea of managing water resources by watershed units first came to prominence at the beginning of the 20th century and was promoted through federal programs and initiatives throughout the early part of that century. Successful river basin management experiments such as the TVA were established but not replicated due to opposition from existing agencies. The idea that water development projects could serve multiple purposes, a precursor of integrated resource management, also arose in the first decades of the 20th century and became the dominant form of water development project for the next 60 years.

The dawn of national environmental consciousness in the 1960s marked a turning point for U.S. water resource management. The introduction of the ecological dimension changed the way water resources were perceived. Pure water development projects were no longer possible, as impacts on living natural resources had to be respected. Water resources were now seen as part of an integrated ecosystem, a watershed, that must be managed instead of exploited. A number of factors, including nonpoint source pollution and its relation to land use, led the U.S. EPA to embrace a philosophy of integrated resource management and look for ways to influence local land-use decisions. In the early 1990s, the U.S. EPA devised then adopted an ecosystem focus and then a watershed management approach. This represented a significant departure from the previous top-down regulatory approach and cast the U.S. EPA in a new role as facilitator and supporter. The approach has been institutionalized and expanded as part of the Clean Water Action Plan. The plan is attempting to overcome one of the major obstacles to effective water resource management by extending integration of water-related programs beyond U.S. EPA, to include all federal agencies with significant water resource oversight responsibilities. There are promising signs, but it is too early to tell whether this collaborative approach can overcome agencies' competing missions and distinct operations.

The importance of the state role in water resource management has increased over the last 20 years as federal power has devolved to the states, and state institutional and water management resources have grown. Massachusetts, Washington, Oregon, and Florida represent the leading edge of state-led watershed initiatives, with many other states also developing watershed approaches and programs.

In recent years the greatest experimentation and innovation in watershed management has occurred within state initiatives at the regional and local level. The 1990s was a period of explosive growth for local watershed initiatives. Hundreds, if not thousands, of such efforts have emerged, each an independent experiment with a variation of the modern watershed approach.

B.3 U.S. Watershed Experience

B.3.1 Introduction

The modern watershed management approach has risen to prominence over the last 10 to 15 years. This approach seeks to manage water and related human and natural resources within a watershed unit. It employs a collaborative, science-based process in which relevant stakeholders develop a mutually acceptable resource management plan that balances competing uses.

Application of the watershed management approach has grown rapidly in the last decade, motivated largely by the failure of the existing system to adequately address more complex and contentious problems relating to water resources, such as water allocation.

One of the most noteworthy aspects of U.S. watershed initiatives is their diversity. Watershed initiatives demonstrate a variety of institutional arrangements. There are nearly as many catalysts for formation, development paths, and organizational arrangements, not to mention scales, as there are watershed initiatives. The diversity of approaches arises in part from differences in presenting issues (e.g., nutrient pollution from agriculture or insufficient in-stream flows to sustain fish) and the scale at which these issues can be addressed, as well as from regional differences in climate, politics, water law, and regulatory framework. Other differences are related to the particular institutions and individuals within a given watershed.

Where the watershed approach has been successfully applied it has created a new paradigm for natural resource management and broken previous bureaucratic gridlock, improving interagency and intra-agency cooperation and public-private collaboration. The watershed approach has not fully proven it can consistently achieve its goal of substantive improvements in one or more aspects of ecosystem health. A number of obstacles have been identified that restrict the growth or potential effectiveness of watershed initiatives. International watershed management experiences can provide important lessons concerning these obstacles.

This section first provides an overview of the types and extent of recent watershed management activity in the United States. This is followed by a discussion of the challenges virtually all watershed initiatives face: pollution management, habitat protection and restoration, and insufficient data and tools.

B.3.2 Nature and Extent of U.S. Watershed Management Activities

The staggering variety of U.S. watershed initiatives makes a comprehensive description of institutional arrangements, processes, and objectives impossible. However, they can be placed into broad categories and illustrated through examples. Watershed initiatives can be found in every state at almost every scale. Each operates within a complex set of climatic, historical, legal, political, and regulatory contexts that can enable and constrain activities. There are programs sponsored by government at all levels and coordinated by local NGOs. They may be catalyzed by an ecological crisis or a desire to protect a pristine resource. While they may define their primary goals broadly or narrowly, environmental protection and restoration is at the heart of most missions. They have received broad support and encouragement from a number of federal, state, and local environmental agencies, largely because they represent the most promising model for addressing some of the thorniest environmental problems, including management of nonpoint source pollution and related land use and habitat protection and restoration, and provide an efficient means for agencies to achieve their objectives. Cross-jurisdictional cooperation and broad stakeholder involvement with collaborative decision making are generally identified as key elements. Most initiatives utilize federal and state agency technical expertise and receive direct and indirect financial support from state and federal agencies.

B.3.2.1 Scale

Just as watersheds come in different scales so too do the initiatives to protect and manage their resources. While there are no comprehensive data on U.S. watershed initiatives, good data on activity in the western states and collections of case studies and reports that include the eastern half of the country indicate that most watershed initiatives work in basins 500 mi^2 or smaller. Some focus on extremely small watersheds. Efforts coordinating activities in watersheds of several thousand square miles are not uncommon, and there are only a small number working in regions tens of thousands or hundreds of thousands of square miles.

To some extent, scale determines the breadth of activity and issue focus of watershed collaborations. Each presenting issue generally has an optimal management scale. Examples of watershed initiatives that work at a diversity of scales and focus on a variety of issues include:

- The Duck Creek Advisory Group works to coordinate water quality and anadromous fish habitat restoration activities in the 1.7-mi^2 Duck Creek Watershed in Juneau, Alaska.
- In the Little Rabbit River Watershed in southwest Michigan, a diverse stakeholder partnership works to mitigate nonpoint source pollution problems in basin's 48 mi^2.
- The Tillamook County Performance Partnership seeks to restore salmon and other habitat in the 570-mi^2 Tillamook Bay Watershed in Oregon.
- The Verde River Watershed Association focuses on preserving the relatively pristine habitat and in stream flows in the 6600-mi^2 Verde River Watershed in Arizona.
- The Chesapeake Bay Program seeks to address nutrient pollution, wetlands loss, and other factors thought to degrade habitat and depress fish populations in the 64,000-mi^2 watershed draining into the Chesapeake Bay. To accomplish these goals, the program coordinates activities in three states and the District of Columbia and enlists the assistance of community-level initiatives working at the sub-watershed level.
- The Colorado River Basin Salinity Control Program focuses on reducing salt loading in the 250,000-mi^2 Colorado River Watershed.

Not surprisingly, scale also influences who participates in the watershed initiative process. Watershed initiatives at smaller scales are more likely to have citizen, landowner, and small business participation. Watershed efforts at larger scales are inherently more complex, encompassing many more political units and other stakeholders. Defining scale according to geographic size is somewhat misleading. Scale in terms of population and relative size of political jurisdictions is equally important. Smaller watershed initiatives in more densely populated watersheds in eastern states with smaller municipal units may face as complex a collection of stakeholders as collaborative efforts in much larger, more sparsely populated basins in western states that have geographically larger towns and counties.

B.3.2.2 Extent of the Watershed Movement

As noted above, there are no comprehensive statistics on watershed management activities in the United States as a whole, though quantitative data on watershed initiative activity in the western United States have recently been published (Kenney et al., 2000). Extensive activity is reported throughout the country with the highest concentration in the Pacific Northwest and California. State efforts to adopt a watershed management approach in

Massachusetts, Washington, Oregon, Idaho, and other states have accelerated activity. For example, there are approximately 150 state-recognized watershed initiatives in Oregon alone.

The data on watershed initiatives in western states is striking. In the area approximately west of the 100th meridian, which bisects North Dakota, South Dakota, Nebraska, Kansas, Oklahoma, and Texas, and including Alaska and Hawaii, researchers identified 346 efforts that featured most of the key characteristics of a watershed approach. Detailed data were collected on approximately one third (118) of these initiatives (from completed survey forms). Data from these surveys supports anecdotal reports of rapid and dramatic growth of the watershed movement in this region (Figure B-1). The graph suggests that the growth of watershed initiatives slowed in the late 1990s. This may not be the case, however, and may reflect deficiencies of the self-selected sample. The data depict responses from the watershed initiatives that returned completed surveys. Follow up with the 228 non-respondents indicated that they tended to have less staff resources and fewer positive results to report. It therefore seems likely that the sample pool contains a disproportionate number of more established initiatives and the apparent slowing in growth may be sample bias. Anecdotal evidence and imperfect data from federal sources such as U.S. EPA's online database suggests that similar, possibly less dramatic growth is occurring in the eastern half of the country.

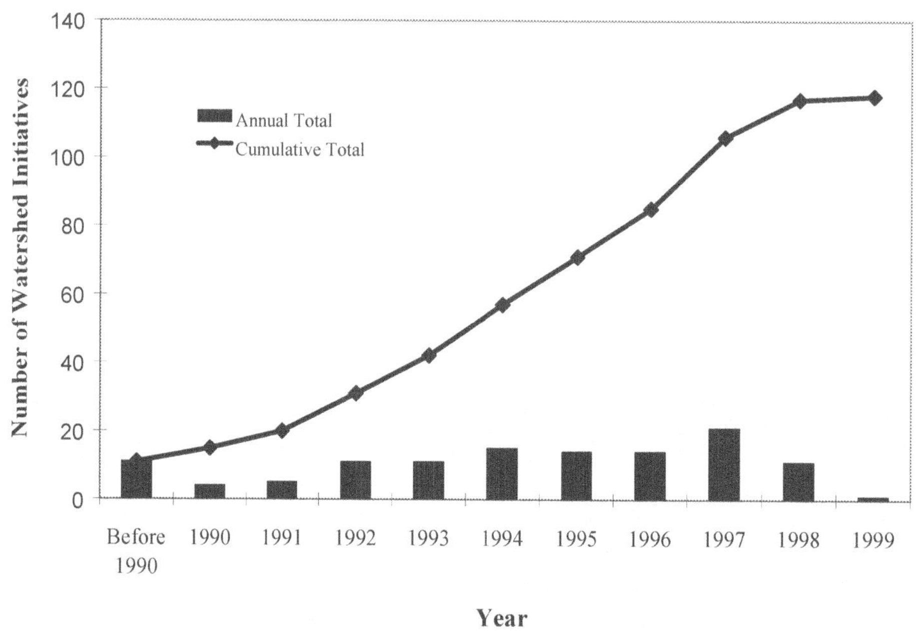

Figure B-1. Growth of Watershed Movement in the Western United States.

B.3.2.3 Climate, Water Law, Regulation, and Politics

Climate, water law, regulatory frameworks, and political conditions all play a role in shaping the institutional arrangements, focus, and modus operandi of specific watershed initiatives.

Climate

The United States encompasses a variety of climates from the arid southwest to the well-watered regions of the northeast and northwest. Climate determines the abundance and dependability of water within a watershed, which significantly influences the focus of watershed initiatives. Moreover, climate creates diverse conditions that support distinct ecosystems, and this diversity must be considered when attempting to generalize lessons learned from individual watershed initiatives.

Localized variation in the quantity, extent, frequency, and extremes of humidity, precipitation, and temperature create relatively unique conditions in each watershed. As a result, no single approach to watershed planning and restoration is possible. Watershed initiatives must all have access to or engage in research and planning specific to each climatic region. Human activities, such as urban and suburban development, agricultural activities, timber harvesting, mining, dams, and levies, have ensured that even watersheds in close geographic proximity and with essentially identical climates can represent unique management problems.

Quantity of precipitation and consistency from year to year probably have the greatest influence on watershed initiatives. The Pacific Northwest, the Great Lakes region, and East Coast states have an abundance of water. Until recently, water quantity concerns were rare in these regions. However, water consumption in these areas is increasing and low-flow conditions and dropping water tables are occurring more frequently. Indeed, Washington's most significant watershed management legislation is intended to help address water allocation conflict issues. Of course, whatever shortages watershed efforts in these areas have to address do not compare to those in the more arid western states, such as Arizona, New Mexico, and Nevada. Water has always been scarce in these regions and is becoming more so. At the same time, western states are experiencing tremendous growth, with projections of a 30% increase in population by 2020. In these states, allocation of water is necessarily a core issue for many watershed initiatives.

Water Rights

Myriad laws and regulations govern the use of water and related activities, such as the modification of river channels. Although they are necessarily intertwined, the legal context can be divided into two parts: 1) federal; and 2) state and local.

Watershed initiatives interested in influencing the quantity of water flowing in a stream or river will need to negotiate the system or systems of water rights within their basin and perhaps beyond. The use of water and related activities, such as the modification of river channels, are governed by numerous, complex systems of laws and regulations.

Water has generally been treated as property, and the nature and allocation of water rights have been governed by state and local law. Because water laws were developed before there was a good understanding of hydrological dynamics, surface water and groundwater have historically been treated as separate entities. In some states, such as Arizona, the law does not recognize any connection between surface water and groundwater. Integrated resource management is difficult where the integrated nature of the hydrology is not recognized in law.

Surface water has been defined relatively consistently from state to state and administered according to three doctrines: riparian rights, appropriation rights, and permitting.

Generally speaking, riparian rights are found in the eastern and Great Lakes states and appropriation rights in the western states. Permitting is a relatively new innovation and generally applied in conjunction with either riparian or appropriation rights.

Riparian law has been adopted by states with a relative abundance of water and only functions effectively under these conditions. Under the riparian system anyone who owns land adjacent to a body of water has rights to "reasonable use" of that water. Users are prohibited from degrading the quality of the water. Water may not be transferred out of the basin so that downstream users benefit from return flow. All users have equal rights. Water rights are generally transferred with the land title, but can sometimes be sold separately. The method for dealing with scarcity is to require all users to diminish consumption equally, a system that is ineffective without considerable enforcement efforts.

Water rights are less frequently an issue in the well watered states, but localized scarcity is increasingly common. For example, on several occasions the Ipswich River in northeastern Massachusetts has been pumped dry in order to satisfy local domestic demand. Attempts by the Ipswich River Watershed Council to rectify this problem are constrained by state water law.

Increasing demand for water and the resulting scarcity in states applying riparian law has led many to overlay permitting systems. Permits tend to restrict the riparian rights. They allow for the allocation of rights to withdraw defined quantities of water. They can be written to allow for additional restrictions and enforcement in times of emergency. Permitting systems may be the most favorable for watershed initiatives in that they allow for rational and coordinated control of water withdrawals by the state government whose interests and objectives are frequently parallel to those of watershed initiatives.

Historically, in the western states access to water was essential for economic survival. The first settlers were farmers, ranchers, and miners. All of these occupations depend on water (the first miners used a hydraulic technique to mine surface deposits trapped in soil). These settlers adopted a water rights system that gave the earliest settlers the greatest security. Under the appropriation rights system, water rights are established by making a public claim to a specific quantity of water for a "beneficial use." The date of the claim establishes seniority, "first in time, first in right," and senior appropriators are permitted to withdraw their full allotment even if that means no water is left for more junior appropriators.

Several features of the appropriation rights system have important implications for the work of watershed initiatives. Protecting habitat in many cases requires increasing in-stream flows and regulating volume. When watershed initiatives in states operating under appropriation rights systems have this objective, they must purchase the necessary water rights. While in theory appropriation water rights can be sold, in practice this is often very difficult.

A second feature of the appropriation rights system prevents the purchase of water rights even if a watershed initiative has a seller and the transaction is practicable. For example, water under the appropriations doctrine can only be appropriated for an accepted "beneficial use," such as domestic, agriculture, or industrial. Many western states have not expanded their definition to include wildlife/habitat protection or in-stream flow. In these states, the water rights cannot be purchased for ecological uses.

Appropriation rights generally also included two other features that further restrict the options of watershed initiatives in these states. The appropriation system operates under a "use it or lose it" principal, such that appropriators who do not use all of their allotment in a given year will forfeit the unused rights. This, of course, encourages consumption, discourages conservation, and creates even less flexibility for apportioning water for fish and other habitat purposes.

Most appropriation rights states have enacted statutes that modify the appropriation system to allow more state control over which uses the limited supply of water is applied. The statutes define a hierarchy of "preferential uses" to be applied when availability is in doubt or to decide between competing uses. Wildlife and recreation are not included within the hierarchy of many states, and when they are included they are often at the bottom of the list.

Politics

States have a range of political philosophies. An east-west division is again observable, with eastern states favoring greater state intervention and more regulation, while western states, excluding those on the Pacific coast, having less.

Regulation and Authority

Federal

The federal government's substantial authority to regulate water-related activities, founded on constitutionally based mandates to oversee navigability, hydroelectric power generation, defense, disposal, and management of federal lands, has spawned 19 federal agencies that oversee some dimension of water management and are directly relevant to watershed management initiatives. The most important of these agencies are the Corps (navigation, flood control, and, more recently, habitat restoration), the Fish and Wildlife Service (species and habitat protection), the U.S. EPA (water quality), the NRCS (flood control, erosion control, and natural resource protection), the BOR (water supply for drinking and irrigation), the Federal Energy Regulatory Commission (FERC; hydroelectric dam construction and power generation), and the National Oceanic and Atmospheric Administration (coastal zone management). The USGS (hydrologic and data collection and analysis) is an important non-regulatory agency.

In the simplest terms, federal agencies offer watershed planning and management initiatives four things: money, permits, knowledge and data, and interactive technical support. U.S. watershed management activities take place within the context of thousands of pages of regulations generated by these agencies. The authority to accomplish one or more of the objectives of a watershed initiative frequently lies with one of these agencies and or a state counterpart. This extensive federal involvement in regulating water and related resources means that participation of one or more federal agencies is extremely common.

All agencies are not relevant to all watershed initiatives. Some are only important in certain geographic areas or for specific presenting issues. For example, the BOR only manages lands and projects west of the Mississippi, and FERC is only relevant on rivers where there are hydropower installations.

Agencies and watershed initiatives have found cooperation mutually beneficial. Watershed initiatives have turned are efficient vehicles for fulfilling agency missions, and agencies have provided watershed initiatives with mechanisms to implement their plans.

Federal legislation has made significant impacts on the activities of many watershed initiatives.

- **Clean Water Act**
  The centerpiece of federal water quality and aquatic habitat protection, the Clean Water Act presents watershed initiatives with a number of tools and also some potential impediments. The act seeks to restore and maintain the chemical, physical, and biological integrity of the nation's water resources.

Three federal environmental laws have had particularly significant influence on U.S. watershed initiatives: the Endangered Species Act (ESA), the Comprehensive Environmental Response, Compensation, and Liability Act (CERCLA), and the Clean Water Act and its amendments. All have been powerful tools and motivators as well as significant obstacles for watershed initiatives. Total maximum daily loads (TMDLs), an emerging facet of the Clean Water Act, are likely to significantly impact watershed efforts throughout the country.

- **Endangered Species Act**
  Endangered species listings and recovery plans have helped catalyze the formation of local and state watershed efforts, heighten recognition of existing initiatives, and prompt state and federal governments to increase available resources. For example, the Southwest Oregon Salmon Recovery Initiative, motivated by the likelihood of the listing of coho salmon as an endangered species, is organized on a watershed basis and is designed around the use of watershed councils as the "backbone of habitat recovery efforts." It has increased the profile and resources available to watershed initiatives in that state. It has been suggested that the threat or fear that a species will be listed as threatened or endangered is more beneficial to watershed efforts than actual listing. Actual listing, it is argued, would place planning and decision-making activities under federal control and thus act as an impediment to local watershed initiatives. This phenomenon has yet to be well documented.

- **Superfund**
  Fear of Superfund designation under CERCLA has had an influence in some watersheds similar to the threat of listing under the ESA. A series of water quality studies documented the extent of degradation of the Animas River in Colorado by more than 2000 abandoned mines and prompted the formation of the Animas River Stakeholder Group. The principal motivation for the formation of the group was fear that U.S. EPA would declare parts of their region Superfund sites, thereby stigmatizing the region and damaging property values and the potential for growth. The Animus River Watershed also exemplifies the ways in which CERCLA can impede the implementation of watershed remediation and restoration plans. The abandoned mines remain the most significant source of pollution in the watershed, but it has been difficult to implement limited remediation of even the worst sites because of the potential that any entity

engaging in such remediation might acquire liability for environmental impacts and be required to complete remediation of the site under CERCLA.

- **Total Maximum Daily Loads**

  TMDLs, a prominent element of the Clean Water Act in recent years, are having a significant influence on many watershed initiatives throughout the country. Section 303 of the Clean Water Act requires states to write plans for meeting TMDLs of a water body's key pollutants, which will bring the water body in line with state water quality standards. Key pollutants can chemical, biological, physical, or temperature. For practical and political reasons U.S. EPA ignored this requirement until court decisions in 1997 forced U.S. EPA to apply the law. More than 24 states have been forced to write implementation plans, and U.S. EPA intends to require all states to do so.

  The impact of TMDLs on the work of watershed partnerships is still uncertain because they have yet to be widely implemented. On the one hand, TMDLs provide U.S. EPA with a sorely needed tool for regulating all nonpoint source pollution. At a glance, they appear to be a rather crude and indirect tool that prevents U.S. EPA from targeting the most significant nonpoint pollution sources. They can also be viewed as giving states complete flexibility in establishing the methods for reaching targets. On the other hand, TMDLs are in some ways a return to top-down regulation. Moreover, experts disagree as to whether it is technically possible to determine appropriate TMDLs of priority pollutants. Critics charge that data are missing, models flawed, and deadlines too tight. States complain that they will create a large administrative burden and restrict local watershed partnerships by determining the ecological issues they should focus on. The final analysis must wait until TMDLs have been widely implemented for a longer period of time.

State

States have generally created sets of regulatory agencies with parallel functions to federal agencies, and these influence watershed initiative activities in a similar fashion to their federal counterparts. In many cases federal agencies have delegated authority to enforce particular laws to the state agency. State governments have a range of responsibilities over water related resources, including monitoring water quality, issuing and managing water quality permits for point source pollution, addressing nonpoint source pollution, regulating water flows, managing fisheries and wildlife habitats, insuring safe drinking water, and dealing with water quantity allocation issues.

State regulations may provide watershed organizations with additional regulatory tools for pursuing their objectives. For example, some states have taken advantage of their license to enact environmental legislation that is more stringent than the federal government and created regulations that empower them to regulate nonpoint-source pollution. The diversity of agency structures makes it difficult to transfer watershed initiative innovations among states and can hinder coordinated management of a watershed that spans two or more states. Since rivers often form state boundaries, there are many such watersheds in the United States.

Government Programs

Federal Programs

Federal watershed management initiatives fall into two categories: broad-based agency programs and mostly unique watershed management experiments within specific river basins or estuaries.

The primary example of a broad based agency program is the U.S. EPA/USDA Clean Water Initiative/Clean Water Action Plan.

The Clean Water Action Plan is founded on the watershed approach U.S. EPA first articulated in 1991, which outlined an integrated resource management program using a collaborative science-based decision-making process that involved an array of stakeholders. The goal of the plan is the protection and restoration of U.S. water resources. It envisions the project as a cooperative effort including multiple federal agencies, states, tribes, local governments, other stakeholders, and citizens. States have been required to carry out unified watershed assessments and identify restoration priorities and action strategies for their watersheds and set up a series of roundtables throughout the country to discuss regional issues that cut across state lines.

The plan integrates all of U.S. EPA's water-related programs within one conceptual framework and is intended to integrate their databases, although they continue to operate functionally as separate entities. Watershed-related programs explicitly brought together under the rubric of the Clean Water Action Plan include Surf Your Watershed, which provides the public with environmental information about their watershed; Adopt a Watershed, designed to help connect citizens with watershed initiatives in their area; and The Watershed Institute, a program designed to provide federal, state, and local agency personnel and others involved with tools related to watershed management.

Existing USDA programs that promote a watershed approach such as the NRCS's Small Watersheds Program are also integrated into the conceptual framework of the action plan. The NRSC has managed this and related programs within a watershed framework since the 1930s. In developing its natural resource conservation flood and erosion control projects, NRCS employs a sophisticated watershed approach that treats natural resources as part of an integrated ecosystem and works with stakeholders in a collaborative "comprehensive resource management" process to develop mutually acceptable science-based action plans. While NRCS planning often takes place within a watershed, the agency will also adjust the planning unit to reflect practical realities, such as political boundaries or socio-economic divisions, so long as it does not compromise the planning process.

While the constituent programs of the Clean Water Action Plan are substantial, it is unclear whether the plan has actually produced more unified and coherent administration of participating agencies organized around a watershed approach.

Other agencies including the Forest Service, the Corps, and the BOR all have planning processes that incorporate some elements of the watershed management approach, but none do so within a watershed unit framework.

In addition to watershed approaches promoted by federal agencies there are also a modest number of experiments for managing one or more water resource issues within large river basins or water resource regions. These represent examples of coordinated management at large scales that span multiple states. Elements of the modern watershed approach are evident in these federal efforts. Although the literature is by no means unequivocal, these institutions are viewed as relatively successful.

Created in 1933, TVA was the federal government's first experiment with water management within a river basin unit for multiple purposes. The TVA is a federal agency governed by three directors appointed by the president. It has broad authority and independence to construct and operate facilities and manage resources to fulfill its missions of power generation, flood control, reforestation, agricultural and industrial development, supporting national defense, and other unspecified purposes. The TVA had an operating budget of roughly $5.5 billion dollars budget in 2002 is the largest power generator in the country. In order to achieve its missions of flood control and navigation the TVA transformed the Tennessee River and its tributaries into one of the most controlled river systems in the world. As the impact of these changes on water quality, aquatic habitats, and other uses have become apparent, the expectations of the valley's residents have changed. In response, the TVA has launched a watershed-based clean water initiative. While state, county, and other federal agency representatives are involved, the TVA designs, directs, and conducts the scientific assessments.

The TVA exhibits a degree of exclusive federal authority and operational and financial independence that has never been replicated. The trend in the last 40 years, and increasingly so during the last two decades, has been toward more equal partnership with the states, as well as greater emphasis on water quality, ecological protection, and restoration as program objectives. These programs include: the Delaware (1961) and Susquehanna (1971) river basin commissions, Northwest Power Planning Council (1980), the Chesapeake Bay Program (1983), the Great Lakes Program (1987), the National Estuary Program (NEP; 1987), and most recently the CALFED Bay-Delta Program (CALFED; 1995) and Everglades restoration effort (2000).

While all of these programs are important in their particular geographic region and can provide important lessons with regard to particular aspects of the watershed approach, the older programs are essentially remnant and not likely to be replicated. However, the CALFED and NEP may serve as models. CALFED is the product of federal agency initiative and was established without congressional action. The NEP is the only one of these programs that relates to more than one basin or water resource region and, therefore, merits particular attention.

- **CALFED Bay-Delta Program**
  CALFED is a partnership of federal and state natural resource agencies established in 1995, following resolution of long-running disagreements between state and federal agencies on an array of issues arising out of inherent conflicts around water quality, quantity, and endangered species. Established via a memorandum of agreement (MOU) the program has now direct authority to regulate, enforce, and make funding decisions. In practice, participating agencies accept its recommendations and act on its behalf. The program has its own staff and borrows staff from participating agencies. The program seeks to meet multiple objectives: providing good water-quality for all beneficial uses, improving and increasing aquatic and terrestrial habitats, reducing the mismatch between

water supplies and beneficial uses, and reducing damage to both human and natural resources from levee failure. The program has undertaken a comprehensive watershed-based resource assessment and analysis and employs an open planning process, including numerous task forces and advisory committees and hundreds of public meetings, designed to include all relevant stakeholders.

- **National Estuary Program**
  The federal government established the NEP in 1987 to identify, protect, and restore nationally significant estuaries in the United States. Inspired by the success of the Chesapeake Bay Program, NEP was an explicit departure from traditional regulatory approaches to environmental protection and incorporated a number of the elements associated with the modern watershed approach. The NEP focuses on maintaining the estuarine ecosystem as a single unified entity, while also accommodating other non-environmental factors. Each program operates independently and seeks to develop multi-stakeholder participation, including representatives from all levels of government, citizens, business leaders, and others. Programs have had mixed success in galvanizing local support and transforming plans into action. Critics have charged that the federal funding structure, which provides extensive support for research and planning, but little for implementation, has contributed to the lack-luster performance of some programs. The NEP is also noteworthy in that it represents the first time the federal government succeeded in replicating a successful basin management model (the Chesapeake Bay Program), although most NEP estuaries are considerably smaller than the Chesapeake. There have been 28 estuary programs established to date.

State Programs

States experiments in integrating and promoting watershed approaches are wide ranging. A number of states have formally adopted a watershed approach, while others have not. Massachusetts, Washington, and Oregon are on the forefront of refocusing their land and resource management programs within a watershed framework. California, Florida, Maryland, Wisconsin, and Minnesota are also leaders. Doing so requires establishing the watershed as the organizational framework for one or more aspect of water resource management: integrating or coordinating formally independent agency policy, management, and regulatory activities relating to point and nonpoint source pollution control, permitting, fish and wildlife conservation, forest management, parks management, soil conservation, agricultural practice, and transportation planning to name a few. Leading states have provided additional and targeted funding for watershed efforts and have relocated agency staff to watersheds and often assigned them responsibility for leading or supporting watershed partnerships. The applicability of successful models to other states is partly dependent on the underlying legal and regulatory framework, as well as climate and political conditions, but individual elements are likely to have broad transferability.

Massachusetts is regarded as having made the most significant institutional changes and introduced a number of innovations. The state began with a somewhat advantageous institutional structure in which all state environmental agencies, including the Departments of Environmental Protection, Environmental Management, and Fisheries and Wildlife, were overseen by a single entity, the Executive Office of Environmental Affairs (EOEA). It has proceeded to restructure the EOEA and its departments to better accommodate watershed-based management. The state created an Office of Watershed Management, which coordinates the Massachusetts Watershed Initiative and the activities of basin teams within the state's 27 USGS cataloging unit scale watersheds. Basin teams are led by full-time team leaders and include federal and state agency personnel, municipal representatives, businesses, local watershed association members, and other relevant stakeholders. The teams strive to develop broad-based consensus decisions regarding basin planning, implementation, and assessment. They also foster local support and capacity for watershed management. The basin management process works on a five year planning cycle with the first three years concentrating on developing a strong watershed partnership and conducting monitoring and assessment. This culminates in year four with the preparation of a watershed action plan, followed by implementation and evaluation of activities. The process synchronizes five previously independent water quality management activities: water quality monitoring and assessment, water withdrawal permitting, nonpoint source pollution control, awarding of water quality related grants, and NPDES wastewater permitting. The role of the state has been defined as facilitation of local problem identification and coordination of implementation activities. Implementation has been enhanced by the creation of an interagency roundtable where senior agency managers set priorities and coordinate resource allocation with input from other watershed stakeholders.

Massachusetts has used grant and similar funding awards criteria in an innovative way to promote the overall watershed approach. Criteria are weighted to heavily favor projects defined on a watershed basis or that advance the state's watershed objectives. When applied to the approximately $200 million SRF that supplies 0% loans, communities became extremely interested in the watershed approach.

Note that despite its success, in 2003 the Massachusetts state budget crisis resulted in the abandonment of the formal watershed initiative. While many involved on the basin teams continue to work together informally, the coordinated state structure and support is no longer in place.

Washington's Department of Ecology (DOE) adopted a Watershed Approach to Water Quality Management in the mid-1990s. Simultaneously, the Department of Natural Resources in Washington began a separate watershed effort. DOE's program cites "lessons learned" that support the assumptions that program design such as "building relationships with partners is essential to nonpoint progress" and "water teams are key to obtaining comprehensive information." A newer statewide initiative enacted an optional process through which local governments and water utilities can choose to come together to form an entity for watershed-wide planning and protection. The local planning groups have a prescribed multi-stakeholder membership. DOE provides funding of up to $75,000 for initial organizing, up to $200,000 for conducting watershed assessments, and up to $250,000 for drawing up a watershed plan. This is a watershed approach in which local governments have more of a leadership role than the state environmental agency. The legislation has given these local entities the power to direct state agencies with regard to state lands by requiring the agencies to adapt rules, regulations and policies for managing state land so that they follow approved local plans. Critics of the Washington legislation charge that it focuses excessively on water quantity issues and does not sufficiently integrate water quality and habitat protection.

Local Programs

Watershed initiatives started at the local level are the most common and exhibit the greatest diversity of size, composition, motivation, objectives, and actions. They are foremost collaborative efforts that are most successful when broad stakeholder participation and effective cross-jurisdictional coordination is achieved. They are considered essential for ultimate success of all watershed management efforts primarily because it is only at the local level, through changes in behavior, land use and development, that issues such as nonpoint source pollution can be addressed.

B.3.3 Characteristics and Key Issues of Watershed Initiatives

B.3.3.1 Motivation

Some watershed efforts have arisen out of crisis or conflict. For example, fish kills and contamination of drinking water wells prompted the county in Long Creek, North Carolina, to organize a citizens' commission. The Nisqually River Watershed protection efforts were sparked by litigation between a municipal wastewater treatment system and the Nisqually tribe about water withdrawals and fishing of endangered species. Others efforts have grown from fear of government intervention, especially via CERCLA or the ESA, as in the case of the Animus River Stakeholders Group and the Oregon Plan for Watersheds and Salmon, respectively. Still others were motivated by the desire to protect a valuable natural resource such as in the Tomorrow-Waupaca watershed in Wisconsin where a local member of Trout Unlimited concerned about erosion and sedimentation in a fishing stream along her property organized a citizens group. Efforts in states such as Massachusetts, Washington, and Oregon have been encouraged by state agencies; others like the Panoche-Silver Creek Watershed in California, by sedimentation and chronic flooding. Others were interested in source water protection like the Eugene Water and Electric Board, which helped launch and fund the McKenzie Watershed Council to protect water consumed by its 200,000 customers. These concerns usually remain a central focus of watershed initiatives, and they almost always include protecting environmental health or environmental restoration.

B.3.3.2 Structure and Scope

Watershed initiatives vary in structure and scope as might be expected given their diverse locations; issues of concern; and historical, legal, and regulatory contexts. For example, the Dungeness River Management Team in Washington comprises a diverse group of 13 citizen and agency stakeholders that share information on water related issues such as salmon restoration, water quality, groundwater and surface water flows, and development and private property rights. This differs markedly from the government-coordinated efforts of Tahoe Regional Planning Agency, which was formed via an interstate compact between California and Nevada and has broad regulatory authority. Paid agency staff coordinate a process that combines input from an Advisory Planning Commission, and collaborative ad hoc and technical working groups to generate and periodically revise a comprehensive regional watershed plan. Many initiatives are coordinated by nonprofit watershed organizations, sometimes with large membership bases. The Charles River Watershed Association in Massachusetts is one such organization. It uses scientific data and modeling to convince local government and industry to control non-point source pollution, conserve water, and restore groundwater.

B.3.3.3 Developing the Watershed Management Plan

A collaborative, consensus-based planning process is the heart of the watershed management process. Stakeholder interests can come into conflict and competing uses must be balanced. Stakeholders enter watershed initiatives with different agendas and, not infrequently, a history of conflict. This is especially true when allocation of water supply is a presenting issue.

Some inherent features of the process are reported to help diffuse tensions and establish an environment where participants can craft the necessary compromises. The common practice of envisioning and setting broad goals at the outset (e.g., return trout to the river, foster better stewardship, prevent resource degradation, ensure sustainability) is a common feature. It allows participants to recognize their shared goals and understand their respective agendas in a new context. Data gathering is an almost universally necessary activity, whether it is formally included in the vision and broad goals adopted by an initiative. Data gathering sometimes can be a task in which stakeholders work together, and may help build personal relationships and a measure of trust, before they enter into the plan development stage, where agendas and conflicts are likely to resurface.

A number of other elements have been associated with watershed initiatives that operate effective collaborative processes: perception of a crisis situation, inclusion of stakeholders, clear rules and formal structure, and communication. Processes seem to work best when the situation is perceived as a crisis, as in the case of the Chesapeake Bay Program where stocks of commercial fish species collapsed. In a crisis, participants agree some action must be taken and that the outcome of retaining the status quo is certain and unacceptable. Inclusion of all stakeholders, in particular those with veto power, is essential and is discussed below.

Rules and formal structures facilitate efficient operation and help establish and sustain trust. Clear rules of order help maintain stakeholder confidence in the equity of the process and enable the process to work smoothly. Formal committees and subcommittees and transparent processes act similarly to maintain a perception of a level playing field, avoid confusion and misunderstanding, and provide for efficient distribution of labor.

Explicit strategies to maintain communication with the organization each stakeholder represents are also reported to be important. Such strategies ensure that the host organization is party to the intermediate decisions and incremental steps in the watershed planning process and is prepared for the final decisions when they come. If communication is not maintained, an organization may reject a plan or decision late in the process, which can kill or cripple the watershed initiative's efforts. Such strategies also ensure that participants accurately represent the position of their respective organizations.

Time is also important. Successful watershed planning processes commonly allow sufficient time to build trust and mutual understanding. As noted above, certain features of the overall watershed management effort help provide the necessary time. Collaborative decision-making is not a process to be rushed.

Management plans developed by watershed initiatives are of variable quality. Some literature suggests that extended planning processes drain momentum and stall the watershed initiative. However, the one rigorous scientific assessment of the effectiveness of watershed initiatives found that better planning processes and more carefully crafted plans correlated with the implementation of more ecologically effective projects. The optimal amount of planning varies with scale and local conditions, but establishing and managing an effective collaborative process appears to be important for successful implementation efforts and indeed reaching the implementation stage in the first place.

B.3.3.4 Implementation Activities

Implementing watershed management plans generally involves stakeholders working to create the on the ground changes identified in their plan and monitoring the impact of those steps. These watershed protection activities often entail a combination of projects: implementing best management practices (BMPs) to filter pesticide and nutrient pollution and reduce sedimentation, conducting public education campaigns such as storm drain stenciling to reduce illegal dumping, purchasing conservation easements or land with critical habitat value, amending local land-use codes to protect headwater catchments from development, and adopting of local guidelines or restrictions on construction, sewage disposal, fertilizer use, blasting, and others. Each watershed initiative needs a unique combination of activities to achieve its ends. However, since many problems are common to multiple watersheds, such as sedimentation or nonpoint source pollution from stormwater runoff, solutions developed in one watershed are often transferable to another. Such sharing occurs, but not always in a systematic and timely fashion.

B.3.3.5 Stakeholder Participation

One of the defining characteristics of the modern watershed approach that is often cited as an essential ingredient for success is broad stakeholder participation. The number of stakeholder groups actively participating in decision making has practical limitations, although watershed initiatives with 50 partner entities, such as the Bronx River Working Group in New York or even the 150 that support activities in the Illinois River watershed, are not unheard of.

Stakeholders vary with project phase, location, and issues. Three quarters of western watershed initiatives surveyed included one or more federal agency representatives; 85% of the time that representative was a member of the NRCS. The local watershed movement is generally considered a synergy of bottom-up grassroots citizen and environmental activists and top-down government agencies, but the survey of western watersheds made the somewhat startling discovery that 40% of groups contained no academic or citizen members and nearly half had no environmental groups as members.

All recent reviews of U.S. watershed management activities highlight that federal and state agency involvement is critical to the success of local watershed initiatives. Their participation is often essential as they are frequently the only or best source of four things: money, permits, knowledge and data, and interactive technical support for watershed initiatives. Federal and state government participation is often explained as a form of enlightened self-interest, in that federal and state agencies have discovered that watershed initiatives offer efficient means of fulfilling their mandates.

Achieving broad stakeholder participation that is perceived as meaningful by participants is a challenging task for large watersheds. The CALFED process, for example, has used a large number of subcommittees and has held countless public meetings. Though resource intensive, it is considered by many to be working well.

B.3.3.6 Cross-jurisdictional Coordination

Given that lack of cross-jurisdictional coordination is one of the main reasons for the absence of effective water and natural resource management in the United States, it is perhaps not surprising that the success of watershed initiatives is at least partly related to how well they achieve cross-jurisdictional coordination. Indeed, the planning and decision-making process of the modern watershed approach addresses this explicitly. Initiatives not only need a range of government agencies to coordinate their objectives, but these agencies are frequently the main agents for implementing agreed actions.

The NRCS as well as state and local soil and water conservation districts, in particular, have been instrumental in facilitating the coordination of all levels of government and multiple federal agencies in local watershed initiatives. Nearly every square mile of U.S. territory was organized into a national, state, or local soil conservation district in the years following the dust bowl. NRCS and soil conservation districts have long histories in their regions and have well developed relationships with local and state authorities, other agencies with related missions operating in the district, and local farmers and land owners.

Cross-jurisdictional cooperation is a challenge at all scales and becomes more complex at larger scales. Larger areas necessarily contain more units of local government. The larger scale river basins encompass multiple states and tribes, often with long histories of conflict over water rights and other water resource issues.

Exluding the obvious need to attract representatives of all relevant jurisdictions early in the planning process, one of the keys to cross-jurisdictional coordination at any scale is early agreement on shared general goals and specific, measurable objectives that express these goals. This is believed to be one of the reasons for the success of the CALFED project. Until 1996, federal and state agencies were not managing the Bay-Delta to any reasonable extent because the agencies disagreed over a number of issues. Agreement in 1996 on a particuar set of water quality standards led the agencies to establish the inherrently cross-jurisdictional CALFED process through which they could resolve differences over other issues.

B.3.3.7 Funding and Technical Support

Application of the modern watershed approach requires a variety of resources, administrative services, office space and equipment, data gathering personnel and materials, scientific and technical analysis, and funding for restoration or other interventions. Watershed initiatives have arrived at a variety of creative solutions for meeting these needs that often combine federal, state, and local agency technical resources. Funding and support often comes from government, in-kind donations, volunteers, corporate and foundation grants, and water and wastewater utilities, to name a few. Watershed initiatives tend to have small staffs and rely heavily on state and federal agencies for technical expertise and support. Federal funding is an important resource for watershed initiatives at all scales, and very few of even the more mature western watershed initiatives operate without some federal support. Nonetheless, federal funding opportunities are generally seen as insufficient and too restricted. While comprehensive data for the states is lacking, the literature indicates that this is also true even in the more advanced states. Many watershed initiatives report that obtaining support for basic operations, overhead, and staff is a particular challenge for younger, community-based watershed efforts.

Watershed initiatives operating in small- and medium-sized regions appear to be relatively lean organizations financially. According to the recent study of western watershed efforts, approximately 27% of surveyed groups operated without any direct funding, 15% with annual budgets of less than $20,000, 26% with budgets between $20,000 and $75,000 and 32% with budgets of more than $75,000. These numbers do not perfectly reflect the financial resources supporting the efforts since respondents reportedly had difficulty accounting for project funding that did not flow through the watershed group.

Federal funding applicable to watershed protection and enhancement activities comes from a relatively limited number of sources: the U.S. EPA's 319 Nonpoint Source Pollution Control Grants program; the Environmental Quality Incentives Program; the Conservation Reserve Program administered by the Farm Service Agency; and the Conservation Reserve Enhancement and Wetland Reserve Programs administered by the NRCS. Approximately $200 million was available through U.S. EPA's 319 Nonpoint Source Pollution Control Grants program in 2000, a five fold increase over 1990 levels, but still at an average of less than $4 million per state, far short of the need.

Most remaining federal funding targets the agricultural sector. The Environmental Quality Incentives Program, administered by the NRCS, directs an additional $200 million for technical, educational, and financial assistance to eligible farmers, ranchers, and foresters to address soil, water, and related natural resource concerns on their lands. In 2000, the federal government spent $1.6 billion on the Conservation Reserve Program, which encourages farmers, ranchers, and foresters to retire highly erodible or environmentally sensitive cropland for 10 years, and provides incentives to leave untilled buffer zones along creeks and ditches. Approximately 3.2 million acres of the statutory limit of the 34.6 million acres permitted to be enrolled in the Conservation Reserve Program have been set aside for the Conservation Reserve Enhancement Program operated by the NRCS. This state-federal partnership offers higher levels of compensation to farmers, ranchers, and foresters in specific geographic areas to address water quality, soil erosion, and wildlife habitat issues related to agricultural use in exchange for permanent or temporary conservation easements. The Wetland Reserve Program is similar in structure to the Conservation Reserve Enhancement Program but focuses on restoring wetlands converted before 1985. It was funded at $165 million dollars in fiscal year 2000.

Specific data regarding state funding were not readily available. The literature suggests that state funds are also an important source for watershed initiatives. However, researchers report that they too are insufficient even in states with strong commitments to supporting watershed-based management activities.

Although there is general agreement that the current level of financial resources available for watershed initiatives is insufficient, some critics charge that the available funds are not being allocated in ways that would most efficiently meet the resource protection and restoration goals of most watershed efforts. The recent rigorous study of the effectiveness of watershed initiatives in the Pacific Northwest noted that while the greatest gains in environmental protection and restoration could be achieved in rural watersheds, urban watershed initiatives were better organized and received a disproportionately large share of the funding. Reviewers of the Colorado Basin Salinity Control Project have pointed to its competitive bid process as a useful model. In the bid process, funding goes to those proposals with the greatest projected salinity reduction per dollar. To avoid duplication and lack of coordination, the Northwest Power Planning Council requires agencies to coordinate their budget submittals and seek public comment before approaching congress, an innovative solution that has proven effective and worth replicating.

B.3.4 Common Challenges

All watershed initiatives face common challenges with regard to pollution management, habitat protection and restoration, and many of other aspects of watershed assessment and planning. In some cases, promising solutions are emerging, while others look likely to persist for some time.

B.3.4.1 Pollution Management

Pollution management issues include integrating point source permitting programs into watershed initiatives, overcoming the historic segregation of point and nonpoint source permitting, and addressing the inherent complexities of nonpoint source pollution. These challenges occur in the shadow of newly prominent TMDL requirements discussed above. Several interesting tools have emerged over the last decade that might assist watershed initiatives in addressing these challenges.

Much of U.S. EPA and most state environmental regulatory agencies devote a significant portion of their resources and energy to issuing and enforcing point source discharge permits, such as the National Pollution Discharge Elimination System (NPDES). In many states, these are large bureaucracies with significant institutional inertia and are not well integrated into watershed planning. Massachusetts is a noteworthy exception. Under the state's Watershed Initiative, NPDES and other permits within a basin have synchronized within the initiative's five-year cycle. Massachusetts' state-wide watershed framework and coordinated planning process, including state-funded basin team leaders, have greatly facilitated the integration of pollution management with other watershed planning and protection efforts.

The TMDL regulations have the potential to assist watershed initiatives in overcoming two of their biggest pollutant management challenges: integration of point source permitting programs in broader watershed initiatives and integration of point and nonpoint source pollution management. Because TMDLs look at the daily load conditions of the waterbody and consider point and nonpoint sources, their implementation virtually requires permitting programs to coordinate with nonpoint source programs. Watershed initiatives are an excellent vehicle for facilitating this coordination and arguably offer the most viable means of actually implementing TMDL plans once developed. However, the focus on meeting TMDL requirements also has the potential to divert attention away from identified priorities.

While watershed initiatives have shown themselves capable of addressing an array of problems, including nonpoint source pollution, habitat degradation, sedimentation, elevated nutrient levels, maintenance of in-stream flows, and reducing agricultural runoff, others have proved more elusive. In particular, the change in hydrodynamics resulting from urban sprawl and increased impervious surface areas continues to be a difficult challenge in many watersheds and reinforces the importance of integrating land-use planning into watershed management efforts. Land-use planning and zoning authority resides at the local municipal level in most states, and zoning changes are notoriously contentious and difficult to enact.

Substantial progress has been made by some watershed initiatives where regional entities have authority for land-use decision making. For example, the Tahoe Regional Planning Agency has used its broad regulatory authority to adopt regional environmental standards, issue permits, and take enforcement actions. It has developed a number of useful tools for addressing development, including an individual parcel evaluation system (IPES), water quality offset requirements for new developments, and a system of transferable development rights (TDRs). The IPES ranks all residential lots in the basin in terms of environmental sensitivity against a number of criteria. Lots with high scores are eligible for building permits. Those with low scores are not. Water quality offset requirements require that all new development projects offset any water quality impacts by 150% with water quality improvements. The first two programs, while unusual and effective, still follow a heavily regulatory model. This is not the case for TDRs. TDRs are part of a system of allocations, transferable development rights and coverage transfers that provide economic incentives and flexibility to encourage the removal of impermeable land cover and development from environmentally sensitive parcels to less sensitive areas.

Market or quasi market-based approaches for addressing water pollution control are relatively rare. Under an effluent trading system a pollution source such as a wastewater treatment plant can sell or trade pollution reduction credits to another source for which an equal reduction in pollution would be more difficult or costly. The Colorado Salinity Control Program's competitive bid system is one example. Tahoe's TDR is another. U.S. EPA has been promoting effluent-trading experiments, and the TMDL process may provide an opportunity for the introduction of effluent trading systems. For example a WWTP might pay a local dairy farmer to build filter strips along ditches and streams to reduce nitrogen, phosphorous, and coliform bacteria loading.

Trading systems have proven efficient at producing cost-effective yet flexible pollution control, particularly for air pollution. However, pollutants generated by nonpoint sources are harder to measure or estimate and it is not at all clear how credits for some items, like sediment, might be distributed. Some pollutants can also be expensive to measure on a routine basis, which may be necessary to ensure the validity of the system. Nonetheless, effluent trading systems may offer watershed initiatives an alternative to the arduous task of promoting voluntary adoption of best management practices and are worthy of consideration.

B.3.4.2 Habitat Protection and Restoration

Habitat protection and the ecological dimension is a fundamental characteristic of most watershed initiatives. It is often expressed in terms of concrete goals for species protection or restoration. Other objectives, such as controlling nonpoint source pollution to meet drinking water standards or erosion control measures to prevent physical deterioration of the landscape, are often necessary to improve physical and chemical conditions in the local aquatic habitat.

Although ignored before the 1960s, habitat protection and restoration has become a major focus of water resource management since the 1980s. Mounting evidence and an accompanying increase in public concern that natural resources are becoming significantly degraded has driven this to prominence. Research illustrates the considerable value of ecological services provided by healthy natural ecosystems and river systems, including filtering of nutrients, pesticides, pathogens, and other pollutants by wetlands; protection of drinking water supplies; nursery and spawning grounds for fish; and flood control.

Many if not most of the common activities watershed initiatives engage in relate to habitat protection (e.g., preserving in-stream flows to keep water temperature within the optimal range for fish and other aquatic organisms, revegetating stream banks to provide shade and increase plant and animal species diversity and abundance, obtaining conservation easements to protect habitat from development and degradation, and restoring wetlands to replace this integral part of aquatic ecosystems and buffer the stream from nonpoint source pollution).

B.3.5 Data and Tools

Science based decision making is one of the pillars of the watershed approach. Data gathering and scientific assessments of the watershed are usually among the first projects U.S. watershed efforts pursue, and they remain central activities. Watershed initiatives use scientific data and tools to:

- Develop an accurate picture of the nature and health of the watershed's ecosystem and hydrology and how both are influenced by human activities;
- Devise conservation and restoration plans; and
- Monitor ongoing changes and the impacts of the initiative's actions in order to practice adaptive management.

Watershed initiatives use one or more of four basic sources of information: watershed-scale inventory and assessment, chemical/biological/hydrological monitoring, computer simulation models, and socioeconomic surveys/analyses with monitoring data.

B.3.5.1 Reliable Monitoring Data

Science-based decision making requires reliable, complete, long-term data to understand the dynamics of the watershed, assess environmental health, develop plausible management plans, and evaluate the impact of actions taken. Unfortunately, reliable, representative monitoring data for the range of relevant biological, chemical, and physical parameters, as well as hydrological and ecological conditions, do not exist in most cases. Many watershed efforts must make ecological management decisions with a limited understanding of the history, nature, and dynamics of their watershed and the impact of human activities.

Despite 30 years of significant investment, U.S. water quality data remain incomplete, fragmented, and often not useful to watershed managers. The focus of most monitoring efforts historically was on effluent quality from point sources. Moreover, different agencies have targeted different parameters and employed different analytical techniques, some of which are no longer considered acceptable. Poor quality control procedures are often an issue. To make matters worse, much of the data that do exist are stored in inaccessible formats.

Fortunately, these defects have been recognized and a national-level effort is underway to develop uniform sampling and analytical protocols, improve field and laboratory quality control procedures, and make data available in standardized formats.

The best sets of high-quality, long-term physical data available to watershed initiatives are hydrologic data from gauging stations, especially those of the USGS. In contrast to the improvement in water quality data, the hydrologic data set is becoming degraded as tighter state and federal budgets have led to the elimination of a large number of gauging stations.

Data on aquatic and riparian habitat, such as stream morphology and stream bank species composition and diversity, vary from watershed to watershed, but in general are sparse. Such data, however, are extremely important for understanding the ecology of the watershed. Some of this data, such as vegetative cover, can be obtained at a relatively crude level via remote sensing, but to construct a detailed picture of a watershed, most data must be collected on the ground in a labor intensive fashion.

Many watershed initiatives have transformed their water quality and habitat data gaps into a means for engaging and involving citizens and volunteers, providing an opportunity to educate children, students, and the general public about the concept of the watershed and the work of the watershed initiative. Common activities include periodic water quality monitoring, transect surveys of the riparian corridor, riffle pool sequence measurements, shellfish bed surveys, and others. The result is that volunteers often gain a greater appreciation for and a personal connection to the natural life of the watershed.

B.3.5.2 Data Accessibility

Historically, the accessibility of data has been problematic. Attempts are underway to remedy this and other defects in national water resource data gathering. The advent of the Internet has provided unimagined opportunities for sharing and accessing data. U.S. EPA, USGS, and the Conservation Technology Information Center (CTIC), a non-profit data and technology information transfer center, have partnered to create the Watershed Information Network designed to educate and inform the public and to assist watershed initiatives, especially new ones. U.S. EPA's portion of this effort is Surf Your Watershed, described as "a service to help you locate, use, and share environmental information about your place." Through this portal visitors can locate their watershed and watershed groups in their area. Watershed initiatives looking for spatial data to use in their Geographic Information Systems (GIS) modeling can access a number of spatial data clearinghouses, such as the Pennsylvania Spatial Data Access System, part of the National Spatial Data Infrastructure effort, which can direct users to other sources of watershed and related data.

B.3.5.3 Scientific Models and Decision-support Tools

GIS are fast becoming ubiquitous tools for the display, modeling, analysis, and interpretation of watershed-related data. GIS software enables users to integrate information within a geographic framework by linking computerized databases to geographic coordinates. GIS maps can combine different layers of data, such as water bodies, wetlands, political boundaries, USGS watershed unit boundaries, topography, dissolved oxygen levels, flow rates, and census data. GIS capabilities and databases significantly enhance watershed research and planning efforts. This is especially true as dynamic computer hydrological and other models have been integrated, enabling the water transport and flow to be simulated and viewed visually.

The rapid pace of adoption and utilization of GIS by watershed initiatives is being outstripped by the growth rate of the complex natural resource simulation models linked to it.

These models have so many adjustable interacting parameters that making a decision is difficult without a systematic method for evaluating the outputs. The National Research Council sees an urgent need to facilitate the development of decision support system software. It envisions a type of expert system that addresses: 1) problem identification; 2) selection of decision criteria; 3) selection of feasible management systems or design alternatives; 4) evaluation of the alternatives by simulation models and/or historic data; and 5) recommendations for a decision.

B.3.6 Effectiveness of the Watershed Approach

Whether watershed initiatives are effective at creating more integrated resource management, implementing plans, and positively impacting ecological resources is of central importance and has not been rigorously assessed. There are many qualitative studies, but only one was identified that uses a rigorous scientific methodology to address some of these questions (Huntington and Sommarstrom, 2000). Watershed initiatives can also be evaluated along other dimensions, such as the degree to which they are able to establish a smoothly functioning multi-stakeholder effort, work to resolve conflicts, or develop a broadly accepted management plan.

Huntington and Sommarstrom independently evaluated the effectiveness of ecological restoration activities and organizational attributes of a random sample of 14 northwestern watershed councils according to strict standardized metrics and criteria. All councils were primarily focused on coho salmon restoration. While emphasizing that the watershed councils fill an otherwise empty niche and their activities generally lead to an improvement over the status quo, the researchers found that many councils:

- Do not consistently plan, design, and implement projects with high ecological restoration values;
- Could not address some of the major ecological stressors such as water allocation, urbanization, and logging, which are only appropriately dealt with at a much larger scale;
- Did not focus on projects that should have been highest priority, for technical and other reasons, including insufficient funding; and
- Had no mechanism for overcoming strong stakeholder opposition.

Huntington and Sommarstrom's conclusion is that government and public expectations that watershed initiatives can effectively address most U.S. water resource management problems are far too high. The narrow geographic focus of this study and the uniform ecological objective of the councils may limit the degree to which the conclusions can be generalized, but the findings are significantly at odds with the dominant view of the potency of the watershed approach.

There is at least anecdotal evidence that contradicts Huntington and Sommarstrom's overall conclusion. Within the U.S. watershed experience there are many examples that illustrate good planning, design, and implementation of projects that have resulted in high restoration value:

- Illinois River Watershed: After restoration activities fish species not seen in the river since 1908 have returned.
- Bigalk Creek, Iowa: Watershed restoration activities led to a six-fold increase in rainbow trout populations since 1992.

- Big Darby Creek, Ohio: A single project has reduced erosion and prevented 400,000 tons of soil from washing into the creek.
- Tomorrow-Waupaca Watershed: Phosphorous and sediment loading to streams has dropped by 21% and 12%, respectively.
- Colorado Salinity Control Project: Delivered the required amount of water to Mexico within the salinity limits established by treaty.
- Tampa Bay National Estuary Project: Facilitated an agreement of local governments to implement 105 projects that are predicted to reduce nutrient loadings, surpass sea grass restoration goals, and restore 1600 ac of habitat.

This anecdotal evidence suggests that Huntington and Sommarstrom's overall conclusion may be excessively pessimistic. Nonetheless, the problems they have identified may be remedied by improved planning, design, and implementation and through technical assistance and application of decision support tools by watershed initiatives.

The absence of other rigorous assessments of the effectiveness of watershed initiatives is an important deficit. Poor assessment of individual watershed initiatives is also a problem. In general, watershed initiatives do not measure and assess their own outcomes and, therefore, cannot apply adaptive management with the level of sophistication that would allow them to improve effectiveness.

B.3.7 Overall Characterization of the U.S. Watershed Experience

The experience described above highlights a number of important findings about efforts to implement a watershed approach in the United States.

- First and foremost, there is growing recognition of the value of a watershed approach, and efforts to apply it are growing at all levels (local, state, regional, and federal). Moreover, this recognition is shared by a wide spectrum of stakeholders, including citizens, government agencies, NGOs, and businesses.
- Lessons, both positive and negative, are not universal. That is, approaches are related to geographic, climactic, legal/regulatory, and political context and, therefore, to some extent are place-specific. The challenge is to identify the key conditions for success and under what circumstances certain approaches or lessons can be transferred and adapted to other watersheds. Various elements seem necessary but not sufficient for successful watershed initiatives. A focused effort and political will seem critical, regardless of other factors.
- Cross-jurisdictional cooperation is virtually always necessary for success, whether it is just among local agencies (municipal planning departments, conservation commissions, and water/wastewater utilities) or many levels of government (municipal, county, state, and federal).
- Early and meaningful multi-stakeholder involvement is important to ensure key perspectives are represented throughout the process and to minimize surprises (and opposition) as processes move towards decision making and project implementation. Time invested up-front in broad involvement will often be saved later in avoided controversy and/or litigation.

- Benefits of an integrated, more comprehensive approach include consideration of a range of issues, not just the immediate presenting issue. This facilitates longer-term success and helps to avoid unforeseen problems.
- It is important to base decisions on good data that is accessible to all stakeholders and tools that are appropriate to the issues and transparent. This helps avoid conflict and provides a shared basis of knowledge and methods for decision making.
- Certain federal regulations are having a profound effect on watershed initiatives, particularly the ESA and TMDLs. The threat of ESA listing is an important driving force for watershed activities in the Pacific Northwest and elsewhere and has focused attention on and increased support for watershed initiatives. TMDLs are an emerging element of the Clean Water Act that have the potential to force greater integration of point and nonpoint source pollution control and better integration of both into watershed initiatives.
- The diversity of institutional structures within watershed initiatives makes it difficult to generalize the relative effectiveness of such structures. However, scale does appear to be important in that larger watershed initiatives tend to have more of a cross-jurisdictional interagency structure and less direct citizen stakeholder participation. Smaller-scale watershed initiatives tend to have much greater citizen participation and be more directly community based. Watershed initiatives in the arid western states are necessarily larger because the watersheds are larger, often thousands of square-miles, whereas watersheds in the wetter eastern states tend to cover smaller areas.
- Independent of scale, certain elements of institutional process and composition are identified as important for effective institutional operation. These include participation of all appropriate stakeholders, including local, state, and federal agencies, to enable cross-jurisdictional coordination and to broker technical assistance and the establishment of formal organizational structures and clear rules for conducting discussion and debate.
- The federal government and many states are developing promising programs to address obstacles to effective implementation of the watershed approach and to support local initiatives. These programs often share several of the following characteristics: formal adoption of a watershed approach, application of agency policy and planning activities according to watershed boundaries, facilitation of interagency coordination and collaboration, and provision of additional funding and technical support for local watershed agencies.
- The Clean Water Action Plan, the multi-agency collaboration to coordinate policy and support implementation of a watershed approach to water resource management, is an attempt to remedy shortcomings in federal policy without significant agency restructuring or legislative reform. The widespread collaboration of different subsets of federal agencies with water resource management responsibilities in hundreds of local watershed initiatives suggests that these agencies are making progress towards overcoming historical conflicts over jurisdictional issues.
- Some state programs have gone farther than the federal government and have established multi-stakeholder, cross-jurisdictional, basin-level initiatives. Certain states coordinate watershed planning activities and support watershed initiatives at smaller scales (Massachusetts), provide new permanent funding sources (Oregon), or provide considerable opportunity for local watershed initiatives to impact state agency watershed activities (Washington).

- Availability of financial and human resources is important for U.S. watershed initiatives. Most operate on a combination of government and private funding and in-kind support from a variety of sources, including federal, state, and local government; businesses, foundations; utilities; and membership fees. Unrestricted funding for staff and overhead expenses is the most difficult to obtain and has impeded organizational growth and continuity in some cases. The inflexibility of federal funding, which many watershed initiatives rely on to some degree, has been identified as limiting funding for core operational expenses. Watershed organizations maintain relatively small staffs and rely heavily on state and federal agencies for technical assistance. Citizen volunteers are frequently important for data gathering and monitoring activities, as well as the implementation of certain types of restoration projects.
- Powerful scientific assessment and planning tools such as GIS are rapidly being developed, disseminated, and adopted by watershed initiatives. Scientific understanding of some watershed processes, such as hydrology, is fairly well understood. Other processes, such as ecology and how it interacts with and is impacted by human activities, are less well understood. Complex modeling tools are evolving rapidly, and watershed planning experts have identified the need for development of decision support systems to use them optimally.
- Scientifically rigorous data on the effectiveness of watershed initiatives is limited. Anecdotal evidence suggests that watershed initiatives are successfully addressing issues of conventional pollution (nutrients and pathogens), sedimentation, and restoration of vegetation in the riparian corridor. However, it appears that watershed approaches have been less effective in dealing with nonpoint source pollution and increasing impervious surfaces within watersheds. The latter reflects in part a more general failure to influence land-use planning, especially urban sprawl and building practices.

B.3.8 Weaknesses or Gaps in U.S. Watershed Initiatives

B.3.8.1 Fragmentation of Authority

While many U.S. watershed initiatives have successfully implemented cross-jurisdictional coordination, fragmentation of authority remains a significant impediment to efficient watershed management. In spite of attempts at systematic coordination, federal and state agencies (water quality regulators, wildlife, public health, land management, forestry, or agriculture agencies) often operate independently to fulfill their missions, which at times may be at cross-purposes. At the local level, drinking and wastewater utilities are often separate from one another, and from parks, planning, public health, and conservation commissions. Moreover, such agencies often operate with very limited budgets and over-extended staff and are not provided the direction or supporting framework to coordinate their activities.

B.3.8.2 Dearth of Watershed Initiatives at the Largest Scale

The absence of large-scale watershed management initiatives in most of the nation's river basins is a serious deficiency. These scales (approximately USGS accounting region and USGS resource region) are appropriate for managing key water quality problems such as sedimentation, salinity, or nutrient loading, and many water allocation problems can only be effectively managed at this scale. In regions where these are the principal problems and there is no large-scale river-basin initiative, the impacts of local watershed efforts on ecological conditions have been extremely limited.

Where large-scale initiatives do exist, such as in the Chesapeake Bay Watershed, they face additional challenges not faced by smaller scale watersheds, including:

- **Effectively involving the full spectrum of stakeholders.** The number of stakeholders affected by the decisions of large-scale watershed initiatives is extremely large, far beyond the optimal number for a collaborative process. Yet the interests and opinions of these stakeholders must be meaningfully included and fairly represented in planning and implementation activities. The relatively few very large watershed initiatives currently operating have developed a variety of solutions, but they have not yet been fully tested.
- **Inequity in cost and benefits between stakeholders in different parts of the watershed.** In larger watersheds it is much more common that the management decisions that provide benefit for all or part of the watershed do not distribute the costs equitably. This imbalance is often seen between headwater and downstream communities. Expensive upstream measures that only marginally change water quality or quantity may be needed to prevent cumulative downstream impacts. Also, larger scale projects generally do not enjoy the same inherent sense of community engendered by smaller scale efforts, making voluntary sacrifices that much more difficult. Outside resources sufficient to compensate a community for bearing an unequal potion of the costs is one potential solution to this problem.

B.3.8.3 Funding Challenges

Less than $1 billion in federal funding ($20 million per state), is allocated for grant programs specifically designed to assist watershed initiatives. Much more money indirectly funds programs through federal agency efforts, but this is still far short of the resource needs of local watershed efforts. States contribute proportionately little funding for watershed initiatives. Other funding sources, such as foundation grants, corporate donations, membership dues, and contributions from utilities, are also available. Use of these funds, however, is often narrowly focused. Thus, important funding needs often go unmet, especially support for watershed organizations' overhead and general operations. There is clearly a need for more total dollars and more flexibility to allow watershed efforts at all levels to implement their highest priority projects. Moreover, funding for watershed initiatives is generally unstable, with few programs having dedicated funding sources, such as from hydropower revenue or water/sewer surcharges. Stability of funding from year to year is extremely important as watershed initiatives can lose the ability to support key staff, resulting in the loss of institutional memory.

B.3.8.4 Lack of Integration of Point and Nonpoint Source Pollution Management

There are generally weak links between state water quality regulatory programs, local or state planning activities, and local watershed partnerships. Pollutant loadings to the watershed and the resulting water quality are the sum of point and nonpoint source pollution. To understand the ecological and water quality impacts of pollution and to develop cost-effective management plans, watershed initiatives need reliable information about both sources. Historically, however, point and nonpoint source pollution management has not been well integrated at the federal or state level or in local watershed initiatives. With broader implementation of TMDLs, this is starting to change. However, watershed planning and programs for the most part are still not well connected to local land-use planning. Thus, watershed protection priorities tend to have little impact on growth management and regulatory decisions such as zoning, building design, and development choices, which profoundly impact nonpoint source pollution and the hydrodynamics of the watershed.

B.3.8.5 Science: Data Gaps, Insufficient Knowledge, and Inaccessibility

- **Absence of high-quality monitoring data.** There is little high-quality, representative monitoring data except for hydrologic data. A standard robust set of biological, chemical, and physical indicators to monitor is lacking. Data on water quality, species diversity and abundance, and habitat conditions are needed throughout the country to identify natural trends and evaluate the relative effect of human activities. This problem is made more acute by the variability of conditions across locations, which limits the ability to extrapolate findings based on data from other areas. Most watersheds do not even have data from more distant watersheds to benchmark local conditions.
- **Hydrologic monitoring data deteriorating.** State and federal funding cuts and the resulting decline in the number of gauging stations is a serious threat to watershed management efforts throughout the United States. The hydrologic data provided by these stations represents is key, as it is generally the highest quality data set available to watershed initiatives and the one that generally offers the longest historical view.
- **Limited knowledge of aquatic ecosystems.** Aquatic ecosystems are not well understood, especially as an integrated system. The current level of understanding is not sufficient to enable general predictions with any precision as to how the structure and function of aquatic ecosystems will respond to human activity within the watershed, including any restoration efforts.
- **Inaccessible scientific information.** Government and other planners and decision makers need to be provided with relevant scientific information in a clear, meaningful, and timely manner. Decision makers do not have access to truly integrated watershed models that satisfy minimum criteria for effectiveness, usefulness, and ease of application. There is a need for more user-friendly models that link GIS and decision support systems, incorporate social and economic sciences, and span a variety of scales. GIS have largely been used to make maps and have generally not been effectively linked to decision support systems and other models.

APPENDIX C

COMPENDIUM OF INTERNATIONAL EXAMPLES

C.1 Introduction

Approaches to watershed management necessarily vary based on the unique geographical, ecological, physical, political, cultural, and economic setting of each watershed. In spite of this diversity, watershed approaches rely on two basic means to work effectively: effective institutional arrangements and relevant instruments. In each example given in the attached compendium the institutional organization and the instruments used in the context of the example watershed are outlined. Specifically, the lead organization and stakeholders are identified, followed by a statement on the formation of the watershed approach. The instruments used to facilitate better management are discussed, with a focus on the results, positive or negative, of integrated watershed management. Finally, potential applications or lessons learned from each watershed that may be applicable to the United States are presented. The examples of watershed management chosen primarily come from developed countries, with the assumption that these watershed lessons are more likely to be applicable in the United States.

The compendium is organized in two major sections, the first section provides examples alphabetically by nation, with a description of relevant national policies and approaches that facilitated watershed approaches, followed by specific examples of watershed management within that nation. The second section describes examples of watershed management involving more than one nation, again in alphabetical order. River basin management across international boundaries is a much more challenging process. Where the countries involved share similar institutions and legal frameworks and cooperate on other issues, such as within the European Union (EU), the process of successfully cooperating on water management has proven more achievable. In regions with traditions of conflicts between the countries, the process is much more difficult and slow. Success in these situations may be just the willingness to discuss water issues, which is the first step in the process of effective watershed management.

For all examples given in the compendium, categorization by institutions or by instruments used is avoided to prevent the misconception that either institutions or instruments alone are sufficient for good management. The evidence from the literature overwhelmingly indicates that both appropriate institutions and instruments *together* are necessary for effective watershed management, and that watershed management entails the management of land and water. The matrix (Table C-1) on the following page gives an overview of the basins.

Table C-1 River Basin Matrix

Table C-1: Issues addressed by watershed in the compendium

National River Basins

COUNTRY	RIVER BASIN	Issues addressed and instruments used: Institutional Coordination	Stakeholder participation	Pollution Management	Ecosystem Services	Instruments Used
Australia	New South Wales	Coord, OrgA, MedC	CapB, LocDM, Lfund	AltP, PNP	EcoR, EcoP	Econ, Reg, Tech, Info/Comm
	Queensland	Coord, OrgA, MedC	LocDM	PNP	EcoP	Econ, Reg, Info/Comm
	Victoria	Coord, OrgA	IncAware	PNP	EcoP	Econ, Reg, Info/Comm
	Murray-Darling Basin	LSM, Coord, OrgA	CapB, IncAware	PNP, AltP	EcoP	Econ, Reg, Tech, Info/Comm
Canada	Ontario	Coord, OrgA	CapB, IncAware, LocDM, Lfund	PNP	EcoP, EcoR	Econ, Reg, Info/Comm
	Grand River	Coord, OrgA	IncAware, LocDM, Lfund	PNP	EcoP	Econ, Reg, Info/Comm
	Fraser River and Estuary	LSM, Coord, OrgA	CpaB, IncAware, LocDM	PNP, AltP	EcoP, EcoR	Econ, Reg, Info/Comm
	Annapolis River	LSM, Coord, OrgA	CapB, IncAware, LocDM, Lfund	PNP, AltP	EcoP, EcoR	Econ, Reg, Tech, Info/Comm
France	Paris	Coord, OrgA, MedC	CapB	PNP, AltP	EcoP	Reg, Tech, Info/Comm
Germany	The Ruhr Basin	OrgA, MedC	LocDM, Lfund	AltP	EcoR	Econ, Reg, Tech, Info/Comm
Japan	Ara and Oppe Rivers	OrgA	IncAware	PNP	EcoR	Reg, Tech, Info/Comm
Mexico	Bahais de Huatulco	Coord	LocDM	PNP	EcoP, EcoR	Reg, Info/Comm
New Zealand	Whaingaroa	OrgA	IncAware, Lfund	AltP,PNP		Info/Comm
South Africa	Working for Water	OrgA, Coord, MedC	CapB, IncAware, LocDM	PNP, AltP	EcoP, EcoR	Reg, Info/Comm
Tanzania	Pagnani Basin	Coord, MedC	CapB, IncAware, Lfund			Reg, Info/Comm
U.K.	Mersey Basin	Coord	IncAware, Lfund, LocDM	AltP, PNP	EcoP	Econ, Reg, Info/Comm

International River Basins

RIVER BASIN	Issues addressed and instruments used				
Aral Sea	LSM, Coord	CapB, IncAware	PNP		Reg, Info/Comm
Danube	LSM, Coord	CapB, IncAware	PNP		Reg, Info/Comm
Mekong	LSM, Coord	IncAware		EcoP	Reg, Info/Comm
Nile	LSM	CapB, IncAware			Info/Comm
Rhine	LSM, Coord, MedC	CapB, IncAware	PNP	EcoP, EcoR	Econ, Reg, Info/Comm

Key:

Institutional Cooperation and Coordination:
Large-scale watershed management **(LSM)**
Coordination of authority at all levels **(Coord)**
Establishing an organization with real authority **(OrgA)**
Mediation capabilities **(MedC)**

Stakeholder participation:
Capacity building **(CapB)**
Increase public awareness **(IncAware)**
Local decision making **(LocDM)**
Local funding of water management **(LFund)**

Pollution Management:
Integrating point and nonpoint source pollution management **(PNP)**
Alternative methods of pollution management (i.e., watershed trading) **(AltP)**

Ecosystem Services:
Ecosystem restoration **(EcoR)**
Ecosystem preservation **(EcoP)**

Instruments:
Economic **(Econ)**
Regulatory **(Reg)**
Information/ Communication **(Info/Comm)**
Technology **(Tech)**

APPENDIX C1

WATERSHED MANAGEMENT IN A NATIONAL CONTEXT

C1.1 Australia

Australia shares much in common with the United States. Both are comparable in size, although Australian has a much smaller population (approximately 20 million). Both have democratic governments with some degree of recognition of states rights. Both began addressing environmental issues in earnest in the early 1970s. During the past 10 years, both nations have placed increased emphasis on ecosystem preservation. For example, Australia adopted the Ecological Sustainable Development Strategy in 1992. Since then, watershed management in Australia has been highly progressive and effective. A broad overview of water management at the national level is given, followed by three examples of watershed management at the state level and one at the multi-state level.

C1.1.1 Institutional Setting:

Australia is a commonwealth consisting of six states and two territories. The commonwealth government has jurisdiction over foreign affairs, defense, and treasury; the states are autonomous in the areas of natural resources and environmental management, albeit with guidance from the commonwealth. This has created a situation in which the administration of water and the environment varies markedly from state to state.

The commonwealth government has three departments that are concerned about water and environment issues: the Department of Primary Industries and Energy (DPIE), the Department of the Environment, Sport, and Territories, and the Department of Health and Family Services. These departments operate to coordinate the efforts of equivalent state agencies and work through ministerial councils that link the federal ministers to their state and territory counterparts. Overall, the Council of Australian Governments (COAG) brings the state premiers and territory chief ministers together with the prime minister to forge policy.

The council has endorsed a comprehensive strategic framework for the efficient and sustainable reform of the Australian water industry, which was to be fully implemented by 2001. Major elements of this framework include:

- Water pricing based on consumption-based pricing and full-cost recovery,
- Reduction or elimination of cross-subsidies,
- Clarification of property rights,
- Allocation of water for the environment,
- Adoption of trading arrangements for water,

- Institutional reform, and
- Public consultation and participation.

The framework demands an explicit allocation of water to the environment, in recognition of a growing understanding of the need to maintain ecological processes. The practical ramifications of this policy are still being tested and debated, amidst tensions between urban, agricultural, power generation, and the environmental uses. Historical patterns of extractions from regulated rivers and the imposition of transport flows in those rivers make it difficult to manage for natural environmental regimes. Also, understanding of how to best regulate flow variations for environmental outcomes is poorly developed. At this stage, efforts are basically focused on implementing a "best-guess" flow pattern, followed by monitoring and refinement. The Murray-Darling Basin Commission (MDBC) introduced a cap on extractions in 1985. Despite controversy over the cap, it has been maintained.

Two federal laws furthered the effectiveness of local watershed management. The Catchment Water Management Act (CWMA) of 1995 created local catchment management boards with responsibility for crucial aspects of planning and management, including flood control, recreational issues, and preserving and promoting environmental quality. Funding of the boards is provided through catchment levies (economic) based on property values, and their ability to borrow money for large capital projects. The application of the CWMA varies widely throughout Australia, with some states (e.g., New South Wales [NSW]) having established legislation, and others (e.g., Western Australia) having no generic legislative powers. Integrated catchment management (ICM) is at the stage of developing strong processes for its long-term implementation. The second important piece of legislation, the Water Resources Act of 1997, emphasized the integration of ecological, environmental, and economic considerations on a geographic basis (regulatory and economic).

Substantial progress has already been achieved in implementation of the water reform framework. All jurisdictions have in place or are planning measures to promote integrated catchment management. Major urban water authorities have also maintained the pace of reform; major changes to institutional arrangements have been made in NSW, Victoria, South Australia, and Western Australia.

Interstate water trading has commenced. The first temporary irrigator-to-irrigator interstate trade between Victoria and NSW occurred late in 1995 under provisions of the Victorian Water (Amendment) Act of 1995. South Australia has agreed in principle to allow the temporary trade of River Murray water across the South Australian border, and NSW and Queensland have begun negotiations on water trading arrangements in the Border Rivers System.

C1.1.2 Ecosystem Health

Heavy regulation of rivers and diversion of runoff, via large in-stream dams and shallow off-stream farm ponds, has stressed Australian river systems. Urban and rural runoff carry sediments, nutrients, and other pollutants and add to the stress, as has the loss of vegetation within the catchment and along riparian corridors.

In response, Australia established a National Landcare Program (NLP) to develop and implement integrated approaches to natural resources management that are efficient, equitable, and consistent with the principles of ecologically sustainable development. The program encourages an approach to the management of natural resources that recognizes that land, water, and living resources are inter-related and cannot be managed separately. The program focuses on long-term productivity and ecological sustainability. NLP contributes to ecologically sustainable development by encouraging conservation of Australia's natural resources, and it underpins the government's strategies for maximizing the contribution of rural industries and regions to the economic and social well being of the Australian community. The commonwealth's focus is on natural resource management issues of national importance.

The NLP program promotes partnerships between the community, industry, and government in the management of natural resources. It provides funds either directly to community groups, for joint commonwealth–state initiatives, or for combined government and community effort, including at a regional level. Through the community component of the NLP, the commonwealth government encourages community groups to responsibly manage and conserve the land vegetation, water, and biological diversity. The DPIE and the Department of the Environment jointly administer the community component of the NLP.

The examples given below include watershed initiatives entirely within a state or territory, as well as a large basin that spans four states.

C1.1.3 New South Wales

NSW is the most populous state in Australia, with over six million people. The Department of Land and Water Conservation administers water and natural resource policy. Environmental protection is the responsibility of the Environment Protection Authority, and the Independent Pricing and Regulatory Tribunal adjudicates the pricing of water services. NSW underwent a major water reform in 1997, which included setting environmental flow objectives for all regulated and unregulated rivers as a management priority, introducing a resource management charge for all categories of rural water users, and referring rural water pricing to the Government Pricing Tribunal for review to ensure consistency with COAG requirements.

C1.1.3.1 Summary

Effective Approaches	Coordination of authority at all levels Creation of an organization with real authority Mediation capabilities Capacity building Local decision-making authority Local funding Integrating point and nonpoint pollution management Alternative methods of pollution management Ecosystem preservation Ecosystem restoration
Lead Organization:	NSW government

Other Stakeholders: Government agencies, communities, farming and fishing industry, indigenous Australian people, nature conservation interests, artists, landscape architects, hydraulic engineers, university students

Timeframe: 1997 to present

C1.1.3.2 Description

The NSW government established water management committees (WMCs) in 1997 to help create management plans for major river catchments throughout the state. These plans provide the basis for water use and trading by industry and communities, river care and other environmental programs for a five-year period. The minister for land and water conservation and agriculture and the minister for the environment approve the WMC plans.

The WMCs have successfully developed and implemented a variety of creative and innovative approaches that rely on government partnership with the community. The committees typically include representatives of all relevant government agencies, the community, the farming and fishing industries, indigenous Australian people, and nature conservation interests. Their approaches include developing a strategic planning framework and process to carry out the necessary rapid research and preparation of the plans, and educating the multi-dimensional representatives, particularly in conflict management, consensus decision making and water-based resource management. Indigenous and other cultural elements are brought into committee activities and planning through art, songs, and storytelling.

For example, a program in Sydney called Restoring the Waters has involved artists, landscape architects, hydraulic engineers, and multicultural communities. The program required replacement of 1970s stormwater management infrastructure with a state-of-the-art wetland system. The first stage involved an important community arts project and the second a sophisticated landscape design for wetlands in collaboration with hydraulic engineers and ecologists.

A similar program for rural communities, the Creative Village program, involves rural communities, students, landscape architects, architects, and artists. Under this initiative, management of the ancient Aboriginal fishery at Brewarrina, NSW, was achieved through the use of creative games as a process for community decision making. Another project looked at new forms of river management to develop innovative small-scale agricultural enterprises. In a coal-mining community subject to flooding and heavily polluted water, an intriguing redesign of the catchment included the urban landscape as art and hydrology.

Of particular note, the Hunter River WMC has created a unique and innovative pollution control program called the Hunter River Salinity Trading Scheme (HRSTS), believed to be the world's first 24-hour online credit exchange for a water emissions trading scheme. HRSTS is the only emissions trading scheme based on real-time environmental conditions rather than modeled predictions. It manages discharges of saline water from coal mines and electricity generators to the Hunter River so that river salinity does not exceed levels that are detrimental to agricultural productivity or environmental quality downstream. This is achieved by:

- Extensive and continuous real time monitoring of environmental conditions and discharges;
- Scheduling saline discharges to complement high river flow rates and low background salinity levels so that salinity targets are not exceeded; and
- Sharing the total allowable discharge according to dischargers' holdings of tradable salinity credits.

By maximizing discharge opportunities in response to river and weather conditions, the scheme also saves the participating discharging industries millions of dollars by reducing the need for large storage dams and desalination programs.

Another innovative program was initiated in 1996 in the South Creek area of the Hawkesbury-Nepean River: a small, self-contained emissions trading scheme (economic). This bubble scheme allows the three participating sewage treatment systems to adjust their individual discharges, provided the total pollutant load limit for the scheme is not exceeded. This enables efforts to reduce pollution to be focused where the costs are lowest. The load limits mandated (regulatory) under the bubble scheme require an 83% reduction in total phosphorus and a 50% reduction in total nitrogen by 2004 compared to a business as usual scenario.

C1.1.3.3 Results/Outcomes

The breadth of the approaches to watershed management in NSW is remarkable. The projects represented above show that the water management committees as well as the innovative programs have been successful in organizing the efforts of multiple stakeholders. The variety of approaches is a result of the diversity of stakeholders in each catchment, who are a major force in laying out the approach for their watershed. The programs address the local context. The programs have also been successful in implementing water management at the grassroots level, particularly for the indigenous people.

C1.1.3.4 Lessons/Applicability to United States

Using committees at the catchment level to create water and environmental management plans for the area is a successful way to establish sustainable management. This example shows that while catchment-level committees are capable of developing successful management plans, these plans are reviewed and accepted or denied at the national level. Having a national review of the plans promotes dialog, coordination, and continuity between the different levels of jurisdiction. Another lesson from this example is to use creativity and cultural elements when trying to inform and recruit local community involvement. Once the information is presented in terms that the local population can understand and identify with, the community will be more willing to participate and accept changes. When the community plays a large role in the development, and actually plays a role in the construction (e.g., displaying art, landscaping), there is more community pride in the project and it will more likely lead to additional projects.

C1.1.4 Queensland

C1.1.4.1 Summary

Effective Approaches:	Coordination of authority at all levels
	Establishment of an organization with real authority
	Mediation capabilities

Local decision making
Integration of point and nonpoint pollution management
Ecosystem preservation

Lead Organization: Queensland government

Other Stakeholders: State agencies, local councils, community participants, including agricultural and industrial interests

Timeframe: 1990 to present

C1.1.4.2 Description

Conflict over floodplain development and related water allocation issues arose in the early 1990s in the Lower Balonne. The agricultural and municipal interests were in conflict over floodplain development in the area. The state government intervened, but the result was an escalation of the conflict because of a lack of community involvement in the decision-making process. The community and local farmers needed a process that allowed them to debate the facts, then negotiate and finally arrive at an outcome together, rather than having one imposed. This resulted in Queensland's first floodplain management plan that allowed the community to provide input on decisions made that directly affected them. It took about eight years to fully formulate the process.

The Queensland government extended the participatory approach through a new policy in the early 1990s, the ICM policy. ICM relies on community-level processes to promote ICM principles in the local community, industry, and government without any statutory powers and only limited finances. The ICM funded the Herbert River Catchment Coordinating Committee (HRCC) in 1993 as a catalytic group. The critical roles and responsibilities for ICM as a partnership between government, industry, and community on natural resource management include supporting community learning, influencing resource use planning processes across the spectrum of political and traditional management boundaries at a catchment scale, providing a community forum for coordination of efforts, networking and social learning, brokering funding opportunities on natural resource management for the catchment, ensuring fairness of process on natural resource management issues, and providing a vehicle for integrating natural resource management information.

ICM principles also supported the Moreton Bay Study, which is an interdisciplinary study of the bay and its major river estuaries. It was initiated to address water quality issues which link sewage and diffuse loading with environmental degradation in the Moreton Region. The study consists of 19 local councils in association with six state agencies and the federal government. It forms a regional component of a national and state program to achieve ecologically sustainable use of the waterways by protecting and enhancing their health, while maintaining economic and social development. The study framework illustrates a unique integrated approach to water quality management whereby scientific research, community participation, and strategy development are done in parallel.

C1.1.4.3 Results/Outcomes

The floodplain management plan in the Lower Balonne Basin has been successful in mediating conflicts between competing users in the floodplain. The success of the management plan is a result of the community driving and owning the process and the outcomes. This collaborative effort resulted in a water quality management strategy, which focuses on the integration of socioeconomic and ecological values of the waterways and has led to significant cost savings in infrastructure by providing a focus on initiatives toward achieving the healthy waterways vision.

In the beginning of the HRCC, there was a lack of cooperation within and between government agencies. There was also slight conflict with sections of government departments who viewed ICM as an initiative of a particular government agency rather than an initiative of the whole. There was little industry awareness and commitment to the environment and resource management beyond immediate production impacts, as well as a lack of incentives for industry to develop environmental objectives. Implementation of the ICM in the Herbert Catchment evolved slowly in a context of substantial change in regard to community values and expectations of natural resource management and resource-use planning processes within the catchment. At times, the process was delayed by a considerable degree of uncertainty and confusion on the intent of the ICM initiative and on what was required or expected of those involved. It also experienced internal and external conflicts between its diverse stakeholders that at times constrained progress. However, ICM has had a substantial influence on natural resource management in the catchment. For example, the HRCC succeeded in raising community and industry awareness of natural resource management issues; had noticeable success influencing land-use planning and development assessment at local government level; and was a voice for an integrated approach to coastal zone management and planning, integrated pest management, and sugar industry expansion in the catchment.

C1.1.4.4 Lessons/Applicability to United States

These examples show the success of integrating scientific research, community participation, and strategy development in water quality management. While collaboration between the three sectors is a key to success, they can continue operating simultaneously to provide momentum and advancement of the project. Integrating the local level with the state and federal government is a key to successfully implementing a water quality management strategy. This jurisdictional integration allows for socioeconomic and ecological values of waterways to be addressed by bringing together the local issues with government standards.

Conflict resolution requires a legitimate, inclusive process for acquiring knowledge, developing shared principles and vision, and negotiating real outcomes (making decisions). This process must be developed and evolved by the participants so that they have confidence that it is genuine. This example is quite applicable to all cross-jurisdictional coordination plans and should be addressed in any U.S. watershed planning effort.

The most important lesson from this example is to inform all participants (government, industry, communities, etc.) of the management plan and be sure they fully understand their role. This is key to avoiding misunderstandings that often lead to conflicts among the participants. By keeping all participants abreast of all activities and changes in the management plan, including what is going on in other sectors, planning and development efforts will have a much higher success rate.

C1.1.5 Victoria

Victoria occupies a relatively small part of the total continental area, but at 4.7 million residents it has the second highest population. Water resources oversight has been consolidated over recent years so that water and wastewater management services are now overseen by five corporate businesses in the metropolitan Melbourne area and 15 independent regional water authorities that supply the rest of the state. The Department of Natural Resources and Environment and the Environment Protection Authority are the primary state regulatory agencies, with the Department of Human Services having an involvement in potable water issues.

C1.1.5.1 Summary

Effective Approaches:	Coordination of authority at all levels Establishment of an organization with real authority Increased public awareness Integration of point and nonpoint pollution management Ecosystem preservation
Lead Organization:	Victoria State Catchment Management Authority
Other Stakeholders:	All levels of government, the community, industries
Timeframe:	Current

C1.1.5.2 Description

The Goulburn-Broken catchment has suffered severe degradation from land clearing, agricultural practices, and water extraction. It ranks as one of the highest catchments in terms of nutrient and salt exports, contributing to the severe decline in the Murray River. The Goulburn-Broken Catchment Management Authority (CMA) is a partnership approach developed to reverse these severe trends and implement sustainable management within the catchment. The authority is constituted under Victorian state legislation and represents all government, community, and industry interests. The authority has established a long-term implementation plan to achieve a 65% reduction in phosphorous export, reverse the rapid growth in salinity discharges, improve stream health, support the return of waterway biodiversity, and maintain the economic and social base of the local community. The CMA together with the catchment community, state and federal government, and other key stakeholders work together to develop, implement, and monitor programs with the aim of producing and passing on to future generations viable and healthy catchment and waterway systems.

C1.1.5.3 Results/Outcomes

Achievement within the catchment has been founded on the integration of individual programs and close relationships with the community. Key achievements include construction of 22 fishways, reduction in phosphorus entering the River Murray, protection of 300 km of riverine frontage, and stabilization of over 150 km of waterway. The Goulburn-Broken catchment community has learned key lessons from the past and is proud of the results achieved.

The community has produced successful program integration, provided a tool for the attainment and adoption of new science, and created a framework for programs to be integrated at strategic planning and implementation levels.

C1.1.5.4 Lessons/Applicability to United States

This example represents a successful framework for integrating strategy, planning, and implementation of projects. Quite often discontinuity exists between these levels of project development, increasing the chance of confusion and eventually project failure. This example also shows the importance of close relations between project developers and communities. Adding the support and participation of the community to a project ensures a successful outcome. Another key lesson from this example is that it takes time to improve the water quality, habitat, and ecosystem health in an area that has experienced severe degradation. Understanding this concept, conveying the concept to the general public, and creating a long-term action plan are keys to successfully retroacting water quality problems of the past.

C1.1.6 Multi-state Management in the Murray-Darling Basin

There is recognition that for problems on the scale of a large river basin, the institutions involved must be appropriately scaled. For example, salinity is very problematic in the Murray-Darling River Basin, which covers an area of approximately 1 million square kilometers (the largest integrated catchment program in the world). To address the problems of the Murray-Darling Basin (MDB), the federal government established the MDB Initiative (MDBI), which formed a consortium among catchment management boards and state and federal officials. Funding for the initiative comes from both the state and federal levels.

C1.1.6.1 Summary

Effective Approaches:	Large-scale watershed management Coordination of authority at all levels Establishment of an organization with real authority Capacity building Increased public awareness Integration of point and nonpoint source pollution programs Alternative methods of pollution management Ecosystem preservation
Lead Organization:	Murray-Darling Basin Ministerial Council
Other Stakeholders:	Murray-Darling Basin Commission (MDBC), community advisory committee, basin-wide community
Timeframe:	1992 to present

C1.1.6.2 Description

On top of the troubling natural conditions of salinity, turbidity, and algae, the MDB's primary problem is the overuse and/or misuse of its land and water. Agriculture is the major land use and economic activity in the basin, followed by forestry and national parks and reserves. Due mostly to the agricultural activity, land has been over-cultivated, over-grazed, and over-cleared resulting in soil erosion, acid soils, native vegetation decline, and increases in weeds and noxious plants. The overuse of water has resulted in major pollution of the basin's surface water and groundwater resources, particularly turbidity and nutrients. The basin has also experienced a decline of native fish species and native vegetation.

The government response was the establishment of the MDB Agreement, which includes the governments of the commonwealth, NSW, Victoria, South Australia, and Queensland. The agreement recognized that existing management arrangements in the basin were not able to properly deal with the basin's emerging problems. It was also recognized that to address the needs of the basin would require more than one government or group of people. It would require a joint effort by all governments and communities. The MDBI was created as a structure to implement this joint effort. The purpose of the agreement and initiative is:

> "To promote and coordinate effective planning and management for the equitable, efficient, and sustainable use of the water, land, and other environmental resources of the Murray-Darling Basin."

Funding for the initiative comes from the state and federal levels. The initiative created three decision-making bodies: a ministerial council, a basin commission, and a community advisory committee. The responsibilities of each organization is as follows:

- **Murray-Darling Basin Ministerial Council**: Ministers from each of the signatory governments make up the council. This council is the decision-making forum for the MDBI.
- **Murray-Darling Basin Commission**: The MDBC gives advice to the ministerial council and carries out the decisions made by the council.
- **Community Advisory Committee**: The committee provides a two-way communication channel between the ministerial council and the community. The committee provides the council with advice from the basin-wide community.

There are many other inter-jurisdictional agreements in the MDB that continue to operate. Continuity is provided because government ministers and officers involved with these other agreements are also involved with the MDBI, although there is no formal link.

The management of the basin has not been problem-free. One lesson learned is that there is a need for a common language in which to hold useful discussions of sustainability and trade-offs between environmental, economic, and social objectives. This should include some agreed measure of river health to use for comparison when discussing these issues. Using water quality and flow regimes are not very useful surrogates for river health as seen in the MDB management because they are drivers that determine the outcomes. The outcomes themselves are biological responses. The best indicators of river health are aquatic invertebrates, fish, and algae.

A sustainable rivers audit is proposed that measures river health based initially on two measures, that coming from the AUSRIVAS models of aquatic invertebrates and a measure of fish populations, such as the Index of Biotic Integrity. Three further measures include an index of water quality, a measure of flow regime, and a habitat assessment. These three elements are drivers of the biological outcomes and serve as useful diagnostics in interpreting the reasons why the biological measures may have changed. The proposal is that the biological measures, along with the water quality and flow index be reported annually for each river valley in the MDB. A more comprehensive sustainability assessment would be carried out initially and thereafter every five years.

The MDBC has established a water market reform working group to fast track the process of advancing water markets and the trading environment. The decision taken by the MDB Ministerial Council in June 1995 to place an immediate moratorium on further diversions of water from the rivers of the basin not only recognizes the importance of striking a balance between consumptive and in-stream (environmental) demands but should also positively assist in driving effective markets for the trading of water entitlements.

C1.1.6.3 Results/Outcomes

One of the outcomes of the MDBI is the Natural Resources Management Strategy (NRMS), a framework for a community-government partnership to develop plans for the integrated management of the basin's natural resources on a catchment basis. The strategy is set forth as follows:

- Outcomes across water, land, other environmental resources, and cultural heritage must be consistent with the principles of ecologically sustainable development;
- The roles of communities and of governments are specified under a catchment coordinating structure; and
- Coordination and financial support for measures to improve the sustainability of natural resources management practices, including communication, education, on-ground works, and evaluation may be obtained through the NRMS.

C1.1.6.4 Lessons/Applicability to United States

The MDBI is an example that covers many issues plaguing the United States. For one, the MDBI is a well-organized and documented example of a large-scale approach to watershed management; crossing over many jurisdictions. The MDBI addresses the integration of authority from local issues to ministerial councils, strategies for increasing awareness and participation among stakeholders, and building capacity. The initiative also addresses point and nonpoint sources of pollution, as well as ecosystem protection.

C1.2 Canada

Under the Constitution Act (1867), the Canadian government established that the ownership and management of water resources are the responsibility of the provinces. Recognizing the decline in the quality of water resources, the Canadian government passed the Canada Water Act in 1970 and created the Department of the Environment (DOE; better known as Environment Canada) in 1971 to provide national guidance for freshwater management. While providing guidance for the provinces on issues of water quality, Environment Canada actively promoted a partnership approach among the various levels of government and private sector interests that contribute to and benefit from the use of the resource.

All of these interests were extensively consulted during the 1984-1985 Inquiry on Federal Water Policy, which conducted Canada-wide hearings toward the development of a federal water policy. Guided by the findings of the inquiry, the government released its policy in 1987, which has since given focus to the water-related activities of all federal departments and which will continue to provide a framework for action in the coming years as it evolves in the light of new issues and concerns. In particular, the federal water policy attempts to balance water uses with the requirements of the many interrelationships within the ecosystem.

The policy takes into account the needs of all Canadians in its overall objective: to encourage the use of freshwater in an efficient and equitable manner consistent with the social, economic, and environmental needs of present and future generations. To manage Canada's water resources, the federal government has defined two goals: to protect and enhance the quality of the water resource; and to promote the wise and efficient management and use of water. The policy stresses that government action is not enough. The public must become aware of the true value of water in their daily lives and use it wisely. The federal government uses five strategies to reach its stated goals, including:

- Water pricing
- Science leadership
- Integrated planning
- Legislation
- Public awareness

The five strategies are broad courses of action that define a supportive yet flexible role for the federal government, one that enables the various federal agencies, other levels of government, and industry to respond to circumstances and challenges. This approach is compatible with the federal structure and the realities of a large and diverse country.

Three examples of the coordination of federal, provincial and local efforts successful in striving for the policy's goals are given below.

C1.2.1 Ontario

C1.2.1.1 Summary

Effective Approaches: Coordination of authority at all levels
Establishment of an organization with real authority
Capacity building
Increased public awareness
Local decision making
Local funding of water management
Integration of point and nonpoint pollution management
Ecosystem preservation
Ecosystem restoration

Lead Organization: Conservation authorities (38 authorities) and municipalities

Other Stakeholders: Local interests, government agencies, businesses, schools

Timeframe: 1990s to present

C1.2.1.2 Description

Ontario has used a nonregulatory approach to watershed management. The province prepared three watershed documents in 1990: *Water Management on a Watershed Basis: Implementing an Ecosystem Approach*; *Subwatershed Planning*; and *Integrating Water Management Objectives into Municipal Planning Documents*. These documents were broadly distributed in combination with a campaign to educate the public on watershed issues. Watershed management projects could be initiated by any organization.

There have been more than 100 projects initiated since 1990. The lead organizations for these projects are typically municipalities or pre-existing conservation authorities. There are 38 conservation authorities in Ontario that promote integrated planning for development. These authorities are organized by local interests and deal with issues like flood control, recreation management, water supply, and water quality. The uniqueness of authorities' management is that they deal with these issues together rather than separately.

The trigger for project initiation has primarily been due to development pressures, but also to augment resource protection and remediation. Funding is entirely at the local level.

C1.2.1.3 Results/Outcomes

More than half of the watershed projects have successfully been completed. The most common projects address aquatic life/fisheries, surface water flows and quality, stream channels/buffers, and wetlands.

C1.2.1.4 Lessons/Applicability to United States

It may be interesting to further research the strategies developed by Ontario. Locally initiated and funded projects have proven to effectively address some of the broader issues of ecosystem preservation and restoration. Public awareness played a key role.

C1.2.2 The Grand River

C1.2.2.1 Summary

Effective Approaches:	Coordination of authority at all levels Establishment of an organization with real authority Local funding Increased public awareness Local decision-making Integration of point and nonpoint source management Ecosystem preservation
Lead Organization:	The Grand River Conservation Authority
Other Stakeholders:	Municipalities, non-government organizations (NGOs), indigenous groups, angling groups, universities, industry, the community
Timeframe:	Began in the 1930s

C1.2.2.2 Description

The Grand River Conservation Authority (GRCA) has achieved outstanding success in restoring the picturesque Grand River in a region that supports a large industrial base and a productive agricultural sector. The municipalities in the Grand River valley worked together in the 1930s to solve common river management problems by forming the Grand River Conservation Commission. This commission, which later became the GRCA, was the first river management and watershed management in Canada. It is the third oldest such agency in the world. The Province of Ontario used the Grand River model as the basis for the Conservation Authorities Act of 1946, which led to the establishment of today's 38 conservation authorities in Ontario. The GRCA is governed by a 26-member board, appointed by municipalities in the Grand River drainage area. The following is a list of the watershed management issues identified in the Grand River Basin:

- Keeping the watershed healthy while accommodating growth;
- Developing a viable tourism industry while protecting the resources on which it is based;
- Improving water quality using a cost-effective balance between cleaning up urban wastewater and controlling rural sources of pollution;
- Managing water supply and water allocation;
- Reducing flood and erosion damages using a mix of structural and non-structural approaches;
- Conserving the natural environment and biodiversity; and
- Conserving heritage and a sense of place.

To address these issues, the GRCA is trying an integrated and continuous watershed approach. One goal is to identify the critical areas where actions will produce the best results and find the appropriate partners to carry out the corresponding activities. The GRCA provides full administrative and technical support to ensure continuous participation. Another goal of the GRCA is to further investigate the capacity of the resources within the basin to determine the best possible path to future self-sufficiency. Throughout implementation of these activities, the GRCA continues to monitor the effectiveness of the actions taking place, organize and maintain an accessible information base, and continuously report on the health of the watershed as well as success and failures within the program.

Funding for the conservation programs is provided by a partnership with the Province of Ontario and the watershed municipalities. Sources of revenue for conservation programs include watershed municipalities, provincial grants, income from conservation area user fees, rental income, hydroelectric production, and the Grand River Foundation.

C1.2.2.3 Results/Outcomes

The GRCA has experienced environmental success while also providing for human needs, a delicate balancing act that communities and governments struggle with all over the world. The GRCA has a well-defined approach to action planning and delivery and has established an effective network of partnerships with 34 municipalities in nine regions and counties, NGOs, indigenous groups, angling groups, universities, industry, and the community.

C1.2.2.4 Lessons/Applicability to United States

The GRCA is an excellent example of a well-organized watershed authority that has succeeded in implementing continuity throughout authoritative levels. The presence of the GRCA provides continuous momentum of watershed programs and requires all participants to be accountable for their actions. The GRCA represents an authoritative structure that not only manages projects, outreach, data, and participants, but also seeks active involvement itself. This example also gives insight into how and where to obtain funding.

C1.2.3 Fraser River and Estuary

C1.2.3.1 Summary

Effective Approaches:	Large-scale watershed management Coordination of authority at all levels Establishment of an organization with real authority Capacity building Increased public awareness Local decision-making Integration of point and nonpoint pollution management Alternative methods of pollution management Ecosystem restoration Ecosystem preservation
Lead Organization:	Environment Canada
Other Stakeholders:	Fisheries and Oceans Canada; Ministry of Environment, Lands and Parks (MELP); Greater Vancouver Regional District; Fraser River Port Authority; North Fraser Port Authority
Timeframe:	1985 to present

C1.2.3.2 Description

The Fraser River Estuary is the social, cultural, recreational and economic heart of British Columbia. With so much at stake to protect the environmental quality of the river and estuary, an efficient and solid management program is a necessity. In 1985, the Fraser River Estuary Management Program (FREMP) was formed to address the public concern about the negative effects of development, rapid population growth, and communication problems between government agencies. The mission of FREMP is "to coordinate the cost-effective management of a living, working Fraser River estuary." The key to the success of the FREMP has been the cross-jurisdictional management framework that is shared between six agencies, thus linking agencies responsible for setting environmental policies to those managing land and water resources. Decisions are made using a consensus-based approach. The actions for the Fraser River Estuary are coordinated by Environment Canada to ensure consistency with policies established for the rest of Canada.

The six participating agencies act within their jurisdiction, but in coordination with the other agencies in the FREMP. One of the participating agencies in the FREMP is the Fisheries and Oceans Canada, which is responsible for all of Canada's water, fish, and habitat. The

Fisheries and Oceans Canada have the only legislation through which an environmental offender may be charged, the Fisheries Act. The MELP is responsible for the management, protection, and enhancement of British Columbia's environment. The MELP is responsible for issuing permits to industries which discharge into the Fraser River Estuary. At the regional level, the Greater Vancouver Regional District (GVRD) is a network of municipalities that manages the sewerage and drainage systems. At the harbor level, the Fraser River Port Authority (FRPA) and the North Fraser Port Authority (NFPA) administer the property within their jurisdictions. The FRPA oversees its harbor and maintains the channel, facilitates international trade through the marine terminal, and encourages environmental stewardship through community involvement.

The NFPA together with the Fisheries and Oceans Canada took part in signing the North Fraser Harbor Environmental Management Plan that facilitates a proactive approach toward port development and environmental management. The plan has established a shoreline classification system; developed a more detailed project review process; created a habitat compensation banking system for industrial, port, commercial, and recreational developments; and established a cooperative management program for fish management and fish habitat.

The Fraser Basin Council (FBC) is a not-for-profit organization formed in 1997 with the vision to develop the Fraser Basin into a place where social well-being is supported by a vibrant economy and sustained by a health environment. Because the council is focused on the basin as a whole it must practice cooperative management or shared decision making in all of its activities. The FBC is a unique governance model, integrating the social, economic, and environmental dimensions of sustainability through cooperative, watershed-based action. Financial contributions to the council have primarily been from the federal, provincial, and local governments. The council is comprised of governmental, non-governmental, private, and civil sectors and facilitates the emergence of common ground in large partnership processes toward sustainable solutions. Senior involvement from each of these sectors enables the council to address multi-jurisdictional issues collectively, and then through the use of specific organizational mandates, implement actions to help resolve those issues. By choice, the council is not an authority and has no legislated powers. It functions as an impartial facilitator, catalyst, jurisdiction and conflict resolution agent, and educator. As a result of the council, more decisions are made at the community level with senior governments providing support and participating whenever possible. The FBC is renowned as a leading example of best practice river basin management. It also assists the emergence of community-based governance models in Russia, Brazil, and the Philippines.

C1.2.3.3 Results/Outcomes

The FREMP has been successful in integrating management from the federal level to the community level. Since its inception, the FREMP has been involved with coordinating decision making on conservation and development in the estuary among more than 30 agencies representing federal, provincial, and local governments and port authorities. Many success stories are also found behind the individual agencies that make up FREMP. For example, the Fisheries Act by Fisheries and Oceans Canada or the North Fraser Harbor Environmental Management Plan of the NFPA with its innovative port management represent activities happening at various jurisdictional levels, but remain interlinked by FREMP.

C1.2.3.4 Lessons/Applicability to United States

The Fraser River Estuary Management Program provides a framework for a highly successful integrated management program. The U.S. may have a significant amount to learn from the various agencies involved in FREMP. The ability for the individual agencies to not only manage jointly, but to also experience success with their individual programs, i.e. Fisheries Act, is an attribute that the U.S. should aim to achieve. In particular, the North Fraser Harbor Environmental Management Plan contains many innovative techniques for managing a harbor that the U.S. may wish to implement. The shoreline rating is used to maintain an economically and environmentally viable harbor by assisting port administrators and habitat managers. The innovative habitat compensation banking system is used to facilitate compensation for industrial, port, commercial, or recreational developments where on-site mitigation or compensation would be impractical or inadequate and where off-site compensation would be considered acceptable.

C1.2.4 Annapolis River

C1.2.4.1 Summary

Effective Approaches:	Large-scale watershed management Coordination of authority at all levels Establishment of an organization with real authority Capacity building Increased public awareness Local decision-making Local funding Alternative methods of pollution management Ecosystem restoration Ecosystem preservation
Lead Organization:	Clean Annapolis River Project
Other Stakeholders:	Government, communities
Timeframe:	1980s to present

C1.2.4.2 Description

Residents of the Annapolis watershed faced declining quality of the resources that sustained the region's economic base. Where central services existed, the sewage treatment plants provided inadequate treatment. In the un-serviced areas, on-site sewage treatment facilities were poorly operating or nonexistent. Waterways were degraded by siltation from forestry, agriculture, and suburban development. These factors were undermining the region's economy. Because it was apparent that the issues were too broad-based and involving divergent interests and the government was not acting, concerned residents concluded that they must take a pro-action stance. That opinion led to the creation of the Clean Annapolis River Project (CARP) and indirectly to the initiation of the Atlantic Coastal Action Program (ACAP).

The establishment of CARP and ACAP can be viewed from several perspectives: the change in some government agencies to facilitate a participatory partnership among all stakeholders as equals; the change in community groups, industry, and resource users to embrace and build on these evolving partnerships; and the willingness and capacity of these often divergent interests to expand the partnerships to an ever-widening circle of participants. Thus, the process was initiated when very different, unrelated interests converged.

In the late 1980s, a regional board of trade made application to the Canada Heritage River program to have the Annapolis River designated as the first heritage river in Canada. Prior to European settlement in 1605, the resources of the region had supported a rich Mi'kmaq culture. The application was rejected due to the environmental degradation that had been created after nearly 400 years of European settlement. This led the regional board of trade to sponsor several public workshops to investigate remedial options. The result was enhanced local awareness of the environmental challenges facing all watershed residents.

Also in the late 1980s, university and government researchers expressed concern about declining water quality in Atlantic Canada. They felt that community based approaches, involving all stakeholders, was the best option to address these distresses. This informal alliance of scientists, the Atlantic Estuaries Cooperative Venture (AECV), sought sites in which they might spark community responses. They choose the Annapolis due to the high level of public awareness arising from the rejected heritage river status. CARP resulted when these interests combined.

These divergent interests combined with other concerns in 1991 to form ACAP, an Atlantic region manifestation of Canada's Green Plan. CARP/AECV and other community-based experiments played a key role in the development of ACAP. The early growing pains of ACAP were at times testy. After considerable discussion, CARP was the first organization invited to participate in ACAP. The initiative was still a government-controlled process in which the participants were expected or required to accept the traditional notion of command and control by the central governmental agency tempered by traditional community input.

To Environment Canada and other institutions, ACAP demonstrated the need to change in order to bring about new partnerships. The dynamics within the ACAP process left no participant unchanged. Environment Canada's role changed from conveyor to facilitator and finally to participant. With initiatives like CARP gaining capacity and demonstrating a license to lead from citizens, the private sector, and all three levels of government, Environment Canada accommodated the community-based leadership. The initiatives and participants continue to change as all gain knowledge, discover interdependencies, and learn to work across boundaries of time, space, and interest.

To Environment Canada, community based does not just refer to place. It refers to place-centered governance initiatives where citizens, all levels of government, and the private sector; social, economic, and environmental interests; and local, regional, and national perspectives come together. The result is redefinition of community, involving all stakeholders. This facilitates shared responsibility for studying issues and developing solutions. While other definitions reduce citizens and communities to participants in government programs, ACAP requires community members to empower one another by collaborating as equals.

The early months were often tense for the CARP and ACAP managers; neither wished to give up their respective realms of ownership. Over time, trust and mutual respect increased allowing each to develop a new relationship that respected the political/institutional opportunities and boundaries that the partners required. All have come to realize that their focus must widen to include other federal agencies as well as other orders of government and expanded stakeholder participation. This is the challenge for the future as ACAP and CARP continue to evolve.

This evolutionary process cannot be attributed to particular individuals. Its success is built on weaving the interests of the diverse partners into new forms of governance. With declining environmental quality and reduced economic options, participants begin to see a shared future that benefits their self-interest. Scientists, farmers, artisans, homemakers, fishers, environmentalists, and dozens of others with differing viewpoints have learned how their counterparts view their separate worlds, and that cooperative action is essential.

C1.2.4.3 Results

While still a work in progress, CARP and the other ACAP sites have many notable environmental, social, and economic achievements. Their focus has evolved from activities to project outcomes that achieve explicit and measurable ecosystem objectives. Like other participants and supporters, Environment Canada is proud to be affiliated with the ACAP sites and their achievements. All participants are particularly pleased with water quality, habitat restoration, environmental conservation, pollution prevention, eco-efficiency, and biodiversity accomplishments. Each has made many direct and indirect contributions to sustainable livelihoods. Equally important, citizens and organizations not part of the traditional power base create and share the outcomes.

Institutional changes have been diverse and profound. Environment Canada has learned to differentiate between community-based initiatives and community participation in government programs. The process has empowered governmental representatives in community-based initiatives to represent all department branches and programs and granted them the financial authority to commit project funding. All participants have a waiting list of citizens and scientists seeking to work with these innovations. Initiatives like CARP are now seen as an opportunity to participate, not manage. Environment Canada has learned how to support a full range of community-based activities, including knowledge generation, capacity building, and actions to address a diversity of issues. Nationally, the department has changed its policy regarding ecosystem initiatives to accommodate community-based approaches.

CARP, along with the other 12 ACAP sites, is involved in the policy development within Environment Canada. The groups participate in quarterly conference calls during which budgetary allocations and policy decisions are jointly made. Having changed the way of doing business, Environment Canada, Atlantic Region is now actively recruiting the participation of other federal and provincial agencies. All groups have been represented at national and international resource management policy development processes.

CARP has developed working partnerships with most other stakeholders. The group has negotiated a letter of agreement with the provincial Department of the Environment. This is an enabling document to facilitate the exchange of human and financial resources. A wide range of academic and research expertise is available to CARP from the institutions in the region and at the national and international levels. Directly and indirectly, CARP receives support from most federal, provincial, and municipal governments; numerous corporations; international agencies;

other environmental groups; First Nations; foundations; and dozens of individuals. Other national governments have even participated.

CARP has been involved in numerous on-the-job training programs that have enhanced the skills of young people. Many of these individuals have built on their experiences and launched careers in the environmental field. These participants have included representatives from all socioeconomic groups.

Through its various programs, such as volunteer water quality monitoring, CARP has afforded residents the opportunity to teach themselves the connection between human activities and environmental quality. They have learned scientific methods and the requirements for reproducible data. This participation has resulted in an enhanced sense of stewardship of the region's environmental resources. Senior research scientists are now seeking the input of CARP and the other 12 ACAP sites.

Due to community efforts, environmental conditions have improved in the Annapolis watershed. Several hectares of salt and fresh water marsh are protected by conservation agreements or donations to CARP. Atlantic salmon are once again spawning in streams that have been restored. Wetlands have been constructed to process agricultural waste. Environmental farm management plans are being adopted. Hedgerows and riparian zones have been replanted. Water conservation and wastewater management programs have resulted reduced sewage discharges. Habitat for waterfowl and other bird species has been enhanced. Local businesses have reduced their waste stream through pollution prevention and waste reduction programs. There is now a high level of community awareness of local environmental conditions. A widely endorsed comprehensive environment management plan for the watershed has been prepared to guide enhancement activities in the watershed.

C1.2.4.4 Lessons Learned

Likely the most important lesson learned from the experiences of ACAP and CARP is that the process works as is evidenced by improved environmental conditions and the participation of all stakeholders. Other, more specific lessons have also been gained by this process.

The government–community partnership is a highly efficient mechanism for the delivery of resource management programs. For every \$1 invested by ACAP in CARP, the other partners have contributed another \$15 of real and in-kind resources.

The initiatives benefit by being holistic at both the institutional and ecological levels. On the institutional plane, this means an inclusive process that openly welcomes the participation of all stakeholders. From an ecological perspective, this means viewing the entire watershed as a single integrated, interdependent whole. It requires the incorporation of traditional knowledge with scientific research.

Each has learned that principle-centered approaches to conflict resolution and conflict avoidance used by community-based initiatives are transferable. Best processes prevail and quickly spread to participating organizations, including government departments. While all community-based initiatives are distinct, successful initiatives share common qualities. They are inclusive, open, transparent, and champions of informed decision making. They combine people and resources in combinations that generate creativity and innovations. They develop solutions that are economically feasible, socially acceptable, and environmentally sound. They align the efforts of public and private landowners, citizens, industry, governments, First Nations, universities, and others in short-, medium-, and long-term actions.

One thing unique about ACAP is the collaboration and collective capacity of participating community-based initiatives. Sites collaborate in addressing larger scale issues, answering larger scale science questions, and serving as resources to one another. Their network enables them to share information, best practices, and lessons learned and to speak with a common voice when necessary. As honest brokers, these initiatives harness and focus a diversity of resources and lever the pubic and political support necessary to address complex issues. While community-based initiatives have exceeded expectations, the best is yet to come. The distinct needs, capacities, and behavior patterns of different community-based resource management initiatives are just beginning to be recognized and accomodated.

C1.3 France

The French established a derivative of the Ruhrverband approach with the creation of River Basin Financing Agencies in 1964, for each of France's six major river basins. Under this approach, basin committees, made up of representatives from state and local authorities and various users, are responsible for preparing and implementing water resource management policies. Associated with each basin is a water agency, under the supervision of the basin committee and the national Ministries of Environment and Finance. The water agency is responsible for the implementation of policies dictated by the basin committees. This structure is illustrated in Figure C1-1.

At the national level, a member of parliament chairs a water committee responsible for giving guidance on water. A 1992 water law established by the parliament emphasized a more integrated approach to water rather than the more sector-oriented logic used in the past. Each basin was required to establish a master plan based on two main principles: 1) progressing from water management to water environment management; and 2) giving priority to collective interests. These plans, developed at the basin level, attempt to achieve balanced water resource management, in the form of ecosystem conservation and management, use reconciliation, and water resource protection and development.

The funding for the water agencies is founded on the application of the "user–polluter pays" principle (Figure C1-2). The water agencies levy water charges on withdrawals and discharges from all users who affect water quality and/or modify flow. The charges to industry are calculated using various parameters appropriate to each sub-basin and the amount of pollution produced by each establishment. Domestic charges are calculated for each community according to permanent and seasonal populations and are collected from individual users based on the quantity of water used as measured by a flow meter.

The rates and charges for the basin are based on the cost of maintaining the system as well as investments in improvements. The water agencies are, therefore, self-sustaining with total charges equal to total budget requirements.

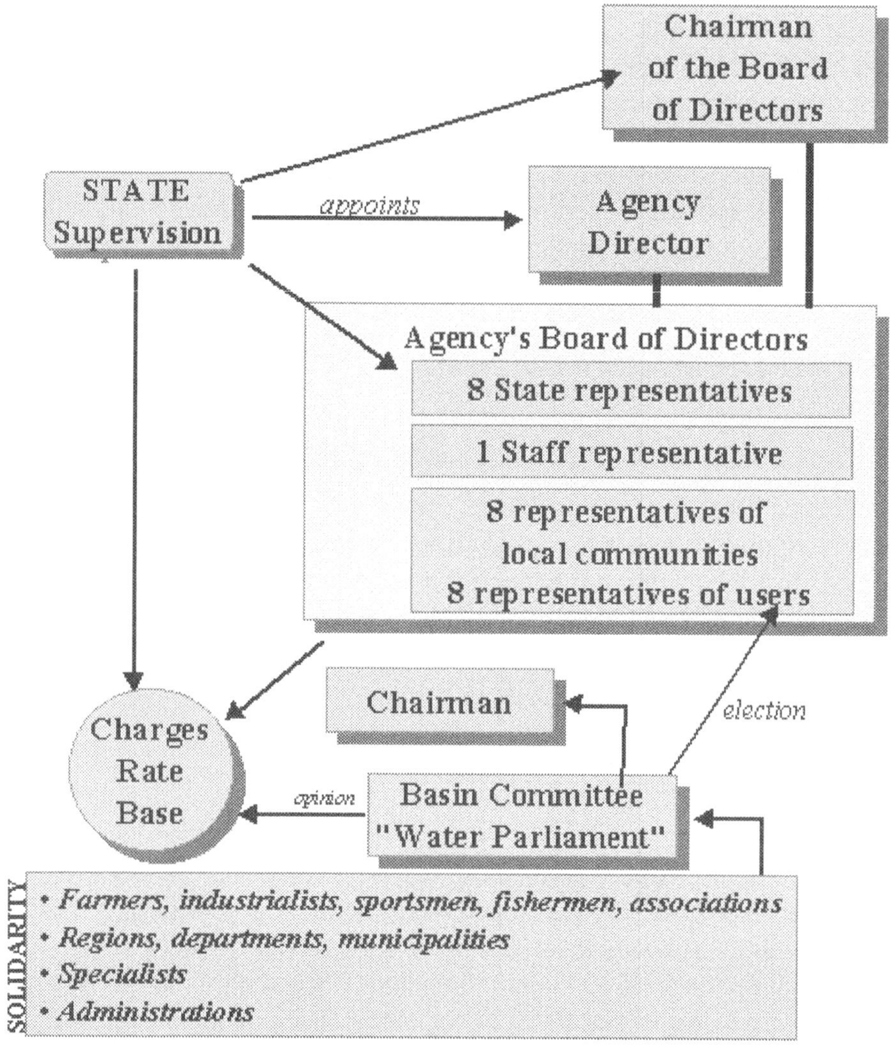

Figure C1-1 Organization of Water Management in France, Organization of the River Basin (International Office for Water, n.d.).

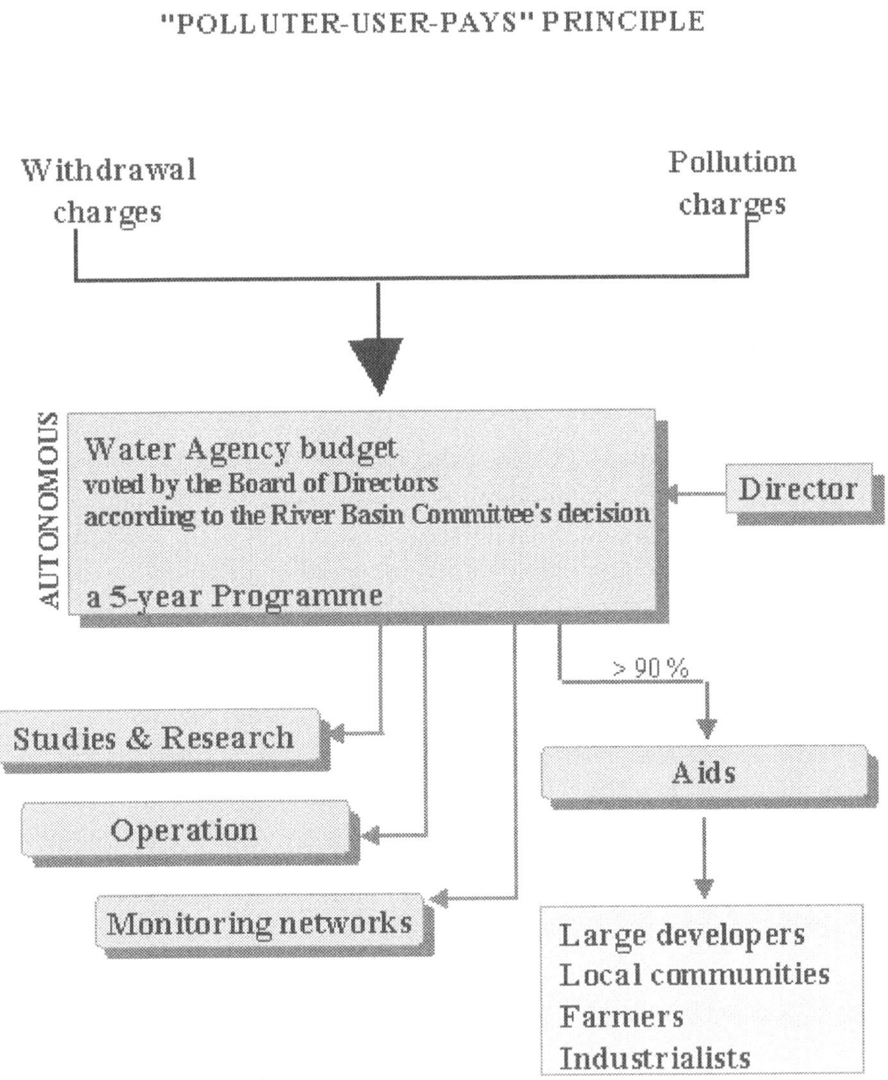

Figure C1-2 "Polluter–User Pays" Principle (International Office for Water, n.d.).

Surface water provides more than 40% of France's freshwater needs. The largest consumers, Paris, Lyons, and Toulouse, basically depend on surface water for their drinking water supply (Reimold, 1998). Upstream industrial sites and multiple uses of the watershed create conflicts regarding water quantity and quality rights. In the past, sector-oriented management was implemented to address these issues. As the resource became more limited, managers began looking for alternative strategies to integrate the needs of water consumers with the preservation of the natural resources within a watershed. France's approach to this task involved setting up legislation and the necessary administrative and technical structures and providing the opportunity for all the partners involved in water management and utilization to act together in the interest of all concerned.

France has a very detailed administrative structure for the management of water resources. This involves relationships between national government officials, six river basin delegates and committees, regional prefects, departmental heads, and local water committees.

- **National Level:** A member of parliament chairs the national water committee. The committee is composed of representatives of the national assembly and the senate and of important institutions and national federations involved. It is consulted on the trends of the national water policy and on drafts of the legislative and regulatory texts.
- **River Basin Level:** France is divided into six major watershed catchments, each have a river basin committee and a water agency. The basin committee, chaired by a local elected official is composed of representatives from the state, local authorities, and various users. The committee's responsibilities include preparing and implementing water resource management policies and supervising the water agency. The Masterplan for Water Development and Management (SDAGE) is prepared and adopted by the basin committee. This masterplan takes into account the main programs decided by public communities and defines, in a general and harmonious manner, the objectives for water quantity and quality as well as the developments and improvements to be undertaken to attain them. The water agency is a financial establishment, responsible for collecting dues from water users, which are then redistributed in the form of loans for assisting the financing of pollution control works and resource protection programs. This is based on the principal that the polluter pays and whoever cleans up receives assistance.
- **Sub-basin Level:** Local water commissions are composed as follows: one half representatives of local communities, one quarter representatives of users, and one quarter state representatives. The local water commissions are set up to prepare and follow up the implementation of the Water Development and Management Scheme (SAGE). SAGE fixes the general objectives for the use, development, and quantitative and qualitative protection of surface and groundwater resources and aquatic ecosystems, as well as the preservation of wetlands. The local water community can be entrusted with the study, completion, and operation of all construction, installation, or equipment of an urgent or general character, aiming at:
    - Developing a basin or part of a hydrographic basin
    - Developing and maintaining a watercourse that is not managed by the state, including accesses
    - Managing water supply
    - Controlling stormwater and run off

France has one of the most impressive water management systems in the world, linking local water commissions to national water committees.

C1.3.1 Paris, France

C1.3.1.1 Summary

Effective Approaches:
Coordination of authority at all levels
Establishment of an organization with real authority
Mediation of disagreements among disparate groups
Capacity building
Alternatives to dealing with pollution management
Integration of point and nonpoint pollution management
Ecosystem preservation

Lead Organization — National Water Committee (national assembly, the senate, and important institutions and national federations)

Other Stakeholders: — River basin committees, water agencies (representatives from the state, local authorities, and various users) and local water commissions (local communities, users, state representatives)

Timeframe: — 1990 to present

C1.3.1.2 Description

Dense urban and agricultural activity in the Paris area is responsible for significant deterioration in water quality. Two parallel tracks have been taken to combat the pollution problems. The first involves improving upon treatment facilities by installing the latest technologies. The Paris area is also involved in a unique plan to address urban wastewater by moving the treatment of wastewater as close as possible to its collection area. This innovative act avoids the transfer of water from upstream of towns in a downstream direction, which in addition to a high transport cost has a significant ecological impact by depriving the whole section of river of water. New technologies allow treatment plants to be built right in the middle of town, sometimes completely underground, while respecting the environment and meeting the most demanding treatment standards.

The parallel track, which is very often a weakness in other management practices, involves significant risk prevention. This approach includes monitoring resources, studying potential contamination risks, diagnosing resource vulnerability, and using crisis management tools after a pollution event, and identification of risk prevention measures.

The Paris area has also set up a program to deal with accidental pollution. By constantly monitoring all pollution sources upstream (over 50 measurements every 24 hours), water managers are kept abreast of any changes in pollutant levels. The Paris area has invested in modeling variations in water quality parameters in the event of accidental pollution, which has prepared the water managers with multiple techniques to avoid any devastating effects from any such event.

C1.3.1.3 Results/Outcomes

The Paris area has two projects that promote the integration of management of point and nonpoint source pollution. One program has brought together research establishments specializing in agronomy and hydraulic modeling, as well as manufacturers of agro-pharmaceutical specialty products. This program is establishing "an acceptable pollutant pressure according to the vulnerability of the environment." This information will be helpful in addressing nonpoint source pollution. The second program is integrating the management of farming practices by bringing farmers together with the Chamber of Agriculture, in association with the fertilizer suppliers and manufacturers and with the cooperatives whose task is to market farm produce. The task of this collective group is to carry out advisory operations on farming practices that will reduce nitrate runoff. The advisory operations address the environmental constraints for farmers, the use of fertilizers and pesticides, and the year-to-year management of cultivated areas.

A technique set forth by the Paris area is to support research into common interest among groups to come up with an initial interest among groups. This interest doesn't have be the one of highest environmental concern, but something that will be successful in getting the groups to work together first and see results of their collaborative effort with the intent of sparking interest in other collaborative projects. This approach was used with farmers in the program of changing farming practices to decrease nitrate runoff. While pesticides are also a major environmental concern, initial research found that farmers would be more willing and interested in addressing the nitrate issue.

C1.3.1.4 Lessons/Applicability to United States

The wastewater treatment facilities presented in this example seem quite innovative and promising. The United States could look into the treatment facilities that Paris is installing in their urban areas and determine if the technology is applicable to the United States. The Paris example also brought up the issue of risk prevention. It may be worth reevaluating the programs that address pollution risk prevention in the United States to compare with programs in Paris. The acceptable pollutant pressure established by the point and nonpoint source pollution management program could act as an acceptable indicator for environment vulnerability. Lastly, the approach of obtaining a common interest among stakeholder groups in Paris to initiate collaboration is an idea that could be quite useful in the United States.

While the detailed administrative structure in France may not be directly transferable to the United States, there are many other lessons and applications to obtain from this example. The administrative and technical structures implemented in France have provided the opportunity for all stakeholders involved to act together, representing the interests of all involved. This is an area where the United States could improve. The main highlight of the France example is the local water communities. Local communities are given full authority over the study, completion, and operation of all construction, installations, or equipment purchases of the majority of water resource projects. While adaptation may be necessary to implement such an application in the United States, the success of this local authority is quite impressive and worth investigating further.

C1.4 Germany

Germany is endowed with relatively large quantities of freshwater resources. By the end of the 1960s, water pollution in Germany began to cause great concern as a result of significant increases in industrial activity and pollution. The federal government in western Germany made water protection a major focal point and initiated a large number of measures designed to improve water quality as quickly as possible and with lasting effect, in both national and international water basins.

The first 10 years of environment policy, from 1970 to 1980, were initially marked by the crisis management to deal with individual situations and some preventative regulatory measures. The goal was to safeguard the environment for human health and welfare, and to protect the soil, water, air, flora, and fauna from negative human impacts. The federal government gained more control over policy, while the states (Länder) were given enforcement authority.

The guideline legislation in the field of water protection is the Federal Water Act of 1986. In order to regulate water quality and water levels, water bodies are considered an integral part of the water balance and should be managed to serve both the common good and, in harmony with this, the use to which individuals put them, so that no avoidable damage takes place. Surface water and groundwater are, as public utilities, subject to an extraordinary public management that leaves the allocation of users' rights to official discretion. The Waste Water Charges Act of 1990 complements the Water Management Act. An annually rising wastewater charge linked to the toxicity of the input wastewater is an effective financial incentive to dispose of wastewater properly. In addition, the Water Boards Act of 1991 created water and soil boards which, as public corporations, implement measures designed to improve water and soil quality. Water legislation promulgated by the Länder goes beyond merely the enforcement of the framework of federal law to determine administrative procedures and regulate issues of private water law. In Germany, the authority to establish water quality objectives in surface water or groundwater rests with the states.

C1.4.1 The Ruhr River

The River Ruhr is a relatively small river, 217 km in length with a mean yearly runoff of $76 m^3/s$ at its mouth. The Ruhr catchment area covers $4488 km^2$.

C1.4.1.1 Summary

Effective Approaches:	Establishment of an organization with real authority Mediation capabilities Local decision making Local funding Alternative methods of pollution management Ecosystem restoration
Lead Organization:	The Ruhrverband
Stakeholders:	Communities, government, business, industry
Timeframe:	1913 to present

C1.4.1.2 Description

One of the earliest examples of successful river basin management was the Ruhr Basin Water Association, also known as the Ruhrverband. The association was founded in 1913 as a self-governing body, but is subject to legal supervision from the state. The Ruhrverband established several key principles. First, all stakeholders (all users and polluters of water including communities, districts, and trade and industrial enterprises) would be members, and policies would be made by a water parliament or assembly of stakeholders, as well as a supervisory board and an executive board. Some positions are elected, some appointed, and some are based on representation from various water users/polluters. Second, economic instruments, such as water charges and pollution fees, would provide the finances for investments and other management activities of the Ruhrverband. Third, water quality objectives were to consider both the costs and the benefits of improvements. Fourth, municipalities along the Ruhr River would maintain the collection of wastewater and distribution of drinking water in their jurisdiction, while the Ruhrverband would manage the major infrastructure, such as the design, construction, and operation of reservoirs and wastewater treatment facilities.

C1.4.1.3 Results

The Ruhrverband continues as the basin water authority in charge of operating a system of reservoirs to protect drinking water supply while minimizing the risk of flooding. In addition, the Ruhrverband is responsible for the water quality management of the Ruhr. In that context, the Ruhrverband plans, constructs, and operates WWTPs, and works to reduce nonpoint sources of pollution to enhance the ecosystem quality of the Ruhr basin. The Ruhrverband has been successful in the challenge of water quality, since the River Ruhr today is one of the cleanest rivers in comparison with those in comparable industrial regions worldwide. Many species of fish, such as trout and grayling, make this their home once more, and kingfishers and other protected birds nest on the riverbanks.

Users of the Ruhr must pay subscription fees to the Ruhrverband association to allow it to accomplish its tasks. The Ruhrverband Act outlines the principles for subscription charges. The statutes and assessment guidelines of the Ruhrverband detail the exact distribution of subscriptions. To calculate subscription fees for water quality management, the dischargers of wastewater are assessed according to the quantity and quality of the wastewater discharged, the expense incurred in removing the pollution and the advantages gained in disposing of it. The subscription fees for water quantity management as well as the part of subscriptions for water quality management, which have to be paid by the water abstracters, are calculated according to the quantity of abstracted water.

C1.4.1.4 Lessons Learned

There are several important lessons from this example:

- Self-administration guarantees participation in the accomplishment of tasks and in the development of guidelines for the distribution of fees. Here it is in the interest of the Ruhrverband to apply the principle of causality, the guiding principle in modern environmental law.
- Since the Ruhrverband is responsible for the entire Ruhr catchment area, it can carry out its tasks according to natural conditions, regardless of community boundaries.
- The organizational structure produces cost-saving effects for the Ruhrverband in design, construction, and operation of its plants.

C1.5 Japan

Physical characteristics as well as cultural and social patterns have made river management an ongoing concern for Japan. The Japanese islands are narrow and mountainous, causing rivers to frequently have sudden and sometimes extreme changes in volume. Japan has historically placed a high priority on damming all of its major rivers, because of a perceived need to stabilize the speed and volume of flowing waters. This is in contrast to the traditional value in Japanese culture of closeness to nature. Massive infrastructure accompanied the construction of dams, resulting in complex irrigation and drainage systems to support rice farming and urban water supply.

A river law enacted in the mid-1960s classified rivers into two categories: A and B. Class A rivers were under central government authority; Class B rivers were under local authority. The central government was to pay two-thirds of the costs of new infrastructure for Class A rivers and half the costs for Class B rivers. Engineering projects included channel straightening, hard-lining banks and beds, grade control work, and the construction and maintenance of dykes and reservoirs.

This approach to river basin management changed in the early 1990s with public outcry over the construction of the Nagara Estuary Barrage. The opponents of the project expressed skepticism about the need for the barrage for flooding issues and a lack of consideration for the effects on river ecology.

The resistance to the heavy engineering approach, in combination with exposure to river management approaches in Germany and Switzerland by central planners led to a close-to-nature movement. In preparation for the next five-year plan on river management in 1992, the Ministry of Construction asked for new perspectives. A set of recommendations was drawn up in 1991 that emphasized the need to create beautiful, relaxing, and ecologically sound river environments as part of any program for flood control. The 1997 five-year plan required stakeholder participation.

The example below illustrates the effectiveness of the close-to-nature movement.

C1.5.1 Neo-Natural River Reconstruction in the Ara and Oppe Rivers

C1.5.1.1 Summary

Effective Approaches: Ecosystem restoration

Lead Organization: Ministry of Construction

Other Stakeholders: Expert scientists, local governments

Timeframe: mid-1980s to present

C1.5.1.2 Description:

The Ara River flows from the northern mountains to Tokyo. Extensive development has lead to ecological deterioration in the river. The Japanese adopted the European approach of neo-natural river restoration to address the degradation issues of the Ara River and its tributaries.

In the early 1980s, Germany was clearing out forests at an accelerated rate to create space for urban development and agricultural land. There was a collapse in the ecological balance of plants and animals in rivers. Rivers once considered clean now were lined with concrete. In 1984, the idea of neo-natural river reconstruction emerged from Germany with the intent to address ecological deterioration in the river systems. Neo-natural river reconstruction consists of the introduction of various species of flora and fauna into a river environment, not with the intent of recreating the organic ecosystem that has deteriorated at the hand of man, but creating a new ecosystem that protects the surviving inhabitants, restores ecological balance, and improves the living situation of animal and plant life. Some refer to these locations of restored nature as biotops (meaning a place for life). Original policy elements of neo-natural river reconstruction, or biotops, include the following:

- A river basin must be regarded as an ecological complex.
- Pre-existing species of flora and fauna should be preserved to the greatest possible extent, and species diversity in the ecosystem should be enhanced.
- The characteristics of each species' habitat should be developed in harmony with the natural dynamic alterations that take place in a river, such as erosion, sedimentation, and floods.
- A balance of natural and human influences according to the condition of the surroundings should develop the characteristic of each river.
- Consistent maintenance of the natural habitat and its unique characteristics is a priority.
- First preserve what remains of the original biostructure.
- Repair and maintenance should be done in harmony with ecological rhythms and cycles.
- Incorporate new agricultural technology.
- Provide sufficient space as dictated by the ecological system requirements.

Canals were dug into the riverbed where there used to be flat barley fields. The canals vary in width and depth and are intentionally crooked to provide a better medium for species variety. The terrain surrounding the river was made uneven with small mounds and gaps, creating shelter and protection for small animals. Kingfishers were provided wooden posts for resting spots, and the banks of the river were gently sloped, allowing for plant growth and fish spawning habitat. The restoration did require the removal of several recreational facilities. In 1994, the Japanese pursued the idea that connecting two or more biotops would create a better environment for a greater number of animals and plants. Using yashiki groves and paddy fields as a stepping-stone network, tablelands and forests are connected.

C1.5.1.3 Results/Outcomes

After only five months of biotop construction, willows were sprouting and kingfishers were breeding. Foxes, raccoon dogs, and various types of ducks now inhabit the area. The Ara River Basin is now home to an extensive wildlife population with a variety of water animals and plants. The basin is also home to some species that are on the Japanese endangered species list. This success is occurring within 40 km of central Tokyo. Other biotops have been installed in other places in Japan. For example, the Oppe River Biotop was created on a site that has been degraded due to past excavations from the local quarry. The natural environment on Oppe River is slowly reviving. While the neo-natural river restoration concept is beginning to spread, the outreach has not been extensive. Valuable information for future projects will come from the monitoring of wildlife in current biotops.

C1.5.1.4 Lessons/Applicability to United States

There are many locations in the United States where river ecology has been degraded to a point where the original natural state cannot be restored (e.g., Kissimmee River, Florida). The approach presented in this example may be applicable to these areas. The neo-natural river restoration approach has been successful in transferring from its initial conception in Germany to full implementation in Japan. While this is an artificial approach and emphasis should be placed on conserving the natural environment and creating a supportive environment for wildlife, the neo-natural river restoration approach is a well-developed tactic for dealing with river ecosystems that have already been degraded.

C1.6 Mexico

The Mexican Constitution of 1917 established that all water resources are held in trust by the state, and only the state can grant the right to use water resources. The implementing agency was the Ministry of Agriculture and Hydraulic Resources. By the 1990s it was clear that water security was a real handicap for economic and social development, and that pollution was becoming a serious problem. In response, Mexico engaged in a fundamental reform of its water sector in 1992, which largely embodied the Dublin and Rio principles. Of particular note were two features: the importance of local participation and formalizing property rights in water. In order to implement the new law, an agency was created, the National Water Commission (CNA). The CNA is the federal agency in charge of the administration, allocation, and custody of Mexico's water resources.

The CNA has initiated a decentralization process that attempts to reform the traditional forms of governing where the central authority has the bulk of the decision and financial resources capacity. One key feature of the decentralization program is increasing users' participation in the decision making of the water industry through basin councils. The councils include representative farmers, peasants, industrialists, and service providers. The basin councils are conceived as a mediator between the CAN, governmental offices, and representatives of the water users within each water basin. Among their duties is the prioritization of water use within their basin. For the municipal and industrial sectors, the CNA controls water quality through the use of fees. Any discharge that is properly treated is exempt from fees.

Important progress has been made in the implementation of irrigation management reforms. The government reorganized irrigation system policies around two basic concepts: a program for transferring their operation to organized users, and a set program aiming to increase the productivity of the agricultural sector. Reform of the irrigation districts has been a remarkable success and an inspiration for reforms in India and Turkey. At the heart of the "irrigation turnover process" was the idea that farmers could form user associations that could manage the irrigation districts. In less than a decade, the results have been impressive. Over 92% of the three million hectares in irrigation districts have been turned over to the user associations representing half a million farmers. With users in control, many good things have happened. Cost recovery has risen from 30% to 80%. In some cases, user associations have mobilized their own funds for rehabilitation and investment. And in some cases, as in the coastal aquifers of Sonora, user associations have evolved into groundwater management agencies with remarkable effect. In Hermosillo, for example, users have voluntarily reduced pumping by about 50%, bringing pumping and recharge into balance. This has been made possible by participation and because those who need water for high-value crops are able to obtain it through transactions in the water market.

C1.6.1 The Bahais de Huatulco

C1.6.1.1 Summary

Effective Approaches:	Coordination of authority at all levels Increased public awareness Local decision making Integration of point and nonpoint source pollution management Ecosystem preservation Ecosystem restoration Sustainable development
Lead organization:	The Centro de Soporte Ecológico (CSE)
Other Stakeholders:	Local stakeholders, indigenous people, tourism industry, government officials
Timeframe:	1990 to present

C1.6.1.2 Description

In 1958, the landscape of the coast of Oaxaca was primarily tropical forest contrasting with the blues of the Pacific Ocean. Less than 5% of tropical forest included traditional fields of corn, beans, and fruit trees; canopies of shade trees covered coffee plants. This isolated area was home to four indigenous ethnic groups.

The introduction of a mega-resort and accompanying infrastructure catapulted the region into the international market, inducing a self-reinforcing cycle of speculation and investment that accelerated the process of impoverishment among the indigenous populations and provoked a growing water shortage. Forty years later, the forest coverage had been reduced by 50%; only 20% resembled its former condition, while the rest suffered from a partial extraction of its timber resources. During the past 15 years the rate of deforestation doubled that of the previous 25 years. These tropical dry forests are among the most fragile ecosystems in the world and are rapidly disappearing.

The indigenous population is now approximately 50,000, living in some 150 subsistence communities widely dispersed over 700,000 ha in the surrounding highlands or in a number of small fishing villages located along the coastline.

In 1992, a local NGO, the CSE, was founded to promote sustainable development. Initially, the CSE concentrated on collaborating with local indigenous communities to rebuild their denuded mountainsides through a long-term program of reforestation. This also included initiatives for landscape management and a modest ecotourism project with quality bungalows to accommodate visitors. The sudden destruction wrought by Hurricane Paulina in October 1997 intensified these problems. Following Paulina, it became clear that a more ambitious program was needed. One of the smaller river basins within the region was selected as a site for a pilot program of intensive reconstruction, involving a series of local works projects to protect river beds and banks, clearing the waterways of debris, and regulating water flows and infiltration to underground aquifers.

Perceiving an opportunity to broaden its program, the CSE initiated specific activities to stem environmental degradation and increase total water availability in the region. To ensure greater water supplies and to stabilize flow rates, large-scale reforestation and river management work was initiated. Another fundamental change was the introduction of the concept of compensating peasant and indigenous communities for the environmental services they offer. These services represent tangible benefits to downstream users and society and are having a major effect on regional planning, contributing to the possibility that these communities continue their conservation and protection activities.

Because of their precarious existence, these communities must be compensated for their collective efforts to increase water production. To be effective, this compensation must include some sort of long-term guarantee for the peasants. The rewards can take several forms:

- Direct payments for their contribution to increasing water supplies, akin to the nascent mechanism of carbon sequestration bonds now marketed to finance some forest conservation efforts in Central America, Bolivia, and Mexico
- Better prices for their goods
- Increased output in existing systems and/or the introduction of new products and services that complement and add value to traditional output

Most people in the region were not even aware of the depths of the impending water crisis. In Huatulco, urban consumers received water for free, and although they complained, the hotels were only charged a fraction of what they would pay at other international locations. In contrast to this lack of consciousness of the water shortage on the coast, the upland communities were acutely aware of the problem.

C1.6.1.3 Results

This reforestation program has already shown great success. The first round of plantations resulted in germination rates exceeding 90%, and replantings brought the effective rate to virtually 100%. The selection of varieties, the techniques used and the anticipated marketing opportunities are creating an extraction profile that will allow the first harvest only five years after the initial effort. Furthermore, the program proved effective from the very beginning as a water production measure; the minor land management works and tree planting have already acted to reduce the speed of water runoff, reducing erosion and visibly increasing water infiltration.

The organizational and funding structure is also innovative. The CSE is a constitutive part of a several local trust funds that channel resources from governmental agencies, the communities, and the private sector into the program. Individual enterprises established by the communities (including a pure water-bottling program in the highlands and the ecotourism project) are coordinated from the CSE. To that end, it has established a formula that attempts to create a solid foundation for future activities. Prices for goods and services must be sufficient not only to cover the direct costs of production, but also to make substantial contributions for further investment in the community and environmental programs in the region as a whole.

The program envisions an eventual charge to the large water users on the coast to cover part of the costs of the regeneration of the river basin. At present, this is difficult because the water system is controlled by the tourism agency, which has been unable either to fulfill its commitments to deliver a high-quality product or to maintain its infrastructure properly. In fact, inadequate water supply is considered to be a major obstacle blocking construction of at least a dozen new hotels in the area. Once the communities in the watershed's upper reaches have demonstrated their willingness and effectiveness in producing water, as well as in reducing the damage from seasonal rains, it is expected that the local authorities will be able charge large-scale users for these environmental services. In the interim, other mechanisms of explicitly integrating the coastal beneficiaries into the program are being explored:

- The hotels are expected to provide some support for the indigenous communities' ecotourism activities, including promotion among their clients and financing of some of the construction.
- Future plans include reserves for native flora and fauna, possibly including areas for large mammals, once common in the zone.
- The communities will shortly begin to develop a dependable capacity to supply fruits and vegetables to the hotels.
- Contracts are under development that will compensate the communities correctly for the real costs of water production, including fair wages for the workers and a charge for the environmental services that are normally not included by the market.

Regular channels are being developed for regional discussions of activities that will increase the overall attractiveness of the area for visitors in a sustainable fashion. This communication network is fundamental to successful collaboration among dissimilar groups and is essential for the long-term consolidation of the CSE agenda.

C1.6.1.4 Lessons Learned

The Huatulco experience illustrates the possibilities of developing a water management program that contributes to sustainable development. A number of original contributions are proving effective. Perhaps most important is the integration of water production and conservation programs into a broader process of diversified, community-based resource management for sustainable development. This is the only way in which the long-term goal of creating an integrated regional program can internalize the true environmental costs; the beneficiaries will participate by paying for the environmental services through the prices of products and services generated in the communities, a concept that will eventually be extended to the use of water itself.

In the meantime, a solid basis is being built to balance downstream-user benefits with obligations. The project (CSE and the trust funds) has moved slowly but deliberately to educate and invite participation from the most important private sector groups; although only one of the hotels has participated to date, the intensity of the water problem will soon require some sort of further collective action. In contrast, several government agencies greeted the project with alacrity since it relieved them of several responsibilities they had been unable to meet adequately. Their contributions have generated temporary employment and supported this unique and efficient integrated regional development program.

Even more important, this program reinforces the vitality of the region's indigenous communities. Since they are the organizational structure through which the program is implemented, not only are their direct environmental management services being rewarded, but their members are also benefiting as they are incorporated into new activities and begin to develop a better understanding of their opportunities in international markets. In this way, they are once again able to define a set of priorities that will ensure their continuity as unique ethnic and/or social groups, while also raising the living standards for their members, now and in the future.

C1.7 New Zealand

New Zealand created river boards in 1868 to deal with flooding and erosion problems. By the late 1980s there were 20 regional boards. Standards were set at the national level, and the regional boards were responsible for meeting these standards, but each board also took into account the issues within their individual watershed (water quality from agricultural runoff, land rights of indigenous people, etc.) In 1989, New Zealand consolidated the regional boards into 16 entities according to watershed boundaries. Combining the local governments into watershed groupings reinforced the idea that the communities most affected by management decisions should be involved and support the management of resource allocation and use in their local area. The new regional councils include technical specialists in the physical science and engineering professions and have sole responsibility over soil conservation, water and air quality, waste disposal, and geothermal resources. Responsibility for coastal resources is shared with the central government, while responsibility for natural hazards, noise, and cultural heritage is shared with the local government.

In 1991, New Zealand passed the Resource Management Act (RMA), which governs the management of natural resources and environments. The RMA is governed by two major principles: sustainable management is the overall objective, and the mechanism for decision making is to move from the central government to the regional and local levels. Under the RMA, each region is required to formulate policy and vision statements to establish the local methods for reaching the goal of sustainable management. Public participation, including that of the indigenous people, is part of the process for formulating these statements. The territorial governments, private resource owners, and commonwealth resource managers are required to observe the policies created by each region. National environmental standards, national policy, water conservation orders, heritage protection orders, and coastal policy remain the responsibility of the Ministry of the Environment (central government).

Implementation of the RMA has mostly been in the hands of local and regional councils, with minimal central governmental support. While the objective is to move from the central government to the regional and local levels, in the initial implementation of the RMA, it would have been useful and more productive if the central government had provided more guidance. The RMA has added more complexity to New Zealand's resource management by adding several levels at which planning is occurring and often the quality varies at these different levels.

The Ministry of the Environment is also responsible for reporting on the national state of the environment. A group at the Ministry of the Environment is developing a core set of national indicators grouped into 14 elements, one of which is a freshwater indicator. They are also developing a river environment classification, which will provide the context for interpreting indicator data.

A downside to this approach in New Zealand is that now resource management agencies are largely autonomous units. This structure has weak incentive for collaboration between agencies and no one central agency that maintains an accessible overview or database on projects and programs undertaken by different agencies. The new government of 1999 intended to pursue a partnership approach to managing the environment and acknowledges the difficulties that agencies have in pursuing collaborative partnerships under the current institutional arrangements.

This example from New Zealand shows that organizing management according to watershed boundaries is a workable method for ensuring local control over water and water-related resources. This will require natural boundaries to blend with previously established administrative boundaries, which may involve aggregation of small administrative units to constitute regions approximating the watersheds. The New Zealand example also shows that responsibilities for water and water-related resources can be logically divided among local, regional, and national authorities.

C1.7.1 Harbor and Catchment Management in Whaingaroa

C1.7.1.1 Summary

Effective Approaches: Increased public awareness

Lead Organization: Ministry of the Environment

Other Stakeholders: Local community

Timeframe: Began in 1995

C1.7.1.2 Description

The community of Whaingaroa, New Zealand, is focusing on integrated catchment management and harbor protection. In 1995 the Ministry for the Environment provided financial support to determine the transferability of an integrated catchment management model from the North American ACAP to Whaingaroa. In addition to working on transferring the ACAP model, Whaingaroa has tried to find different avenues that would result in more stakeholder participation and involvement in water-related management issues. Whaingaroa has invested in research on farming practices in the area. The research has been ongoing for the past few years. Its goal is to convince farmers to change their non-sustainable practices.

C1.7.1.3 Results/Outcomes

Due to lack of local preparation, the application of the ACAP resulted in some "alienation within the community." Further investigation determined that not all principles of the ACAP model are locally applicable within the New Zealand community. There appears to be "a clash of cultures within the catchment." For example, the area is experiencing rapid population growth and the "fundamental ecological nature of the catchment has not been taken into account." There are many local residents eager to help protect the area, but there is little coordination of these efforts. There seem to be communication difficulties in transferring the knowledge from the research agencies to the community groups (i.e., farmers). National strategies on biodiversity management and sustainable water management recently released should provide some of the drivers for local communities to require change. Individual property rights and the potential adverse effects on rural profit margins would seem to be two of the many reasons why rural land use changes are difficult to initiate without regulation.

C1.7.1.4 Lessons/Applicability to United States

When thinking about applying a model from a different country, it is important to prepare not only the regulation authorities, but also the local community. It is also important to investigate the applicability of the model to the local area with great detail, consider future growth, and determine the longevity the applied model will have on the local community. This example shows that coordination plays a key role in stakeholder involvement and participation. There may be plenty of stakeholders willing to participate, but if there is no organization, the volunteers will lose interest and eventually doubt the success of the program. This example also shows that good communication and a similar language used between researchers and the local community are key elements to successfully motivating the local community to change practices.

C1.8 South Africa

The majority of users in South Africa (SA), as well as the government, had regarded water as a free good that the government must provide. The SA government had made little attempt to recover the real costs of water supply from the beneficiaries, and consumers were required to pay only a nominal tariff that was not consumption related. The lack of recovery of even the water supply and sanitation service operations and maintenance costs has all the ramifications that have been experienced elsewhere in the world, including:

- Failure by the community to identify with the water system because it is state owned;
- Inappropriate system design resulting from lack of consultation with the community;
- Increasing the national operations and maintenance cost burden, which reduces funds for new projects; and
- Ultimately, wasted water.

A major shift in governance occurred in SA after the election in 1994, and the subsequent introduction of the constitution in 1996. The constitution guarantees the right to life, life with dignity, access to sufficient water, and an environment that is not harmful to health or well being. In response to these objectives, the newly appointed minister of the Department of Water Affairs and Forestry (DWAF), Kader Asmal, pioneered major reforms in the water legislation, which culminated with the National Water Act of 1998.

The water act of 1998 made public recognition of the value of water a priority and established a reserve of water to ensure that basic human and ecological needs are met before water is allocated for commercial or industrial uses. The act imposed a sliding cost scale to allow the poor to pay what they can afford. Higher water tariffs for those who can afford them provide financing to cover operating and maintenance costs and capital costs. The water act further acknowledged the duty to ensure neighboring countries have equitable access to shared rivers.

Responsibility for water supply in SA is divided among central, regional, and local authorities and nonprofit bulk-supply authorities (water boards), with the central government's DWAF managing the overall policy framework. DWAF is responsible for operating many of the country's major dams, setting policy, issuing forestry permits (based on estimated water use and runoff reduction), and coordinating long-term water-resources development. The water act established democratically controlled water boards, based on the concept of ICMs. These water boards are called CMAs and represent each major river. The purpose of the CMA is to place water resources management at the local level and involve local communities within the framework of the national water resources strategy. The CMA investigates and advises on the protection, use, development, and conservation of the management area, guided by a sustainable catchment management plan (CMP). The CMP takes into consideration the state of the catchment and develops management strategies that will guide the use, conservation, control, and protection of the water resources within the catchment.

Each catchment holds catchment management forums to allow for further decentralization of responsibilities and decision making to the local level. The CMP's function is to encourage voluntary participation in ICM by providing a forum for stakeholders, organizations, and various tiers of government to come together and discuss the opportunities and problems in their catchment and lobby for action.

CMAs may be further divided into water user associations (WUA), largely created to be inclusive of pre-existing irrigation associations. WUAs have the right to participate in the election/appointment of members to the CMA.

The water act created a formal process for conflict resolution, including the creation of a water tribunal comprised of persons appointed by the minister but nominated through a public process.

C1.8.1 Working for Water Program

C1.8.1.1 Summary

Effective Approaches:

Coordination of authority at all levels
Establishment of an organization with real authority
Mediation of disagreements among disparate groups
Capacity building
Increased public awareness and involvement
Alternatives to dealing with pollution management
Integration of point and nonpoint source pollution management
Ecosystem restoration

Lead Organization: Department of Water Affairs and Forestry

Other stakeholders: Other government agencies and communities

Timing: 1999 to present

C1.8.1.2 Description

The South African government's Working for Water (WFW) program uses labor-intensive methods for clearing invasive alien plants from water catchment areas and river courses. Although not formally a watershed management program, the scope of the WFW's work on invasive species is based on the effects on water resources by watershed unit. The aim of the program is to recover water lost to, and prevent the further spread of, invasive alien plants. The program also helps to conserve biological diversity and ecological integrity, while creating employment and training opportunities within the local communities. The problem posed by invading alien trees is that they use considerably more water than natural vegetation and, as a result, dramatically reduce groundwater levels and the flow of water in rivers.

The National Water Act provides the WFW program with broad legal authority. For example, WFW may develop criteria by which landowners can be charged for the stream-flow reduction of invading alien plants. This is effectively a user-pays charge that will help to fund the clearing of the invading alien plants in the affected catchment. A second facet of the legislation relates to catchment management, where water users are charged in order to raise the portion of funds necessary to ensure that catchments are kept clear of invading alien plants. The argument here is that it is in their self-interest to fund the work, as the costs of not doing the work would be far greater. It also relates to the requirements regarding to the water reserve (i.e., ensuring that there is sufficient water for ecological functioning). The third major aspect relates to the section on pollution, where it is possible to charge landowners with invasive alien plants for the seed pollution on the land of others. Although this power is not implemented at this stage (and certainly not retrospectively), it remains a powerful option for control of invasive species in the future.

The WFW program has been funded principally from the SA Government Rural Development Program and poverty alleviation funds, and through water tariffs from water users. WFW recognizes that in order to succeed there is a need for greater support from all of the beneficiaries. The project at Walker Bay, near Hermanus, is an example of how WFW's funding is undergoing a major shift, where the landowners pay for half of the clearing costs. They have also agreed to pay the full maintenance costs of keeping the area clear.

C1.8.1.3 Results

The WFW program invests in the most marginalized sectors of SA society, the unemployed, while enhancing ecological integrity. Involvement of the unemployed sector further ensures an awareness of the importance of sustainable use of resources and promotes the development of sustainable economic benefits from land and water resources. There are currently 300 projects in SA, spread through the nine provinces. Projects are selected on the basis of the need for follow up of areas previously cleared, the impact on regional water resources, the extent and distribution of alien plants, the level of poverty and unemployment, sustainability, capacity, and the potential for institutional partnerships.

In fact, one of the program's key successes has been the development of partnerships at an interdepartmental level and other key stakeholders outside government. These partnerships have worked well. Partnerships are beneficial in that expertise across line functions is maximized, the duplication of work is avoided, and, in many instances, funding has been secured. Interagency cooperation also facilitates the development of more effective policies and legislation.

The formation of partnerships has not been without certain challenges. In the past, it has been difficult to get agencies to cooperate, primarily due to a lack of political support. To meet these challenges, the program has spent a great deal of effort to secure political endorsement, to invite representatives to the national steering committee, and to consider the exchange of staff from co-operating agencies.

Through a government partnership with the Department of Land Affairs and Agriculture (DLAA), the DLAA has funded WFW projects in communities involved in land reform initiatives. The partnership grew out of a suggestion to resolve a crisis in a land restitution case in the Western Cape. In 1998, the DLAA agreed to fund a number of partnership projects that were win-win for the agencies and also of benefit to the local communities.

Another very strong partnership is developing between the program and the Department of Welfare and Population Development, with jointly run initiatives on reproductive health, employment of the victims of crime and of ex-offenders, and the t-shirt production by community sewing groups. The department also provides valuable guidance in the welfare-related efforts of the program.

C1.8.1.4 Lessons

The WFW program's success is largely due to the combination of coordinated efforts between national level departments and local communities and emphasis on public education and involvement. This example illustrates the importance of approaching watershed management from a holistic view, one that includes social as well as environmental concerns, and makes clear to the public that environmental impacts have real economic consequences.

C1.9 Tanzania

Tanzania is endowed with abundant surface water and groundwater sources. However, water resources development in the country was managed following sectoral, regional, and district interests of the central government. The government agencies involved in the planning and management of water resources did so without consulting or involving water users/stakeholders. This resulted in the following problems:

- **Conflicts**: The limited amount of water for a growing set of demands resulted in serious conflicts between upstream farmers and downstream users and between agricultural and hydropower interests.
- **Pollution**: Serious pollution of the water sources took place at the catchment area where population and industrial densities are high, affecting drinking water interests and the food and beverage industry.
- **Inefficient Water Use**: Traditionally, irrigation furrows were also used for domestic purposes, using an extensive network (approximately 1800 km) of canals. The canals were mostly unlined and had no facilities to regulate water flow. The efficiency of the traditional furrows was estimated to be as low as 20%.

- **Deforestation**: The increasing population in the catchment led to an increased demand for energy, construction materials, and land for agricultural production and livestock. This resulted in the destruction of forests, which in turn affected river flows as some of the springs dried up.

The strategy to improve conditions in the country rested primarily in the decentralization of water resources management. The principal legislation governing water resources in Tanzania is the Water Utilization Act No. 42 of 1974 with amendments in 1981, 1989, and 1991. The 1991 amendment divided the country into nine hydrological areas, with the analysis, evaluation, and management made at the basin level. In pursuing this strategy the government established basin water offices in Pangani (1991) and Rufiji (1993) as a first step. Pangani was identified to need immediate attention because of serious user conflicts and deterioration of resources due to misuse and lack of comprehensive planning.

C1.9.1 The Pangani Basin

C1.9.1.1 Summary

Effective Approaches:	Coordination of authority at all levels Local funding Mediation of disagreements among disparate groups Capacity building Increased public awareness
Lead Organization:	United Republic of Tanzania government
Other Stakeholders:	Public, private interests, NGOs, women's organizations, the district, village communities
Timeframe:	1991 to present

C1.9.1.2 Description

The Pangani River Basin has a catchment area of 42,200 km^2. In 1991, the government adopted a river basin as a planning unit in hopes to decentralize water resource management. Water boards were formed at two main levels, the Central Water Board (CWB) and the Pangani Basin Water Board (PBWB). The CWB is set at the national level with 15 members appointed by the minister for water affairs. The CWB acts as the principal advisor on national water issues and is given full executive power on pollution issues. The PBWB consists of six members from the public, private interests, NGOs, and women's organizations, again appointed by the minister for water affairs. With the cooperation of the government and regional authorities, the CWB and PBWB coordinate water resource management and water pollution efforts at the basin level. The two boards also promote participatory planning and involvement of stakeholders in decision making at the district or village level. The boards aim to educate and train the communities on the need for management, protection, and conservation of water resources. The CWB and PBWB encourage and support communities to form their own WUAs and to make the communities more aware of the factors contributing to the reduced flow in the river and the direct impact those factors have on their socioeconomic livelihood.

C1.9.1.3 Results/Outcomes

At district to village levels the WUAs promote participatory planning and systematic involvement of stakeholders in decision making. At these stages, the communities are educated and trained on the need for management, protection, and conservation of water resources. Moreover, they are made aware of the factors contributing to the reduced flows of the Pangani River that affect their socioeconomic livelihood and are encouraged to form WUAs. Amazingly, the communities and users/stakeholders are responding positively to their best capacities whenever their cooperation is needed. As a result, the Pangani Basin Water Office has achieved a lot toward improving the situation in the basin.

- All water withdrawals from the Pangani River Basin have been registered, mapped, and stored in the database.
- Control gates and canal lining have been installed, thus improving efficiency and improving flow levels.
- Reduced conflicts and efficient water use has evolved as a result of WUAs.
- Through education programs, there is an increasing awareness of the value of water. Stakeholders/users now have to apply for water rights, and are charged in two ways:
    1. Application fees: cover the costs of water rights application forms.
    2. Water user fees: paid every year for the total amount of water used. These funds have enabled the office to operate without any budget allocation from the central government.
- Some pollution has been reduced.
- A water quality monitoring program has been initiated.

C1.9.1.4 Lessons/Applicability to United States

This example reinforces the need for local user involvement. For example, the WUAs promoted the use of an irrigation timetable created by irrigation users that not only reduced conflict among the irrigation users, but also increased efficiency of water use.

C1.10 United Kingdom

Until recently, Great Britain managed its water and watersheds through river authorities that had responsibility for managing individual drainage basins, with jurisdictions defined by watershed boundaries. The river authorities controlled all aspects of the water cycle within their areas. The Water Act of 1989 privatized the river authorities, resulting in 10 water companies that effectively act as monopolies for water distribution, abstraction, and treatment, as well as the collection, treatment, and disposal of wastewater.

The federal government does retain some control over water management through the Environment Agency (EA), which is responsible for the protection of the water environment and management of competing demands, primarily through a system of licensing abstraction and wastewater discharges. With respect to water planning, EA has input into and provides final approval of the formulation of each water company's water resources strategy. The Office of Water Services (OFWAT) is the economic regulator of the water industry.

The water companies are required to use cost benefit analysis (CBA), including monetary valuation of non-market environmental and social impacts, and must consider supply augmentation and demand management in planning.

C1.10.1 The Mersey Basin Campaign, Northwest England

C1.10.1.1 Summary

Effective Approaches:	Coordination of authority at all levels Increased public awareness Local decision making Local funding Alternatives to dealing with pollution management Integration of point and nonpoint source pollution management Ecosystem preservation
Lead Organization:	The Mersey Basin Campaign (MBC)
Other Stakeholders:	Public, private, and volunteer organizations
Timeframe:	1985 to present

C1.10.1.2 Description

The Mersey Basin covers almost 5000 km^2 and supported a strong fishing industry until the late 18th century. In the early 19th century, textiles, chemicals, and paper industries experienced rapid expansion in the Mersey Basin. Many industries discharged their untreated effluent directly into rivers, polluting the water for human and aquatic life. Conditions improved with the introduction of national regulations to control wastewater discharge during the 1950s and 1960s; however, the river remained heavily polluted in 1980.

Due to strong public support, the national government initiated a partnership approach to improving conditions in the Mersey Basin, with the result of the creation of the MBC in 1985. The Department of Transport Environment and Regions is providing core funding for the MBC through 2010. The goal of the MBC is to promote, manage, and support a flexible partnership between organizations and individuals, and to empower communities to act for themselves. The MBC has enjoyed the backing, expertise, and resources of its partners, which include local authorities, such as United Utilities (formerly North West Water, Ltd.); businesses; voluntary organizations; and government-sponsored agencies, including the Environment Agency.

The aims of the campaign are as follows:

- To improve water quality so that all rivers, streams, and canals within the MBC area are clean enough to support fish by 2010.
- To stimulate the strategic development of attractive, sustainable waterside environments for wildlife, recreation, heritage, housing, business, and tourism.
- To encourage communities in all sectors to value their watershed environment.

A notable element of the MBC is the involvement of businesses. In 1987 ICI, Shell, and Unilever agreed to support campaign-related projects. With the increasing number of businesses actively supporting the campaign, the Mersey Basin Business Foundation was launched in January 1992. The original three members of the foundation have increased to more than 20 and cover a wide range of business sectors. The principle aim of the business foundation is to accelerate the progress of the MBC by encouraging greater business participation in partnership with the public and voluntary sectors, contributing to the strategic development of the campaign, and providing sponsorship.

C1.10.1.3 Results/Outcomes

Already, many projects have been implemented with strong support from the Department of Transport Environment and Regions, EA, and from other campaign partners. The main project themes include grant schemes for voluntary groups, campaign publicity and promotion, environmental education for schools, water quality monitoring systems, encouraging sponsorship for River Valley Initiatives (RVIs), and award schemes and competitions, together with help in kind through various services. The RVIs give an important local focus for delivery of campaign objectives. The MBC was awarded the first International Riverprize for Best River Management Initiative at the 2nd International River Symposium in Brisbane, Australia.

However, more than 20% of the waterways in the basin still cannot support fish. To address this, the campaign has focused on land reclamation and is working with the North West Development Agency and key campaign partners to develop a strategic program of watershed restoration. In addition, the MBC is promoting best practices in farm and rural businesses, as well as looking for methods to reduce the adverse effects of runoff from more urban and developed areas.

The campaign's current projects include Water Watch, which targets litter in and around rivers and canals; Stream Care, which supports voluntary action to clean up and care for stream and rivers; Water Detectives, which involves schools; and the Mersey Basin Weekend, which is a weekend of conservation and recreational activities.

The MBC has played the role of consultant on several key environmental policy documents and legislation, including the *EC White Paper on Environmental Liability*; the *EC Communication on proposed changes to the EC Bathing Water Directive*, and the *North West Regional Assembly Draft Regional Planning Guidance*.

C1.10.1.4 Lessons/Applicability to United States

Wide-ranging partnerships play an important role in breaking down the barriers between the public, private, and voluntary sectors and in creating new ways of approaching conventional problems. The MBC has pioneered this principle with particular focus on the watershed and the involvement and education of the public. Partners contribute according to their own capabilities while benefiting from the campaign's strategic vision and networking capacity.

APPENDIX C2

WATERSHED MANAGEMENT IN AN INTERNATIONAL CONTEXT

C2.1 Aral Sea

C2.1.1 Summary

Effective Approaches:	Large-scale watershed management Coordination of authority at all levels Capacity building Increased stakeholder awareness and involvement Integration of point and nonpoint source pollution management
Lead Organization:	Interstate Council for the Aral Sea (ICAS)
Other Stakeholders:	International donor community, local experts (senior specialists from the government, technical institutes, the academies of science)
Timeframe:	1994 to present

C2.1.2 Description

The Aral Sea Basin lies within Kazakstan, Kyrgyz Republic, Tajikistan, Turkmenistan, and Uzbekistan. A sizable portion of Central Asia's agriculture is dependent on irrigation. Irrigated agriculture in the region predates by millennia the era of tsarist conquests of the 18th and 19th centuries. What has become new about irrigation in the region, however, is the huge amount of water diverted from the region's major rivers and, in turn, the large proportion of arable land devoted to cotton production. In the 1950s, the Syr and Amu Darya rivers were diverted into the area's steppes and deserts primarily to irrigate cotton and rice. This diversion in combination with new reservoirs reduced flows into the Aral Sea by 30 km^3 to 50 km^3 per year. Beginning in the late 1970s, no water from the Syrdarya reached the Aral Sea, and the Amudarya supplied only a minimal and ever-decreasing volume.

Since 1960, the sea has lost 75% of its water volume. The Aral Sea has been split in two, the Little Aral and the Big Aral. This loss in volume has also increased the salinity in the sea over three times, from brackish (10 g salt/L) to hypersaline (40 g salt/L).

The Soviet leadership's response to decreased crop yields due to increasing salinity was to use more agrochemicals and increase irrigation. The increased use of pesticides further contaminated groundwater and surface water supplies. Further, municipal and industrial waste was discharged into rivers with little or no treatment, even though a primary use of the water downstream was drinking water.

The receding shoreline left a bed of drying toxic sediments that contained pesticides and fertilizers, which are prone to wind dispersal. Dust storms from the dry seabed were observed by satellite as early as 1975. In 1981, storms covering a radius of 300 km were observed 90 days per year. The combination of exposure to agricultural chemicals and poor drinking water has had a staggering effect on public health in the Aral Sea Delta. Almost all women and children in Karakalpakstan are anemic. Airborne salts and particulates have contributed to dramatic increases in respiratory and eye diseases. An increase in esophageal cancers is directly attributable to water contamination. Other health effects include high infant mortality (82 per 1000 births in Uzbekistan), gastrointestinal problems, viral hepatitis, contaminated breastmilk, and reduction of life expectancy by 20 years compared to the Commonwealth of Independent States.

The region had contained a wealth of wildlife, with 24 species of fish, many mammals, and more than 170 species of birds. But with the loss of vegetation, almost all of the species in the Aral Sea have either migrated or died. The losses lead to the collapse of the fishing industry that once employed 60,000 people. Fishing villages are now located more than 20 km from the sea.

Awareness of the potential degradation surrounding the Aral Sea drawdown was widespread, even in the 1950s and 1960s—a time when policy makers had a blind faith in the use of technological fixes to overcome obstacles in the paths of economic development. It was also a time when the Soviet government did not allow organized dissent. Articles about the risks of degradation appeared in Soviet journals, at least from the 1960s, and were translated into other languages. However, even the most ardent advocates of preserving the Aral underestimated the range, rate of change, and intensity of the degradation that subsequently transpired.

The serious environmental degradation of the Aral Sea Basin has resulted in a commitment to begin corrective activities in each of the basin states. This commitment of state governments has lead to mutual agreement regarding program activities, even where a country's individual priorities vary greatly. The Aral Sea Basin Program was established in 1994 with support from the international donor community, including the World Bank and the Global Environment Fund.

The program has four objectives: to stabilize the environment of the Aral Sea Basin, to rehabilitate the disaster zone around the sea, to improve the management of international waters of the Aral Sea Basin, and to build the capacity of the regional institutions to plan and implement the above programs. The cornerstones of the Aral Sea Basin Program are the regional institutions established by the heads of state of the five countries. These institutions are ICAS, with 25 high-level representatives from the five states; the executive committee of the ICAS, which is the organizational committee for developing and implementing policies and programs; and the International Fund for the Aral Sea (IFAS), which aids in financing the program.

As a result of the Aral Sea Basin Program, eight major sub-programs have evolved. The first is the Regional Water Resources Management Strategy, which is developing a common strategy of rational water use in the Aral Sea Basin while taking into account regional and national goals. The second program, Hydrometeorological Services, aims to establish a single entity throughout the region and also develop a unified environmental monitoring system. This program includes the development of an electronic database that would contain data on socioeconomic consequences of water and land and the environmental status of the Aral Sea region. The third program, Water Quality Management, will establish a water quality monitoring system and pollution sources accounting system. This project will identify main pollution sources in surface water and groundwater and develop strategies for protecting the resource and reducing the pollution contribution from these sources. The fourth program, Wetland Restoration, aims to restore wetlands in the region using artificial measures and hopes to revive specie populations and the fishing industry. The aim of the fifth project, Clean Water, Sanitation/Health, is to provide clean drinking water to the region, and improve sanitation measures. The sixth program, Integrated Land and Water Management in the Upper Watersheds, will conduct demonstrations, socioeconomic studies, field trials, and physical interventions designed to promote sustainable land and water management practices, improve water quality, regulate river flows, and increase living standards of people in these areas. The seventh program, Operational Water Resources Management, aims to restore and renew technical equipment and monitoring/management techniques. The last major project, Capacity Building, invests in office equipment (i.e., computers) and staff training.

C2.1.3 Results/Outcomes

Unfortunately, after almost 10 years of planning and funding, not much has been done to increase the Central Asian standard of living or to rehabilitate the sea. Failures are largely attributed to lack of attention given to local needs and concerns, or involvement of local NGOs. The international donor community typically does not use local specialists to implement projects. A local joke in the communities of Central Asia is that "if every foreigner who visited the region brought one bucket of water with him, the sea would have been full and the problem would be solved."

In defense of the low success rate of projects is the reality of the length of time necessary to restore the sea. Even if the river flow into the sea were restored to that of 1960, it would take an estimated 200 years for the sea to recover.

C2.1.4 Lessons/Applicability to United States

One lesson from the Aral Sea is the need to focus on the needs and concerns of local people and to involve local experts who have knowledge of the ethnic, cultural, and natural environment specific to their area. This decentralized approach will not be easy to accomplish because of the strong traditions of these countries for centralized planning with no citizen input and little oversight.

Achieving sustainable water use in the Aral Sea will require the following changes, based on the concepts from the 1987 Bruntland report:

- *Accountability*. The various agencies in the region that are responsible for managing the region's resources must be made directly responsible and accountable for ensuring that appropriate and sustainable water use strategies are developed and carried out.
- *Risk*. There must be an independent, complementary capacity to assess risks that is free from political influences (to the extent possible). This would include scientific and technical peer reviews that would add credibility to the projects.

- *Choices*. The rights, roles, and participation of all stakeholders affected by development and planning, decision making, and project implementation must be elaborated and expanded.
- *Investments*. Financial institutions and governments must invest in sustainable resource programs in order to advance projects.
- *Education and empowerment*. An educated public (knowledgeable about ecological and water resource issues) is a necessity to obtaining useful stakeholder involvement.
- *Access*. Distributing data, reports, information on the latest technology available, etc., to decision-makers and local stakeholders (e.g., farmers) is essential to educating the stakeholders and ensuring good communication between various jurisdictional levels. For example, once a farmer can see the direct impacts his/her farming practice has on the resources in the basin, the farmer may feel like part of the system/organization and be more likely to adjust his/her practices to accommodate necessary improvements.

Although the U.S. situation appears to be quite different politically and socially, the concepts given above are directly applicable to the United States.

C2.2 Danube River Basin

C2.2.1 Summary

Effective Approaches:	Large-scale watershed management Coordination of authority at all levels Capacity building Increased stakeholder awareness and involvement
Lead Organization:	Riparian countries
Other Stakeholders:	Riparian country counterparts, local donors, international donor community
Timeframe:	1991 to 2000

C2.2.2 Description

The Danube River basin is spread over 16 countries in southern Europe. The river flows for a distance of 2857 km and drains an area of 817,000 km^2. Land use in this large basin is highly diversified, including a range of agricultural practices, forestry, mining, natural areas, settlements, and industries. The critical interdependence of upstream and downstream users in the Danube can be seen at all levels of the basin. In addition, there are important linkages with the Danube River, its delta (the second largest natural wetland in Europe), and the environmental quality of the Black Sea (considered to be a "World Heritage" site).

Among the transboundary issues of the Danube River Basin are the following:

- Ecosystem deterioration;

- The quality of the water: pollutants, wastewater, agricultural practices, and toxic substances;
- The competition for water among users/nations;
- River navigation, including dams and regulations; and
- Depletion of fisheries.

Historically, cooperation in the Danube tended to be based on bilateral agreements between neighboring countries. Following the World War II, a basin-wide agreement was made for navigation. Even though the problems in the Danube were broadly recognized, further agreements were difficult to achieve due to the different political systems and economic structures among the 16 nations.

Following the political change in Central and Eastern Europe (CEE), the concept of an environmental program for the Danube was initiated in Sofia in 1991 by the Global Environment Fund (GEF), which is part of the United Nations Development Programme (UNDP). The Environmental Programme for the Danube River Basin (EPDRB) began operating in 1992, and was directed by task force comprised of representatives from each country, NGOs, and donor agencies. Preliminary work by the EPDRB led to the development of the Convention and Cooperation for the Protection and Sustainable Use of the Danube River in June 1994. All of the Danube countries signed the convention by 1995.

Two priority issues were agreed upon by the countries: the pollution of water resources intended for drinking water supplies and the degradation of riverine ecosystem sand water quality by eutrophication. Three specific goals were also agreed upon to address the priority issues: to work towards sustainable and equitable water management, to control hazards from accidental spills and floods, and to reduce the pollution load to the Black Sea. Inter-ministerial committees were formed in each country to provide input into the EPDRB, and to promote a dialog at the sub-national level with local stakeholders affected by any on-the-ground implementation. With input from the committees, a strategic action plan (SAP) was prepared in 1995, followed by the development of a strategic implementation plan (SIP) in 1996, which supported an array of projects in the region, primarily at the sub-basin level. The SAP was subsequently revised in 1999, transferring the leadership of the task force from donors to the countries of the Danube.

C2.2.3 Results/Outcomes

One of the GEF–Danube projects has developed a way "to detect and reduce transboundary pollution sources in the Danube River tributaries in Romania, Slovakia, and Hungary, and to minimize the effects of this pollution downstream." As part of this project, support is being directed toward transboundary accident and early warning systems and monitoring and laboratory information management. The underlying goal of this project is to "reduce emission of toxic compounds and pathogens into international waters in areas where potential pathways of human exposure cross national boundaries." The pollution from three tributary basins, the Hornad/Hernad, Bodrog/Upper Tisza, and Cris/Koros, contribute to the deterioration in the quality of water used downstream for drinking, irrigation, fishing, and recreation in Slovakia, Hungary, and Romania. The core of this project is to reduce direct and indirect discharges of toxins and pathogens from industries and municipal water treatment systems, and develop the capability to monitor and respond to periodic spikes of contamination.

A series of monitoring stations placed at the boarders of Ukraine/Slovakia, Slovakia/Hungary, and Romania/Hungary will allow for early detection of water quality degradation and alert users downstream to cease withdrawals until the problem has been eradicated. Eventually, this early detection monitoring system will be dispersed throughout the 13 countries through which the Danube flows. There will be a central Principal International Alert Center (PIAC), located in Vienna. The PIAC will receive warnings from all monitoring systems and then alert the affected areas. At the same time that the monitoring systems are being installed, a large effort will be made to update municipal treatment facilities with the latest technologies and also to improve runoff management aiming to control pollution at its source. Public awareness will be a major priority with the involvement and cooperation of local and national water agencies and NGOs.

Funding will ultimately be covered by the Danube nations, with some support from international sources for specific projects in the near term.

It will be years before there is a balanced sharing of rights and responsibilities in the basin, but the program is placing a strong emphasis on capacity building for weaker nations. The EPDRB maintains a Web site to inform the public and to provide a forum for discussions.

C2.2.4 Lessons/Applicability to United States

The Danube example demonstrates the importance of consultation with stakeholders. The iterative national consultations provided in the structure of the EPDRB facilitated the final agreements at the multi-national level for acceptance of the SAP and SIP. A clear structure of shared decision making is essential with the involvement of so many nations. The equal representation of nations on the task force allowed them to play the role of mediator in the discussion of the SAP and SIP.

The gradient of political, legal, and technical capabilities along the Danube highlighted the need to level the playing field. Early efforts focused on nonstructural measures, such as institutional strengthening and capacity building, in order to provide more voice and a more educated voice to weaker nations.

C2.3 Mekong River Basin

C2.3.1 Summary

Effective Approaches:	Coordination of authority at all levels Increased stakeholder awareness and involvement Ecosystem preservation Fisheries program
Lead Organization:	Mekong River Basin Commission (MRC)
Other Stakeholders:	Government economic planners, fishermen, the public
Timeframe:	Began in 1995

C2.3.2 Description

The Mekong River, one of the world's largest rivers, may be considered Southeast Asia's most substantial single natural resource. The river, with a total length of 4800 km, starts in the Tibetan plateau in Yunnan province of the People's Republic of China (PRC), flows through six countries (PRC, Myanmar, Lao PDR, Thailand, Cambodia, and Viet Nam) and empties into the sea, via two main branches, the Mekong and Bassac. The average yearly river flow is large, at 475 billion cubic meters. Only a small portion is used for water supply, irrigation, and hydropower generation. The Lower Mekong Basin is home to a population of 55 million people, increasing at an annual rate of 2%.

The MRC riparian countries, except Thailand, are classified as low-income nations with GDP between USD 250 and USD 400 per capita. The growth rate before the recent economic crisis was high, ranging from 5% to 8%. Signs of recovery are emerging and with rapidly increasing population pressure, the requirements for water, food, and energy are going to increase.

The MRC was formed in 1995 to enable the signatory countries (Cambodia, Laos, Thailand, and Viet Nam) to cooperate in a constructive and mutually beneficial manner for sustainable development, utilization, conservation, and management of the Mekong River Basin water and related resources. Three years later, in October 1998, the MRC ministerial council retained and extended these objectives when it adopted as the MRC mission; to promote and coordinate sustainable management and development of water-related resources for the four countries' mutual agencies involved in funding general development in the basin on a bilateral basis. Consequently, to achieve focus on the river-basin-related aspects of this mandate, the strategic thrust of MRC lies in three main directions, specifically covered in the 1995 agreement:

- Developing rules for the cooperative utilization of water through the water utilization program
- Basin development planning, facilitating cooperative sustainable development of water related resources
- Environmental monitoring

The commission in principle now only undertakes development activities to fulfill the mandate of MRC as a river basin commission rather than as a general development agency.

The MRC Fisheries Program is one such major development activity on the commission's program that illustrates the opportunities and challenges ahead of the rapidly increasing population in the basin. Close to 13,000 different fish species have now been identified in the Lower Mekong Basin. The high fish production in the basin is closely linked with seasonal flooding. In 1998, Tonle Sap (a very large freshwater lake/floodplain fishery in Cambodia) was included in the World Network of Biosphere Reserves of the United Nations Educational, Scientific and Cultural organization, and it has been proposed as a World Heritage Area. Many internationally funded development projects are focusing on Tonle Sap and its preservation and sustainable use, but competing economic development challenges these efforts. Historically, lack of knowledge seems to present one of the greatest dangers to the inland fisheries. Loss of habitat and obstacles to migrations are the most important threats to the fisheries sector. No individual country in the Mekong sub-region will be able to manage their fisheries resources alone. The MRC Fisheries Program was created to address these issues. Cooperation among the four countries is crucial for the success of the management plan. Another crucial factor for success is that information created by the program must be brought out to stakeholders, ranging from government economic planners to fishermen and the public and further explained/presented in a form they can understand.

C2.3.3 Results/Outcomes

A Technical Advisory Body on Fisheries Management has been founded recently with high-level members from the related national agencies for fisheries and the planners of the national Mekong communities. This advisory board will oversee the cooperation and interact in the development of fisheries management systems.

C2.3.4 Lessons/Applicability to United States

Lessons learned from this example include the recognition that fisheries need to be managed at the basin level, not the local level. There needs to be a joint effort of all parties in the basin to achieve sustainable management for fisheries. This example also shows the importance of communication between the government planners and the local fishing community. The local community needs to fully understand the processes taking place in order to appreciate and participate in sustainable practices.

C2.4 The Nile River Basin

C2.4.1 Summary

Effective Approaches:	Large-scale watershed management Capacity building Increased stakeholder awareness and involvement
Lead Organization:	The Nile Basin Initiative: Burundi, Democratic Republic of Congo, Egypt, Ethiopia, Kenya, Rwanda, Sudan, Tanzania, Uganda
Other Stakeholders:	Local and sub-basin participants
Timeframe:	Began in 1992

C2.4.2 Description

The Nile Basin Initiative (NBI) promotes a participatory approach from the ground up and encourages decision making from the lowest possible level. NBI aims to integrate management basin-wide from the local level up to the sub-basin and national levels. Projects supported by the NBI include hydropower development and interconnection, irrigation and drainage, environmental management, river regulation, drought and flood control, and water use efficiency improvements. The organizational structure includes a council of ministers (water affairs ministers of the Nile basin states), which governs the NBI and is supported by the Nile Technical Advisory Committee (senior officials from the various countries). NBI seeks financial support from the international community. In order to implement plans and policies on the ground, NBI created the Shared Vision and Nile River Basin Strategic Action Program. The Shared Vision program is based on a cooperative framework supported by confidence building and stakeholder involvement; socioeconomic, environmental, and sectoral analyses; development and investment planning; and applied training. A set of guidelines have evolved from the cooperative framework for the action programs which address planning at the sub basin level, ensuring equal consideration of all stakeholders involved and promoting communication among participants across jurisdictions.

C2.4.3 Results/Outcomes

The NBI has yet to implement projects, but has given a forum for discussion about possible joint management of the basin.

C2.4.4 Lessons/Applicability to United States

This example is applicable to large-scale watershed approaches or more complex systems in the United States (e.g., the Colorado River). This is an example on how to organize an initiative and what issues one should consider when working with cross-jurisdictional sub-basins.

C2.5 The Rhine

The Rhine River Basin is an example of divergent interests on water quality—with Switzerland, France and the Federal Republic of Germany—resulting in the pollution of the downstream drinking water supplies in Germany and the Netherlands.

C2.5.1 Summary

Effective Approaches:	Large-scale watershed management Coordination of authority at all levels Capacity building Mediation of disagreements among disparate groups Increased stakeholder awareness and involvement Integration of point and nonpoint source pollution management Ecosystem restoration Ecosystem preservation
Lead Organization:	International Commission for the Protection of the Rhine
Other stakeholders:	National governments
Timeframe:	1958 to present

C2.5.2 Description

International cooperation on the Rhine began in the early 19th century, with the formation of the Central Commission for the Navigation of the Rhine, which ensured freedom to navigate the Rhine across country borders. At that time, more than half a million Atlantic salmon made their annual migration to the Rhine. Fifty other species of fish could also be found in the Rhine. But heavy industrial pollution and the construction of large dams through the early 20th century eliminated salmon in 1958. The ability to use the river as a drinking water source had become increasingly difficult. The Rhine had the nickname of the "sewer of Europe."

An International Commission for Protection of the Rhine (ICPR) was founded during this period to study the problems of the Rhine's degradation and to propose actions to solve them. An international measurement network was set up, with common analysis methods. Although initially tedious to organize, the common basis for objective assessment of the water quality proved invaluable in establishing trust among the nations involved. The ICPR did make steady improvements in water quality during the first decades of its work, primarily using 'end-of-pipe' techniques that treated wastewater, rather than preventing the pollution. Yet by 1975, only 29 of the more than 50 fish species remained, with no return of salmon. The Dutch water authority published a report in 1985 that shifted attention from end-of-pipe approaches to looking at the ecosystems of the receiving water bodies and what was necessary to support these ecosystems. Efforts to improve the Rhine continued. Unfortunately, a serious chemical spill in 1986 further deteriorated the Rhine ecosystem. More than 500,000 fish died, and sheep that drank Rhine water were also killed. The Netherlands drinking water supply was contaminated. Some scientists predicted that the river would be biologically dead for 20 years.

The catastrophic spill catalyzed popular support and political will to create a 15-year Rhine Action Plan, under the authority of ICPR. The plan covered water quantity and quality, the shipping trade, and included the exchange of data and joint research. Public outreach played a large role in the continuing public support of the plan. The first two phases of the plan have been completed: identifying and eliminating the main sources of pollution with a more active program of monitoring, including pollution patrols; and imposing large fines on polluters. By the early 1990s not only did the Rhine see the return of salmon, but also other species that had disappeared, including sea trout, flounder, river lamprey, bullhead and perch, returned.

C2.5.3 Results

In 1999, a third phase was agreed upon to broaden the scope of the plan. The result was the acceptance by the commission of a 40-year action plan in 2000. The plan includes measures on flood management and habitat protection and restoration in the alluvial zone around the river's banks. The flood management goal is to restore as far as possible the natural course of the river. The emphasis is on planning around the water system itself, rather than trying to control the water. This is a major shift, particularly for the Dutch, who have been building dikes for 1000 years.

C2.5.4 Lessons

Transboundary cooperation is a time-consuming process, made by small increments. Cooperation requires trust. The decisions of individual countries must be left to each country. Common legal frameworks help the process of cooperation. Common monitoring infrastructure contributes to mutual trust.

The Rhine example illustrates the importance of societal awareness, based on the involvement and education of affected stakeholders. Great emphasis has been placed on creating broad public knowledge and understanding of the importance of preserving the water ecosystem. Disasters with broad impacts may be good opportunities to promote transboundary cooperation. The media focus on fish mortality and contamination of drinking water provided the impetus for the fundamental shift in water management in the Rhine basin. Because of strong public support, charges for water and sewerage are met with little resistance. One NGO, the International Association of Waterworks in the Rhine River Basin (IAWR/RIWA) uses data from its own monitoring network to negotiate for better treatment directly with industrial polluters, the governments, and the Rhine Commission. Certain requirements IAWR made regarding water quality to be suitable for drinking water production were adopted by the commission and are now part of the commission's quality objectives.

APPENDIX D

WORKSHOP AGENDA AND PARTICIPANTS

D.1 Meeting Agenda

GLOBAL LESSONS FOR WATERSHED MANAGEMENT IN THE U.S.

Water Environment Research Foundation Project 00-WSM-5

WERF Project Manager: Margaret Stewart

Contractor: Tellus Institute

Contact: James Goldstein jgold@tellus.org

Agenda for 10/14/01 Project Workshop

Georgia World Congress Center
Atlanta, GA

8:30 – 9:00 Introductions and Project Overview
(Tellus)

9:00 – 10:15 Status of Existing Watershed Management Practices in the U.S.
(Panel: Tellus, U.S. EPA, Other)

10:15 – 10:30 Break

10:30 – 12:00 Case Study Presentations and Discussion
The Rhine Basin, Harm Oterdoom
The Mersey Basin, Louise Hopkins

12:00 – 1:30 Lunch

1:30 – 3:00 Case Study Presentations and Discussion
The Fraser Basin, David Marshall
The Grand River Basin, Lorrie Minshall

3:00 – 3:15 Break

3:15 – 5:00 Roundtable Discussions of Preliminary Findings and Key Questions
(All, facilitated by Tellus with short presentation on preliminary findings and structured questions for discussion)

D.2 Workshop Participants

GLOBAL LESSONS FOR WATERSHED MANAGEMENT IN THE U.S.

Panel from International Case Studies:

Rhine: **Harm Oterdoom**
Secretary General
International Rhine Committee
harm.oterdoom@iksr.de
IKSR
D-560002 Koblenz, Germany
Tel: (+49) 261 12495 Fax: (+49) 261 36572

Mersey: **Louise Hopkins**
Deputy Chief Executive
Mersey Basin Campaign
business@merseybasin.org.uk
Tel: +44 (0) 161 242 8200

Fraser: **David Marshall**
Executive Director
Fraser Basin Council
1st Floor, 470 Granville St., Vancouver, BC V6C 1V5
dmarshall@fraserbasin.bc.ca
Tel: (604)488-5357 Fax: (604) 488-5351

Grand: **Lorrie Minshall, P. Eng.**
Manager of Watershed Resources Planning
Grand River Conservation Authority
lminshall@grandriver.ca
Tel: (519) 621-2763 ext. 231 Fax: (519) 621-4945

U.S. Watershed Experts:

Stephen Born
Department of Urban and Regional Planning
Institute for Environmental Studies, University of Wisconsin-Madison
smborn@facstaff.wisc.edu

Mark T. Imperial, Ph.D.
Department of Political Science
University of North Carolina at Wilmington, Political Science Dept.
601 S. College Rd UNC-Wilmington, Wilmington, NC 28403-5607
Phone: (910) 962- 3220 Fax: (910) 962-3286
imperialm@uncw.edu

Andrew Purkey
Executive Director, Oregon Water Trust
(503)480-0694
Andrew@owt.org

Douglas Kenney
Natural Resources Law Center
University of Colorado School of Law
Campus Box 401, Boulder, CO 80309-0401
303-492-1296
Douglas.Kenney@Colorado.Edu

Paul Sabatier
Director, Watershed Partnership Project
Department of Environmental Science and Policy
University of California-Davis, Davis, CA 95616
pasabatier@ucdavis.edu
wdleach@ucdavis.edu

Patti Grace-Jarrett
Water Quality/Quantity Administrator
Louisville/Jefferson County Metropolitan Sewer District
700 West Liberty Street, Louisville, KY 40203
502-540-6145
grace@msdlouky.org

Stephen Estes-Smargiassi
Director of Planning
Massachusetts Water Resources Authority
Charlestown Navy Yard
100 First Ave., Boston, MA 02129

Tel: (617) 788-4303
Fax: (617) 788-4888
smargias@mwra.state.ma.us

Eugene Stakhiv, Ph.D.
Chief, Policy and Special Studies Div.
U.S. Army Corps of Engineers
Institute for Water Resources
7701 Telegraph Road, Alexandria, VA 22315-3868
Eugene.Z.Stakhiv@WRC01.USACE.ARMY.MIL

Billy G. Turner
President
Columbus Waterworks
1421 Veterans Parkway, Columbus, GA 31902-1600
Tel: (706) 649-3430 Fax: (706) 327-3845
bturner@cwwga.org

WERF Project Steering Committee Members:

Robin L. Autenrieth, Chair
Department of Civil Engineering
Texas A&M University
College Station, TX 77843-3136
979-845-3593 – Office 979-845-6156 or 862-1542 - Fax
r-autenrieth@tamu.edu

Dan Cloak, P.E.
Principal
Dan Cloak Environmental Consulting
1824 Prince Street, Berkeley, CA 94703-2462
(510) 540-5618 – Office (208) 279-1806 - Fax
dan@dancloak.com

Ronald D. Neufeld, P.E., Ph.D.
University of Pittsburgh
Department of Civil and Environmental Engineering
939- Benedum Engineering Hall, Pittsburgh, PA 15261
(412) 624-9874 – Office (412) 624-0135 - Fax
neufeld@civ.pitt.edu

Gary S. Bowen
Toronto and Region Conservation Authority
Watershed Specialist, Duffins-Carruthers
Watershed Management Division
5 Shoreham Drive, Downsview, Ontario M3N 1S4

(416) 661-6600, ext. 5385 – Office (416) 661-6898 - Fax
boweng@trca.on.ca

Donald Brady, Ph.D. (substituting for Jim Pendergast)
Chief, Watershed Branch
Office of Water
U.S. EPA Headquarters
Ariel Rios Building, 1200 Pennsylvania Avenue NW
Mail Code: 4503F, Washington, DC 20460
(202) 260-1261 – Office
brady.Donald@epa.gov

Drew McAvoy (substituting for Tom Feijtel, Ph.D.)
The Procter and Gamble Company
Environmental Science Dept.
PO Box 538707, Cincinnati, OH 45253
513-627-1176—phone 513-627-1208—fax
feijtelt@pg.com

William (Bill) Booty, Ph.D.
National Water Research Institute
Aquatic Ecosystems Impacts Branch
867 Lake Shore Road, Burlington, Ontario, Canada
L7R 4A6
(905) 336-4718 – Office (905) 336-4400– Fax
Bill.Booty@cciw.ca

WERF Project Manager:

Margaret Stewart
Water Environment Research Foundation
635 Slaters Lane, Alexandria, VA 22314-1177
(703) 684-2470 - Ext. 7147- Office (703) 299-0742 - Fax
mstewart@wef.org

Tellus Institute

Kenneth Strzepek, Ph.D. (PI)
Senior Associate
Tellus Institute
11 Arlington Street, Boston, MA 02116
(303) 530-3818—Office (303) 530-4428—Fax
strzepek@colorado.edu

James Goldstein (Co-PI)
Senior Research Director

Tellus Institute
11 Arlington Street, Boston, MA 02116
(617) 266-5400—Office (617) 266-8303—fax
jgold@tellus.org

Annette Huber-Lee, Ph.D.
Senior Scientist
Tellus Institute
11 Arlington Street, Boston, MA 02116
(617) 266-5400—Office (617) 266-8303—fax
ahuberlee@tellus.org

APPENDIX E

ADDITIONAL RESOURCES

The following resources were useful in the preparation of this report, although they are not specifically cited in the text.

Adler, R.W.; Straube, M. 2000. Lessons from Large Watershed Programs: A Comparison of the Colorado River Basin Salinity Control Program with the San Francisco Bay-Delta Program, Central South Florida Project, and the Chesapeake Bay Program. National Academy of Public Administration, Washington, D.C.

Aral Sea Basin Project Programme 6: Integrated Land and Water Management in Upper Watersheds. (n.d.) www.hcg.helsinki.fi/projects/aralsea2.html. Helsinki Consulting Group.

Armstrong, H. (n.d.) *River Management Through Creative Design with Communities: The Creative Village Program.*

Aso, Y. River Bureau Ministry of Construction, Japan. (n.d.) Bio-Diversity Reconstruction. http://www.mlit.go.jp/river/english/bio-d1.html .

Barkin, D.; Batt, R.; DeWalt, B. 1990. *Food vs. feed: The global substitution of grains in production*. Boulder, CO: Lynne Rienner Publications.

Barkin, D.; Pailles, C. 1998. Water as an Instrument for Sustainable Regional Development. *Arid Lands Newsletter* [On-Line Journal], *Fall/Winter, 1998.* Retrieved from http://ag.arizona.edu/OALS/ALN/aln44/barkinfinal.html.

Barkin, D. 1998. *Wealth, Poverty and Sustainable Development*. Mexico City: Editorial Jus, Centro de Ecología y Desarrollo, and Centro Lindavista.

Barkin, D. 2002. *The Production of water in Mexico.* Managing a Sacred Gift: Changing Water Management Strategies in Mexico. (S. Whiteford, Ed.). Center for U.S.-Mexico Studies, San Diego, CA.

Bellamy, J.; McDonald, G. (n.d.) *Lessons from the Herbert River Catchment: Community-based Integrated Catchment Management.*

Berry, R. (n.d.) *Herding Cats—Policy and Legislative Trends in Water Management in New Zealand.*

Booty, W.G.; Crowe, A.S.; Piggott, A.; deLoe, R.; Bowen, G.S.; Holysh, S.; Svensson, T.; Beatty, B.; Kauffman, G. (ed.). 1998. Proceedings of the Groundwater in a Watershed Context Symposium, Canada Centre for Inland Waters, Burlington, Ontario, Dec 2-4, 1998, Canadian Water Resources Association.

Born, S.M.; Genskow, K.D. 1999. Exploring the Watershed Approach: Critical Dimensions of State-Local Partnerships; The Four Corners Watershed Innovators Initiative Final Report. River Network, Portland, OR.

Born, S.M.; Genskow, K.D. 2000. *The Watershed Approach: An Empirical Assessment of Innovation in Environmental Management.* National Academy of Public Administration, Washington, D.C. Includes case studies of:
- Nisqually River Watershed in Washington
- Dungeness River Watershed in Washington
- Black Earth Creek Watershed in Wisconsin
- Tomorrow-Waupaca River Watershed in Wisconsin
- Upper Little Tennessee River Watershed in North Carolina
- Long Creek Watershed in North Carolina

Burrard Inlet Environmental Action Program (BIEAP) and Fraser River Estuary Management Program (FREMP). (n.d.) www.bieapfremp.org.

Butcher, J.B.; Water Environment Research Foundation Staff. 1998. *Watershed-Scale Ecological Risk Assessment.* Water Environment Research Foundation, Alexandria, VA.

Campagnolo, I.V. (n.d.) 21st Century: Age of Alliances, Watersheds and You: A case study of the Fraser Basin Council. Fraser Basin Council.

Canadian Water Resources Association, the Ministry of Environment and Energy, Environment Canada—National Water Research Institute. *Proceedings of Watershed Management Symposium, Defining: State of the Art, State of the Knowledge, State of the Practice, Burlington, Ontario, Canada, 6-8 December 1995.* Burlington, Ontario, Canada: Canada Centre for Inland Waters.

Cantwell, M.; Day, J.C. 1998. Citizens initiated river basin planning: the Salmon Watershed example. *Environments* 25:2-3:80-90.

Chess, C.; Gibson, G. 2000. *The Navesink Watershed Management Effort.* National Academy of Public Administration, Washington, D.C.

Chichester, P. 1997. Resurrection on the Rhine. *International Wildlife* [On-Line Journal], *Sept/Oct, 1997.* Retrieved from http://www.nwf.org/internationalwildlife/salmon.html.

Clark, J. 1997. Watershed Partnerships: A Strategic Guide for Local Conservation Efforts in the West. Prepared for the Western Governors' Association. Retrieved from http://www.westgov.org/wga/publicat/wsweb.htm.

Clean Water Action Plan. Signatories: DOA, DOC, DOD, DOE, DOI, DOT, EPA, TVA, and USPS. (n.d.). Federal Multi-Agency Source Water Agreement: Integration of State, Tribal and Local Drinking Water Source Assessment and Protection Initiatives within a Watershed Framework. http://www.cleanwater.gov/swa/.

Clean Water Action Plan. Signatories: Army Corps of Engineers, DOA, DOC, DOD, DOE, DOI, EPA, TVA. 2000. Unified Federal Policy for a Watershed Approach to Federal Land and Resource Management. Retrieved on April 5, 2001 from http://cleanwater.gov/ufp/.

Clean Water and Watershed Restoration Accomplishments. (n.d.) Website: www.cleanwater.gov/progress/keyact.html. U.S. Department of Agriculture et al.

Conservation Ontario. 2001. *The Importance of Watershed Management in Protecting Ontario's Drinking Water Supplies*. Conservation Ontario, Newmarket, Ontario.

Conservation Ontario, Ministry of Natural Resources and Housing. 2001. *Memorandum of Understanding on Procedures to Address Conservation Authority Delegated Responsibility.*

Conservation Technology Information Center. 2000. *Bridge builder: A guide for watershed partnerships*. Conservation Technology Information Center, Purdue University, West Lafayette, IN.

Conservation Technology Information Center, Purdue University. 2000. Building local partnerships: A guide for watershed partnerships. Retrieved on May 30, 2001, from http://www.ctic.purdue.edu/KYW/Brochures/BuildingLocal.html.

Council of Ministers of Water Affairs of the Nile Basin States. (n.d.) Policy Guidelines for the Nile River Basin. Retrieved from http://www.nilebasin.org/Documents/TACPolicy.html.

Crandall, J. 1999. Rehabilitation of the 'Small' Aral Sea. *Restoration and Reclamation Review* [Online Journal], *4.* Retrieved from http://www.hort.agri.umn.edu/h5015/rrr.htm.

Credit Valley Conservation. 1997. *Facing Change: A Business Plan for Watershed Management.* Credit Valley Conservation, Meadowvale, Ontario, Canada.

Cullen, P. (n.d.) *Learning from the Mistakes of the Murray.*

Cuming, P. (n.d.) *Making River Plans—A Working Partnership Approach.*

Dainton, J.; O'Kane, R.; Pell, J.; Sheed, W.K. (n.d.) Goulburn Broken Catchment Management Authority, Victoria, Australia.

De Ruyter Van Steveninck, E.D.; Admiraal, W.; van Zanten, B. 1990. Changes in Plankton Communities in Regulated Reaches of the Lower River Rhine. *Regulated Rivers: Research & Management* 5:67-75.

Dennison, W.; Abal, E. (n.d.) *Moreton Bay Study: An integrated research/management program to understand and reduce impacts of nutrients, sediments and toxicants on an Australian Estuary.*

Don Watershed Regeneration Council. 2000. *A Time for Bold Steps: The Don Watershed Report Card 2000.* Toronto and Region Conservation Authority, Downsview, Ontario, Canada.

Duda, A.; El-Ashry, M. 2000. Addressing the Global Water and Environmental Crises through Integrated Approaches to the Management of Land, Water and Ecological Resources. *Water International* 25:1:115-126.

Ehlers, L.J.; Pfeffer, M.J.; O'Melea, C.R. 2000. Making Watershed Management Work. *Environmental Science and Technology* 34:21:464A-471A.

Fraser Basin Council. (n.d.) Website: www.fraserbasin.bc.ca.

Freemuth, J. 2000. "Science, Politics and Watersheds: Thoughts on Their Integration." In *Proceedings of the Seventh Biennial Watershed Management Council Conference, Boise, Idaho, 19-23 October,1998*, Centers for Water and Wildland Resources, Watershed Management Council.

Gippel, C. (n.d.) *Australia's Environmental Flow Initiative: A Major National Research Collaboration to Promote Excellence in River Management.*

Glantz, M.H. 1999. The dying Aral Sea. Public Affairs Section. *Work in Progress: A Review of Research Activities of the United Nations University* 15:2:12-17.

Gordon, J.A.; Jones, T.T. 1998. *Monitoring and evaluation of selected rural watershed councils in the continental United States.* Buffalo River Stewardship Foundation, Fayetteville, AR.

Grabbe, J.; Boully, L. (n.d.) *Conflict Resolution and Floodplain Management Planning in the Lower Balonne.*

Grand River Conservation Authority. 1997. *Grand Actions: The Grand Strategy Newsletter* 2:4.

Grand River Conservation Authority. 1997. *Grand Actions: The Grand Strategy Newsletter* 2:5.

Grand River Conservation Authority. 2001. Grand River Conservation Authority. Retrieved in Sept 2001 from http://www.grandriver.ca/.

Grand River Conservation Authority. 1997. *State of the Grand River Watershed: Focus on Watershed Issues 1996-1997*. Grand River Conservation Authority, Cambridge, Ontario.

Grand River Conservation Authority. (n.d.) *Text for Corporate Brochure*. pp. 1-9.

Gregg, F.; Clark, J. 1998 *The federal role in watershed partnerships*. Western Governors Association, Denver, CO.

Handley, J.; Wood, R.; Kidd, S. 1999. Sustainable Development and Institutional Design: The Example of the Mersey Basin Campaign. *Journal of Environmental Planning and Management* 42:3:341-354.

The Healthy Waterways Trust. c.2001. "Breathing New Life: Canal Oxygenation." Retrieved from: http://www.merseybasin.org.uk/programs/landmark_projects/default.

Hennessey, T.; Imperial, M.T. 2000. *Rhode Island's Salt Ponds: Using a Special Area Management Plan to Improve Watershed Governance.* National Academy of Public Administration, Washington, D.C.

Hiltunen, M. 1998. *Environmental Development Cooperation Opportunities-Kazakstan, Kyrgyz Republic, Turkmenistan, Uzbekistan.* Finnish Environment Institute, Helsinki, Finland. http://global.finland.fi/english/services/publications/envalku.html.

Hooper, B.P. 1997. Improving watershed management using an innovative integrated resources management approach in Australia. *Journal of Professional Issues in Engineering Education and Practice*, ASCE, 123:2:57-61.

Hopkins, L. 2001. Personal Communication.

Humber Watershed Alliance. 2000. *A Report Card on the Health of the Humber River Watershed.*Toronto and Region Conservation Authority, Downsview, Ontario, Canada.

Imperial, M.T.; Hennessey, T. 2000. *Environmental Governance in Watersheds: The Importance of Collaboration to Institutional Performance*. National Academy of Public Administration, Washington, D.C.

Imperial, M.T.; Hennessey, T. 2000. *The Tampa Bay Estuary Program: Developing and Implementing an Interlocal Agreement.* National Academy of Public Administration, Washington, D.C.

Imperial, M.T.; Hennessey, T. 2000. *The Tillamook Bay National Estuary Program: Using a Performance Partnership to Implement a CCMP.* National Academy of Public Administration, Washington, D.C.

Imperial, M.T. 2000. *Delaware Inland Bays Estuary Program: Using a Nonprofit Organization to Implement a CCMP*. National Academy of Public Administration, Washington, D.C.

Imperial, M.T.; McGee, S.; Hennessey, T. 2000. *The Narragansett Bay Estuary Program: Using a State Water Quality Agency to Implement a CCMP*. National Academy of Public Administration, Washington, D.C.

International Cooperation for River Basin Management in the Danube River Basin. 2000. Component 2: (Tisza. River Basin): Identify joint investment projects in waste water treatment facilities in cross-border areas between neighbouring municipalities. Retrieved from http://www.rec.org/DanubePCU/flyers/coop_2.html.

International Development Research Center. 2002. Managing Natural Resources—Africa and the Middle East (MNR-AME) People, Land, and Water (PLW). Retrieved from http://www.idrc.ca/plaw/05e-projects3.html.

Jones, T. 1999. The Mersey Basin Campaign. *Water Science Technology* 40:10:131-136.

Kauneckis, D., Koziol, L.; Imperial, M. 2000. *Tahoe Regional Planning Agency: The Evolution of Collaboration.* National Academy of Public Administration, Washington, D.C.

Kenney, D.S. 2000. *Arguing about consensus: Examining the case against western watershed initiatives and other collaborative groups active in natural resources management.* Natural Resources Law Center, University of Colorado School of Law, Boulder, CO.

Kenney, D.S. 1997. *Resource Management at the Watershed Level: An Assessment of the Changing Federal Role in the Emerging Era of Community-Based Watershed Management; A Report to the Western Water Policy Review Advisory Commission.* Natural Resources Law Center, University of Colorado School of Law, Boulder, CO.

Kenney, D.S. 2000. *Assessing the Effectiveness of Watershed Initiatives: The Current State of Knowledge.* Natural Resources Law Center, University of Colorado School of Law, Boulder, CO.

Kenney, D.S. 1999. Historical and sociopolitical context of the western watersheds movement. *Journal of the American Water Resources Association* 35:3:493-503.

Kier, W. 1995. *Watershed restoration: A guide for citizen involvement in California.* NOAA Coastal Ocean Program, Decision Analysis Series No. 8, US Department of Commerce, Washington, DC.

Kristensen, J. 2000. "The Mighty Mekong and its Basin, Opportunities and Challenges." Prepared for the *Third International River Management Symposium, Brisbane, Australia, 6-8 September, 2000.*

Makule, D.E. 1998. *The Pangani Basin Water Resources Management.* World Bank/WBI's CBNRM Initiative. Retrieved from http://srdis.ciesin.org/cases/tanzania-019.html.

Marshall, D. 1998. Watershed management in British Columbia: The Fraser Basin experience. *Environments* 25:2-3:64-79.

Massachusetts Executive Office of Environmental Affairs and The Natural Resources Conservation Service. 1998. *Massachusetts Unified Watershed Assessments, Restoration Priorities, and Restoration Action Strategies.* Massachusetts Executive Office of Environmental Affairs, Boston, MA. Website: http://www.epa.gov/owow/uwa/ma.pdf.

McCulley, P. 1996. *Silenced Rivers: The ecology and politics of large dams.* London: Zed Books.

McGinnis, M.V.; Woolley, J.T. 2000. *Changing California: From watersheds to healthy watersheds: A characterization of California watershed organizations and activities.* The Ocean and Coastal Policy Center, Marine Science Institute, University of California Santa Barbara, Santa Barbara, CA.

McKinney, D.C. 1996. Sustainable Water Management in the Aral Sea Basin. *Water Resource Update* 102:15-30.

McMahon, P.; Postle, M. 2000. Environmental valuation and water resources planning in England and Wales. *Water Policy* 2:397-421.

Mersey Basin Campaign. 2000. *Progress Report 1985-2000.*

Mersey Basin Campaign. 2001. *The Next Decade.*

Mersey Basin Campaign. 2001. Retrieved on August 13, 2001 from www.merseybasin.org.uk.

Milich, L. 1998. Managing transboundary resources: lessons from river-basin accords. *Environment* 40:8:10-15, 35-41.

Ministry of Environment and Energy, Ministry of Natural Resources. 1993. *Integrating Water Management Objectives into Municipal Planning Documents*. Ministry of Environment and Energy, Ministry of Natural Resources, Ontario, Canada.

Ministry of Environment and Energy, Ministry of Natural Resources. 1993. *Water management on a Watershed Basis: Implementing an Ecosystem Approach*. Ministry of Environment and Energy, Ministry of Natural Resources, Ontario, Canada.

Ministry of Environment and Energy, Ministry of Natural Resources. 1993. *Subwatershed Planning*. Ministry of Environment and Energy, Ministry of Natural Resources, Ontario, Canada.

Ministry of Environment and Energy, Ministry of Natural Resources, Ministry of Municipal Affairs and Housing, Ministry of Agriculture, Food and Rural Affairs, Association of Municipalities of Ontario, Association of Conservation Authorities of Ontario. 1997. *Watershed Planning Initiative, Final Report*. Ministry of Environment and Energy, Ontario, Canada.

Mullen, M.W.; Allison, B.E. 1999. Stakeholder involvement and social capital: Keys to watershed management success in Alabama. *Journal of the American Water Resources Association* 35:3:655-662.

Murray-Darling Basin Initiative. (n.d.) www.mdbc.gov.au.

Murray-Darling Basin Ministerial Council. 2000. *Review of the Operation of the Cap: Draft Overview Report of the Cap Project Board.* Murray-Darling Basin Commission, Canberra, Australia.

National Academy of Public Administration. 2000. *environment.gov: Transforming Environmental Protection for the 21st Century.* National Academy of Public Administration, Washington, D.C.

National Showcase Watersheds. 2002. www.epa.gov/owow/showcase/projects.html. U.S. Environmental Protection Agency, Office of Water, Oceans, and Watersheds. Includes:

- Duck Creek Watershed, AK
- Big Nance Creek Watershed, AL
- Gila River Corridor Recovery Project, AZ/NM
- Suwannee River Watershed, GA/FL
- Bear Creek Watershed, IA
- Sun River Basin, MT
- Blackfoot Watershed, MT
- Carson River Watershed, NV
- McCoy Creek Watershed, OR
- Lititz Run Watershed Alliance, PA
- White River Partnership, VT
- Duwamish-Green River Watershed, WA

National Research Council, Committee on Watershed Management. 1999. *New Strategies for America's Watersheds.* Washington, D.C.: National Academy Press.

Nile Basin Initiative: Background. (n.d.) www.nilebasin.org/nbibackground.htm.

Nile Basin Initiative Program, Strategic Action Program. 2000. http://www.nilebasin.org/nbiprocess.htm. Nile Basin Initiative.

Ollila, P.; Uitto, J.; Crepin, C.; Duda, A. 2000. *Multicountry Project Arrangements: Report of a Thematic Review*, Monitoring and Evaluation Working Paper Number 3. Global Environment Facility, Washington, D.C.

Olsen, C.H.; Garcia, M.W. 1996. Water Quality 2000—Watershed Program Criteria. In *Proceedings of Watershed '96, Baltimore, Maryland, 8-12 June 1996* (35-39). Washington, D.C.: U.S. Environmental Protection Agency.

Ontario Ministry of Natural Resources and Ministry of Environment and Energy. 1993. *Water Management on a Watershed Basis: Implementing an Ecosystem Approach.* Ontario Ministry of Natural Resources and Ministry of Environment and Energy, Ontario, Canada.

Ontario Ministry of Natural Resources and Ministry of Environment and Energy. 1993. *Integrating Water Management Objectives into Municipal Planning Documents.* Ontario Ministry of Natural Resources and Ministry of Environment and Energy, Ontario, Canada.

Ontario Ministry of Natural Resources and Ministry of Environment and Energy. 1993. *Subwatershed Planning.* Ontario Ministry of Natural Resources and Ministry of Environment and Energy, Ontario, Canada.

Ontario Ministry of Natural Resources and Ministry of Environment and Energy. 1993. *Watershed Planning Workshop, Proceedings.*Ontario Ministry of Natural Resources and Ministry of Environment and Energy, Ontario, Canada.

Ontario Ministry of Natural Resources and Ministry of Environment and Energy. c1997. *Inventory of Watershed Management Projects in Ontario 1990-1995.* Ontario Ministry of Natural Resources and Ministry of Environment and Energy, Ontario, Canada.

Ontario Regulation 149: Policies, Procedures and Guidelines for the Fill, Construction and Alteration of Waterways Regulations. (n.d.)

Oregon Watershed Enhancement Board. (n.d.) www.oweb.state.or.us.

Pacific Fisheries Think Tank. 1997. The Fraser Basin Council. In *The Pacific Fisheries Think Tank Workshop Record #2: Exploring Cooperative Management in Fisheries, 21-22 April 1997.* Vancouver, British Columbia, Canada: The Pacific Fisheries Think Tank.

Rieke, B.; Kenney, D.S. 1997. *Resource management at the watershed level: An assessment of the changing federal role in the emerging era of community-based watershed management.* Western Water Policy Review Advisory Commission, Denver, and Natural Resources Law Center, University of Colorado School of Law, Boulder, CO.

Riley, A. L. 1998. *Restoring Streams in Cities: A Guide for Planners, Policymakers, and Citizens.* Washington, D.C.: Island Press.

Robbins, R.W.; Glicker, J.L.; Bloem, D.M.; Niss, B.M. 1991. *Effective Watershed Management for Surface Water Supplies*. American Water Works Association Research Foundation, Denver, CO.

Rogers, P. 1993. *America's Water: Federal Roles and Responsibilities, A Twentieth Century Fund Book.*Cambridge, MA: The MIT Press.

Romm, J. 1997. The Pursuit of Watershed Stability: Lessons from Science, Policy and Practice. In *Proceedings of the Sixth Biennial Management Council Conference, Watershed Management Council, Lake Tahoe, California/Nevada, 23-25 October 1996* (153-156), Davis, CA: Centers for Water and Wildland Resources, Water Resources Center.

Scurlock, M.; Curtis, J. 2000. Maximizing the Effectiveness of Watershed Councils: Policy Recommendations from Pacific Rivers Council and Trout Unlimited. Pacific Rivers Council, Eugene, OR.

Shikaze, S.G.; Crowe, A.S. 1998. Impact of flooding and flood mitigation on groundwater quality, p. 389-396. *In* Booty, W.G., Crowe, A.S., Piggott, A., deLoe, R., Bowen, G.S., Holysh, S., Svensson, T., Beatty, B. and Kauffman, G. (ed.), Proceedings of the Conference Groundwater in a Watershed Context, Dec 2-4, 1998, Burlington, Ontario.

Sippel, P. 2000. Controlling the Nagara: changing Approaches to water management in Japan. *WaterPolicy* 2:283-297.

Slaughter, C. (Ed.) 2000. *Proceedings of the Seventh Biennial Watershed Management Council Conference: Western Watersheds: Science, Sense, Strategies, Boise, Idaho, 19-23 October 1998,* Centers for Water and Wildland Resources, Water Resources Center, University of California, Riverside, CA.

Sommarstrom, S. (Ed.) 1997. *Proceedings of the Sixth Biennial Management Council Conference: What is Watershed Stability?, Lake Tahoe, California/Nevada, 23-25 October 1996,* Centers for Water and Wildland Resources, Water Resources Center, University of California, Riverside, CA.

South Africa, Department of Water Affairs and Forestry. (n.d.) Working for Water Programme. Retrieved from http://www-dwaf.pwv.gov.za/wfw/.

South Africa, Department of Water Affairs and Forestry. (n.d.) Website: http://www-dwaf.pwv.gov.za/.

State of Oregon. 1997. *Oregon Coastal Salmon Restoration Initiative*. Submitted to National Marine Fisheries Services, Salem, OR.

State of Washington. (n.d.) Chapter 90.82 RCW Watershed Planning Sections. http://www.mrsc.org/mc/rcw/rcw%20%2090%20%20title/rcw%20%2090%20chapter.htm.

Stoks, P. (n.d.) Personal Communication.

Tsukui, T. 2000. *Ara River Biotop: Nature Revival at the Hand of Man.* Foundation of River & Basin Integrated Communications, Japan. Rivers and Japan. No. 15/Febuary 2000.

Thomas, C.W. 1999. Linking public agencies with community-based watershed organizations: Lessons from California. *Policy Studies Journal* 27:3:544-564.

Toronto and Region Conservation Authority. 2001. *2001 Business Plan of the Toronto and Region Conservation Authority*. Toronto and Region Conservation Authority, Downsview, Ontario, Canada.

Toronto and Region Conservation Authority. 2000. *Annual Report 2000: Making a Living City Region.* Toronto and Region Conservation Authority, Downsview, Ontario, Canada.

United Nations Environment Programme. 1983. *Rain and storm water harvesting in rural areas.* Dublin: Tycooly International Publishing Ltd.

U.S. Agency for International Development. Global Environment Facility Project: Transboundary Pollution Reduction in three Danube Tributary Basins. (n.d.) Website: www.ee-environment.net/docs/dan-gef.shtml.

U.S. Department of Agriculture et al. 2000. Watershed Success Stories: Applying the Principles and Spirit of the Clean Water Action Plan. Retrieved on November 9, 2000 from www.cleanwater.gov/success/watershed.pdf.

U.S. Department of Agriculture et al. 2000. Watershed Reinvention Opportunities. www.cleanwater.gov/reinvent/opportunities.html.

U.S. Environmental Protection Agency. 1996. *Top 10 watershed lessons learned, EPA 840-F-97-001*, U.S. Environmental Protection Agency, Office of Water, Washington, D.C.

U.S. Environmental Protection Agency, Office of Water. 1996. *Environmental Indicators of Water Quality in the United States.* (EPA 841-96-002).

U.S. Environmental Protection Agency and U.S. Department of Agriculture. 1998. *Clean Water Action Plan: Restoring and Protecting America's Waters, EPA 840-R-98-001,* U.S. Environmental Protection Agency, Washington, D.C.

U.S. Environmental Protection Agency, Office of Water, Office of Wetlands, Oceans, and Watersheds. 2001. *Protecting and Restoring America's Watersheds: Status, Trends and Initiatives in Watershed Management, EPA 840-R-00-001*, U.S. Environmental Protection Agency, Washington, D.C.

U.S. Environmental Protection Agency, Office of Water. 1999. *Protecting Sources of Drinking Water: Selected Case Studies in Watershed Management, EPA 816-R-98-019*, U.S. Environmental Protection Agency, Washington, D.C.

Final Framework for Unified Watershed Assessment, Restoration, Priorities, and Restoration Action Strategies. 1998. http://www.cleanwater.gov/uwafinal/cont.html. U.S. Environmental Protection Agency.

About the National Estuary Program. (n.d.) http://www.epa.gov/owow/estuaries/about2.htm. U.S. Environmental Protection Agency.

U.S. Executive Office of the President. 2000. *Budget of the United States Government, Fiscal Year 2001.* Washington, D.C.: U.S. Government Printing Office.

Van der Kleij, W.; Dekker, R.H.; Kersten, H.; De Wit, J.A.W. 1991. Water management of the River Rhine: past, present and future. *European Water Pollution Control* 1:9-18.

Van Roon, M. (n.d.) *Abstract: Managing Harbor and Catchments at Whaingaroa, New Zealand.*

Veale, B. 2001. *A Brief History of Conservation Authorities in Ontario*. pp. 1-4.

Wallis, R.L.; Robinson, S.J. 1991. Integrated catchment management: The Western Australia experience. *Environment* 33:10:31-33.

Watershed Planning Implementation Project Management Committee (Ministry of Environment and Energy, Ministry of Natural Resources, Ministry of Municipal Affairs and Housing, Ministry of Agriculture, Food and Rural Affairs, Association of Municipalities of Ontario, Association of Conservation Authorities of Ontario). 1997. *An Evaluation of Watershed Management in Ontario, Final Report*. Ministry of Environment and Energy, Ontario, Canada.

Watershed Planning Initiative Coordination Resources and Effectiveness Task Group. 1997. *Final Report.* Submitted to the Watershed Planning Implementation Project Management Committee, Ontario Ministry of Natural Resources, Canada.

Watershed Planning Initiative Relevance and Responsiveness Task Group. 1997. *Final Report.* Submitted to the Watershed Planning Implementation Project Management Committee. Ontario Ministry of Natural Resources, Canada.

Watershed Planning Initiative Science and Technology Task Group. 1997. *Final Report.* Submitted to the Watershed Planning Implementation Project Management Committee. Ontario Ministry of Natural Resources, Canada.

Wells, A. 1994. "Up and Doing: a brief history of the Murray Valley Development League, now the Murray Darling Association, from 1944 to 1994." Murray Darling Association, Albury.
Western Water Policy Review Advisory Commission. (1998) *Water in the West: The Challenge for the Next Century*. www.den.doi.gov/wwprac/reports/final.htm.

Who We Are and What We Do. (n.d.) www.merseybasin.org.uk/who_we_are/. Mersey Basin Campaign.

Woolley, J.T.; McGinnis, M.V. 1999. The politics of watershed policymaking. *Policy Studies Journal* 27:3:578-594.

World Meteorological Organization. (n.d.) Retrieved from www.wmo.ch/web/Press/WDW99.html and rhine.html.

Zinn, J. 2000. Conservation Reserve Program: Status and Current Issues. [On-line Report] *Congressional Research Service Report to Congress 97-673*. Retrieved from www.cnie.org/nlc/ag-65.html.

REFERENCES

Blackmore, D.J.; Goss, K.F.; Newman, R.J.; Powell J. 1999. Salinity Management in the Murray Darling Basin. Presented at the Australian Academy of Technological Sciences and Engineering Symposium on the Spirit of the Snowy—Fifty Years On, Melbourne, Victoria, Australia, Nov 1999.

British Columbia Ministry of Environment, Lands and Parks and Environment Canada. 1993. State of the environment report for British Columbia. Vancouver. 127 pp.

Butler, R.W. 1992. Abundance, distribution and conservation of birds in the vicinity of Boundary Bay, British Columbia. Delta, B.C.: Environment Canada, Canadian Wildlife Service. Technical Report No. 155.

Butler, R.W.; Canning R.J. 1989. Distribution of birds in the intertidal portion of the Fraser River delta. Environment Canada, Pacific and Yukon Region, Canadian Wildlife Service. Delta, B.C. Technical Report No. 93.

Chenowith, J.L.; Malano H.M. 2001. Decision Making in Multi-jurisdictional River Basins: A Case Study of the Murray-Darling Basin. *Water International* 26:3:301-313.

Clements, J.; Creager, C.; Beach A.; Butcher, J.; Marcus, M.; Schueler, T. 1996. *Framework for a Watershed Management Program, Project 93-IRM-4*, Water Environment Research Foundation, Alexandria, VA.

Conservation Ontario. 2000. Retrieved in Sept 2001 from http://www.conservation-ontario.on.ca.

Crabb, P. 1997. *Murray-Darling Basin Resources*. Murray Darling Basin Commission, Canberra, Australia.

Crabb, P. 2001. Australia's Murray-Darling Basin Initiative—Correcting the Record. Research and Technical Note in *Water International* 26:3:301-313.

De Ruyter Van Steveninck, E.D.; Admiraal, W.; Breebaart, L.; Tubbing, G.M.J.; van Zanten, B. 1992. Plankton in the River Rhine: structural and functional changes observed during downstream transport. *Journal of Plankton Res*. 14: 1351-1368.

Dieperink, C. 2000. Successful International Cooperation in the Rhine Catchment Area. *Water International* 25:3:347-355.

Dorcey, T. 1997. School of Community and Regional Planning at the University of British Columbia. Integrated Planning Framework. Retrieved from http://www.interchange.ubc.ca/dorcey/innovations/intplan.html.

Finlayson, B.L.; McMahon, T.A. 1988. *Australia vs. the World: A Comparative Analysis of Streamflow Characteristics*. Fluvial Geomorphology of Australia. (R.F. Warner, Ed.), pp. 17-39. Sydney, Australia: Academic Press.

Fuchs, U.; Statzner, B. 1990. Time scales for the recovery potential of river communities after restoration: lessons to be learned from smaller streams. *Regulated Rivers: Research & Management* 5:77-87.

Greater Vancouver Regional District, Quality Control Division. 1992. Iona deep sea outfall 1991. Environmental monitoring program. Summary report, June. Vancouver: 35 pp.

Gleick, P. 2001. Making Every Drop Count. *Scientific American* Feb 18:38-33.

Global Water Partnership. 2000. *Integrated Water Resource Management*. Global Water Vision, The Hague, The Netherlands.

Government of Ontario, Canada. 1990. *Conservation Authorities Act* (R.S.O. 1990, c. C-27). Toronto, Ontario, Canada: Publications Ontario.

Grand River Conservation Authority. (n.d.) Who pays for our programs. http://www.grandriver.ca/index/document.cfm?Sec=12&Sub1=52.

Hamilton, G. 1996. Phantom of the Fraser. *Canadian Geographic* 116:Jul/Aug:56-64.

Handley, J.; Wood, R.; Kidd, S. 1999. Sustainable Development and Institutional Design: The Example of the Mersey Basin Campaign. *Journal of Environmental Planning and Management* 42:3:341-354.

Hooper, B.P.; McDonald, G.T.; Mitchell, B. 1999. Facilitating integrated resource and environmental management: Australian and Canadian perspectives. *Journal of Environmental Planning and Management* 42:5:747-766.

Huisman, P.; de Jong, J.; Wieriks, K. 2000. Transboundary cooperation in shared river basins: experiences from the Rhine, Meuse and North Sea. *Water Policy* 2:83-97.

Huntington, C.W.; Sommarstrom, S. 2000. An Evaluation of Selected Watershed Councils in the Pacific Northwest and Northern California, Prepared for the Pacific Rivers Council and Trout Unlimited. http://www.pacrivers.org/Publications/council.html.

International Office for Water. (n.d.). Organization of Water Management in France. http://www.oieau.fr/anglais/gest_eau/part_a.htm.

International Office for Water. (n.d.). Polluter-User-Pays Principle. http://www.oieau.fr/anglais/gest_eau/intro.htm

Jongman, R. 1992. Vegetation, River Management and Land Use in the Dutch Rhine Floodplains. *Regulated Rivers: Research & Management* 7:279-289.

Kenney, D.S.; McAllister, S.T.; Caile, W.H.; Peckham, J.S. 2000. *The New Watershed Source Book: A Directory and Review of Watershed Initiatives in the Western United States.* Natural Resources Law Center, University of Colorado School of Law, Boulder, CO.

Leeuwen, R. van. 1992. *De actieradius van Both Ends in het Noorden.* Amsterdam: Both Ends, cited in International Regime Development: Lessons from the Rhine Catchment Area, by Carel Dieperink in *TDRI Quarterly Review,* Vol. 12 No. 3 September 1997, pp. 27-35.

Lelek, A.; Kohler, C. 1990. Restoration of fish communities of the Rhine River two years after a heavy pollution wave. *Regulated Rivers: Research & Management* 5:57-66.

Marmulla, G.; Ingendahl, D. 1996. Preliminary results of a ratio telemetry study of returning Atlantic salmon (Salmon *salar*) and sea trout (Salmon *trutta trutta*) in River Sieg, tributary of River Rhine in Germany. In E. Baras, J.C. Philippart (eds), *Underwater Biotelemetry, Proceedings of the First Conference and Workshop on Fish Telemetry in Europe, Liège, Belgium 4-6 April 1995* (pp. 109-117). University of Liège, Belgium.

Marshall, D. 1998. Watershed Management in British Columbia: The Fraser Basin Experience. *Environments* 25:2-3:64-79.

Minshall, L. 2001. Personal communication with Lorrie Minshall of the Grand River Conservation Authority, Sept 24, 2001.

Noort, W.J. van. 1987. Actievoeren: strategieën en tactieken. In *Protest en pressie, Een systematische analyse van collectieve actie,* W.J. van Noort e.a. Assen: Van Gorcum., cited in International Regime Development: Lessons from the Rhine Catchment Area, by Carel Dieperink in *TDRI Quarterly Review,* Vol. 12 No. 3 September 1997, pp. 27-35.

Pigram, J.J. 2000. Towards Upstream-downstream Hydrosolidarity: Australia's Murray-Darling River Basin. *Water International* 25:2:222-226.

Pomeroy, W.M. 1995. The Fraser River Basin—Towards Sustainability. *Water Science & Technology* 31:8:33-39.

Postel, S. 2001. Growing More Food with Less Water. *Scientific American* February 18:34-41.

Postel, S. 1999. *Pillar of Sand: Can the Irrigation Miracle Last?* Washington, D.C.: Worldwatch Institute.

Reimold, R. J., Ed. 1998. *Watershed Management—Practices, Policies and Coordination.* New York: McGraw-Hill.

Rhoades, R.E. 1998. *Participatory Watershed Research and Management: Where the Shadow Falls*. International Institute for Environment and Development, London. Available at: http://www.iied.org/agri/gatekeepers/gk81.html.

Rogers, P.; Priscoli, J. 1993. Problems of Management of Shared River Basins: Concepts, Principles and Issues. Presented at the *Planning Meeting for a Consultative Forum on Management of International Freshwater Resources Basins,* 11-13 November 1993, Geneva, Switzerland.

Ruchay, D. 1995. Living with Water: Rhine River Basin Management. *Water Science & Technology* 31:8:27-32.

Schulte-Wulwer-Leidig, A. 1992. International Commission for the Protection of the Rhine against Pollution—The integrated ecosystem approach for the Rhine. *European Water Pollution Control* 2:3:37-41.

Schulte-Wulwer-Leidig, A.; Van Dijk, G.M.; Marteijn, E.C.L. 1995. Ecological rehabilitation of the River Rhine: plans, progress and perspectives. *Regulated Rivers: Research & Management* 11: 377-388.

Smith, T. 1997. A Grand River Success Story. *Grand Actions: The Grand Strategy Newsletter* 2:5:1-2.

Stakhiv, E. 1996. Return to the Future: Watershed Planning—The Quest for a New Paradigm. In *Proceedings of Watershed '96, Baltimore, Maryland, 8-12 June 1996* (246-249), Washington, D.C.: U.S. Environmental Protection Agency.

United Nations. 1977. Report of the United Nations Water Conference, Mar del Plata, Argentina, 14-25 March 1977. New York: United Nations.

Van den Brink, F.W.B.; Van Katwijk, M.M.; Van der Velde, G. 1994; Impact of hydrology on phyto- and zooplankton community composition in floodplain lakes along the Lower Rhine and Meuse. *Journal of Plankton Res.* 16:351-373.

Van Dijk, G.M.; Marteijn, E.C.L.; Schulte-Wulwer-Leidig, A. 1995. Ecological rehabilitation of the River Rhine: plans, progress and perspectives. *Regulated Rivers: Research & Management* 11:377-388.

Weber, U. June 2000. The "Miracle" Of The Rhine. *Unesco Courier* [On-Line Journal]. Retrieved on xx from http://www.unesco.org/courier/2000_06/uk/planet.htm.

Wood, R.; Handley, J. 1999. Urban Waterfront Regeneration in the Mersey Basin, North West England. *Journal of Environmental Planning & Management* 42:4:565-580.

World Water Forum. 2000. World Water Vision commission report: A Water Secure World—A vision for water, life, and the environment. World Water Commission, The Netherlands.

WASTEWATER UTILITY

Alabama
Montgomery Water Works & Sanitary Sewer Board

Alaska
Anchorage Water & Wastewater Utility

Arizona
Gila Resources
Glendale, City of, Utilities Department
Mesa, City of
Peoria, City of
Pima County Wastewater Management

Arkansas
Little Rock Wastewater Utility

California
Calaveras County Water District
Central Contra Costa Sanitary District
Contra Costa Water District
Crestline Sanitation District
Delta Diablo Sanitation District
Dublin San Ramon Services District
East Bay Dischargers Authority
East Bay Municipal Utility District
El Dorado Irrigation District
Fairfield-Suisun Sewer District
Fresno Department of Public Utilities
Irvine Ranch Water District
Las Virgenes Municipal Water District
Livermore, City of
Lodi, City of
Los Angeles, City of
Los Angeles County, Sanitation Districts of
Napa Sanitation District
Orange County Sanitation District
Palo Alto, City of
Riverside, City of
Sacramento Regional County Sanitation District
San Diego Metropolitan Wastewater Department, City of
San Francisco, City & County of
San Jose, City of
Santa Barbara, City of
Santa Rosa, City of
South Bayside System Authority
South Coast Water District
South Orange County Wastewater Authority
Sunnyvale, City of
Union Sanitary District
West Valley Sanitation District

Colorado
Boulder, City of
Colorado Springs Utilities
Littleton/Englewood Water Pollution Control Plant
Metro Wastewater Reclamation District, Denver

Connecticut
The Mattabasset District
New Haven, City of, WPCA

District of Columbia
District of Columbia Water & Sewer Authority

Florida
Broward, County of
Fort Lauderdale, City of
Gainesville Regional Utilities
JEA
Miami-Dade Water & Sewer Authority
Orange County Utilities Department
Orlando, City of
Reedy Creek Improvement District
Seminole County Environmental Services
St. Petersburg, City of
Stuart Public Utilities
Tallahassee, City of
Tampa, City of
Toho Water Authority
West Palm Beach, City of

Georgia
Atlanta Department of Watershed Management
Augusta, City of
Clayton County Water Authority
Cobb County Water System
Columbus Water Works
Fulton County
Gwinnett County Department of Public Utilities
Savannah, City of

Hawaii
Honolulu, City and County of

Illinois
American Bottoms Wastewater Treatment Plant
Greater Peoria Sanitary District
Kankakee River Metropolitan Agency
Metropolitan Water Reclamation District of Greater Chicago
North Shore Sanitary District
Wheaton Sanitary District

Iowa
Ames, City of
Cedar Rapids Wastewater Facility
Des Moines Metro Wastewater Reclamation Authority
Iowa City, City of

Kansas
Johnson County Unified Wastewater Districts
Unified Government of Wyandotte County/Kansas City, City of

Kentucky
Louisville & Jefferson County Metropolitan Sewer District

Louisiana
Sewerage & Water Board of New Orleans

Maine
Bangor, City of
Portland Water District

Maryland
Anne Arundel County Bureau of Utility Operations
Howard County Department of Public Works
Washington Suburban Sanitary Commission

Massachusetts
Boston Water & Sewer Commission
Upper Blackstone Water Pollution Abatement District

Michigan
Ann Arbor, City of
Detroit, City of
Holland Board of Public Works
Lansing, City of
Owosso Mid-Shiawassee County WWTP
Saginaw, City of
Wayne County Department of Environment
Wyoming, City of

Minnesota
Rochester, City of
Western Lake Superior Sanitary District

Missouri
Independence, City of
Kansas City Missouri Water Services Department
Little Blue Valley Sewer District

Nebraska
Lincoln Wastewater System

Nevada
Henderson, City of

New Jersey
Bergen County Utilities Authority
Passaic Valley Sewerage Commissioners

New York
New York City Department of Environmental Protection
Rockland County Solid Waste Management Authority/Sewer District

North Carolina
Charlotte/Mecklenburg Utilities
Durham, City of
Metropolitan Sewerage District of Buncombe County
Orange Water & Sewer Authority

Ohio
Akron, City of
Butler County Department of Environmental Services
Columbus, City of
Metropolitan Sewer District of Greater Cincinnati
Northeast Ohio Regional Sewer District
Summit, County of

Oklahoma
Tulsa, City of

Oregon
Clean Water Services
Eugene/Springfield Water Pollution Control
Water Environment Services

Pennsylvania
Philadelphia, City of
University Area Joint Authority, State College

South Carolina
Charleston Commissioners of Public Works
Mount Pleasant Waterworks & Sewer Commission
Spartanburg Sanitary Sewer District

Tennessee
Cleveland, City of
Knoxville Utilities Board
Murfreesboro Water & Sewer Department
Nashville Metro Water Services

Texas
Austin, City of
Dallas Water Utilities
Denton, City of
El Paso Water Utilities
Fort Worth, City of
Gulf Coast Waste Disposal Authority
Houston, City of
San Antonio Water System
Trinity River Authority

WERF SUBSCRIBERS

Utah
Salt Lake City Corporation

Virginia
Alexandria Sanitation Authority
Arlington, County of
Fairfax County Virginia
Hampton Roads Sanitation District
Henrico, County of
Hopewell Regional Wastewater Treatment Facility
Loudoun County Sanitation Authority
Lynchburg Regional WWTP
Prince William County Service Authority
Richmond, City of
Rivanna Water & Sewer Authority

Washington
Edmonds, City of
Everett, City of
King County Department of Natural Resources
Seattle Public Utilities
Sunnyside, Port of
Yakima, City of

Wisconsin
Green Bay Metro Sewerage District
Madison Metropolitan Sewerage District
Milwaukee Metropolitan Sewerage District
Racine, City of
Sheboygan Regional Wastewater Treatment
Wausau Water Works

Australia
South Australian Water Corp.
Sydney Water Corp.
Water Corp. of Western Australia

Canada
Greater Vancouver Regional District
Toronto, City of, Ontario
Winnipeg, City of, Manitoba

Mexico
Servicios de Agua y Drenaje de Monterrey, I.P.D.

New Zealand
Watercare Services Limited

Singapore
Singapore Public Utilities Board

United Kingdom
Yorkshire Water Services Limited

STORMWATER UTILITY

California
Los Angeles, City of, Department of Public Works
Monterey, City of
San Francisco, City & County of
Santa Rosa, City of
Sunnyvale, City of

Colorado
Boulder, City of

Georgia
Griffin, City of

Iowa
Cedar Rapids Wastewater Facility
Des Moines Metro Wastewater Reclamation Authority

Kansas
Overland Park, City of

Kentucky
Louisville & Jefferson County Metropolitan Sewer District

Maine
Portland Water District

Minnesota
Western Lake Superior Sanitary District

North Carolina
Charlotte, City of, Stormwater Services

Pennsylvania
Philadelphia, City of

Tennessee
Chattanooga Storm Water Management

Washington
Seattle Public Utilities

STATE

Arkansas Department of Environmental Quality
Fresno Metropolitan Flood Control District, Calif.
Urban Drainage & Flood Control District, Colo.

CORPORATE

ADS Environmental Services
The ADVENT Group Inc.
Alan Plummer & Associates
Alden Research Laboratory
Aqua-Aerobic Systems, Inc.
Aquateam–Norwegian Water Technology Centre A/S
BaySaver Inc.
BioVir Laboratories, Inc.
Black & Veatch
Boyle Engineering Corporation
Brown & Caldwell
Burns & McDonnell
CABE Associates Inc.
The Cadmus Group
Camp Dresser & McKee Inc.
Carollo Engineers Inc.
Carpenter Environmental Associates Inc.
CDS Technologies Inc.
Chemtrac Systems Inc.
CH2M HILL
Damon S. Williams Associates, LLC
David L. Sheridan, P.C.
Dewling Associates, Inc.
Earth Tech Inc.
Eco-Matrix
Ecovation
Environmental Engineers International
EMA Inc.
The Eshelman Company Inc.
Finkbeiner, Pettis, & Strout Inc.
Freese & Nichols, Inc.
ftn Associates Inc.
Fuss & O'Neill, Inc.
Gannett Fleming Inc.
Golder Associates Inc.
Greeley and Hansen LLC
Hazen & Sawyer, P.C.
HDR Engineering Inc.
HNTB Corporation
HydroQual Inc.
Infilco Degremont Inc.
Ingersoll-Rand Energy Systems
Insituform Technologies Inc.
Institute for Environmental Technology & Industry, Korea
Jacobson Helgoth Consultants Inc.
Jason Consultants Inc.
Jordan, Jones, & Goulding Inc.
KCI Technologies Inc.
Kelly & Weaver, P.C.
Kennedy/Jenks Consultants
Komline Sanderson Engineering Corporation
Lawler, Matusky & Skelly Engineers, LLP
Limno-Tech Inc.
Lombardo Associates Inc.
Malcolm Pirnie Inc.
McKim & Creed
MEC Analytical Systems Inc.
Metcalf & Eddy Inc.
MWH
New England Organics
Odor & Corrosion Technology Consultants Inc. (OCTC)
Oswald Green, LLC
PA Government Services Inc.
Parametrix Inc.
Parsons
Post, Buckley, Schuh & Jernigan
R&D Engineering/Conestoga Rover & Associates
The RETEC Group
RMC, Inc. (Raines, Melton & Carella)
R.M. Towill Corporation
Ross & Associates Ltd.
Royce Technologies
SAIC Maritime Technical Group
Short Elliott Hendrickson, Inc.
Stantec Consulting Group, Inc.
Stormwater Management, Inc.
Synagro Technologies, Inc.
Tetra Tech Inc.
Trojan Technologies Inc.
URS Corporation
Veolia Water NATC
Wade-Trim Inc.
Weston Solutions Inc.
Woodard & Curran
WRc/D&B, LLC
WWETCO, LLC
Zoeller Pump Company

INDUSTRY

American Electric Power
ChevronTexaco Energy Research & Technology Company
The Coca-Cola Company
Dow Chemical Company
DuPont Company
Eastman Kodak Company
Eli Lilly & Company
Merck & Company Inc.
ONDEO Services
Procter & Gamble Company
PSEG Services Corp.
RWE Thames Water Plc
Severn Trent Services Inc.
United Water Services LLC

Note: List as of 7/1/04

WERF Product Order Form

As a benefit of joining the Water Environment Research Foundation, subscribers are entitled to receive one complimentary copy of all final reports and other products. Additional copies are available at cost (usually $10). To order your complimentary copy of a report, please write "free" in the unit price column. WERF keeps track of all orders. If the charge differs from what is shown here, we will call to confirm the total before processing.

Name _______________ Title _______________

Organization _______________

Address _______________

City _______________ State _______ Zip Code _______ Country _______

Phone _______________ Fax _______________ Email _______________

Stock #	Product	Quantity	Unit Price	Total
			Postage & Handling	
			VA Residents Add 4.5% Sales Tax	
			Canadian Residents Add 7% GST	
			TOTAL	

Method of Payment:

(All orders must be prepaid.)

❑ Check or Money Order Enclosed

❑ Visa ❑ Mastercard ❑ American Express

Account No. _______________ Exp. Date _______

Signature _______________

Shipping & Handling:

Amount of Order	United States	Canada & Mexico	All Others
Up to but not more than:	Add:	Add:	Add:
$20.00	$5.00*	$8.00	50% of amount
30.00	5.50	8.00	40% of amount
40.00	6.00	8.00	
50.00	6.50	14.00	
60.00	7.00	14.00	
80.00	8.00	14.00	
100.00	10.00	21.00	
150.00	12.50	28.00	
200.00	15.00	35.00	
More than $200.00	Add 20% of order	Add 20% of order	

*minimum amount for all orders

Note: Please make checks payable to the Water Environment Research Foundation.

To Order (Subscribers Only):

Log on to www.werf.org and click on the "Product Catalog."

Phone: (703) 684-2470
Fax: (703) 299-0742.

WERF
Attn: Subscriber Services
635 Slaters Lane
Alexandria, VA 22314-1177

To Order (Non-Subscribers):

Non-subscribers may be able to order WERF publications either through WEF (www.wef.org) or IWAP (www.iwapublishing.com). Visit WERF's website at www.werf.org for details.

IWA Publishing's authorised EU representative for General Product Safety Regulations is Diane D'Arras, 15 rue Duret, 75116 Paris, France, e-mail: safety@iwap.co.uk.

Printed and bound by CPI Group (UK) Ltd, Croydon, CR0 4YY
12/05/2026
02108868-0007